Athletic Enhancement, Human Nature and Ethics

INTERNATIONAL LIBRARY OF ETHICS, LAW, AND THE NEW MEDICINE

VOLUME 52

For further volumes:
http://www.springer.com/series/6224

Jan Tolleneer • Sigrid Sterckx • Pieter Bonte
Editors

Athletic Enhancement, Human Nature and Ethics

Threats and Opportunities of Doping Technologies

Springer

Editors
Jan Tolleneer
Faculty of Kinesiology
and Rehabilitation Sciences
Department of Kinesiology
KU Leuven University, Leuven
Belgium

Sigrid Sterckx
Faculty of Arts and Philosophy
Department of Philosophy
and Moral Sciences
Ghent University, Ghent
Belgium

Pieter Bonte
Faculty of Arts and Philosophy
Department of Philosophy
and Moral Sciences
Bioethics Institute Ghent
Ghent University, Ghent
Belgium

Advisory Committee Members
Andreas De Block
Institute of Philosophy
Centre for Logic and Analytical Philosophy
KU Leuven University, Leuven
Belgium

Paul Schotsmans
Centre for Biomedical Ethics and Law
KU Leuven University, Leuven
Belgium

ISSN 1567-8008
ISBN 978-94-007-5100-2 ISBN 978-94-007-5101-9 (eBook)
DOI 10.1007/978-94-007-5101-9
Springer Dordrecht Heidelberg New York London

Library of Congress Control Number: 2012953100

Printed on acid-free paper

Springer is part of Springer Science+Business Media (www.springer.com)

Preface

Technology and Sport, Meanings and Realities

Read this book only if you want your cherished ideas to be challenged and your intellectual horizons to be expanded. Whatever your convictions about sport, human nature, and enhancement technologies, you are sure to find arguments here that will puzzle and possibly offend you, as well as ones that provide fresh insights into the assumptions underpinning your own beliefs. Your views may change, or not, but you will have a far sharper understanding of the range of positions once you've engaged with the ideas presented in this book.

Along the way you will witness concepts such as human nature and the natural, as they are deployed in the enhancement debate, get a good cuffing, but also some sympathetic and sophisticated defenses. Misleading labels such as "bioconservative" and "techno-progressive" make their appearance, but for the most part the authors in this volume attempt to explain what they mean by them rather than using these terms as a lazy, dismissive shorthand.

An exercise I recommend is to take careful note of how different authors use or define the idea of sport. Is their focus exclusively or predominantly on elite sport, or do they include the multitudes of people around the world—likely in the hundreds of millions if not more—who engage in some form of sporting activity? Is their concern limited to formal competitions or does it encompass informal and non-competitive athletic endeavors? Do they provide insights into or at least a clear presentation of their own views on the purpose or meaning of sport? You may find, as I did, a wide range of ideas from sport as joyful play to sport as a method to establish hierarchies of genetic superiority. You will have to decide which portraits of sport are most compelling, which capture most fully your own experience as athlete, parent, spectator, or critic.

One tension, inescapable in a scholarly volume of this type, is between abstract theoretical analysis and immersion in the particulars of whatever practices or institutions are under investigation. You will find a spectrum of approaches in this book, some chapters tilting heavily towards the theoretical, some striving mightily to sustain the tension, and a few with a strong empirical bent. Each approach has its virtues, though I confess to a preference for analyses that embody a deep understanding of the realities of sport—born no doubt from my experience at The Hastings Center where the mantra "good ethics begins with good facts" is a fundamental precept. A key word in that mantra is "begins": the hard work of ethical analysis is profoundly informed by what we learn about the underlying realities, but the search for insight and practical wisdom can be difficult and challenging.

Imagine we are talking not about sport but rather about music. Scholars can legitimately analyze music as sequences of sounds of particular frequencies and intervals. They can represent those sounds by symbols written according to the notational conventions of whatever system they employ.

Representations such as these can reveal interesting patterns. But they may not be the best way to capture all the richness of the music—the experience of playing, singing, dancing or listening; how some music is joyful, even transporting and rapturous. Like sport, one's experience of certain forms of music such as jazz, classical and many other genres, is enhanced the more one participates in and understands it. Music can also be unsettling and challenging—think 12-tone—and earn its value that way.

Of course, music can also be utterly dull, uninteresting, boring, listlessly performed or electronically produced. Autotune is an enhancement technology of sorts in that it corrects a singer's errors of intonation. It can help make a pop star out of someone barely able to carry a tune. (The skills that matter most here appear to be in marketing, public relations, and hairstyling rather than musicality.)

A formal analytical structure may struggle to capture such nuances, but it is in such details that we are likely to find what makes particular forms of music, particular performances, important and moving in our lives. Such theoretical structures may also have difficulty explaining why anyone should care about music. What is it that motivates us to play, sing, dance, and listen? If we desire to understand what gives music its meaning and human significance, we are well advised to attend to the experiences of performers and audiences.

The same is true of sport. One of the first things one learns from a close look is the limit of sport as a test-bed for how cultures receive new technologies. Sport can be quite perverse. As much as we welcome in most of our activities technologies that make us more efficient, or that correct our errors (such as spell-check, with its occasionally hilarious results), sport often prohibits them, from golf balls that fly overly far and straight to motors on bicycles in the Tour de France. We can only make sense of such perversity if we grasp what meanings and values people find in that sphere of human endeavor we call sport.

Technologies that sport may choose to prohibit as unwanted enhancements may be quite acceptable, even ethically preferable, in other spheres. Imagine a drug that steadies one's hand. It may be banned in sports such as archery and biathlon; but if it improved surgeons' ability to perform delicate operations without notable risk to the surgeon's health we might well expect that good surgeons would no more think of operating without taking this drug than they would not bother sterilizing their instruments. We need to look to the meaning and values within each sphere, along with the particulars of the candidate enhancement technology, if we wish to understand how to deal wisely and fairly with each purported enhancement.

A second, equally crucial insight one learns from attending carefully to the realities of competitive sport is the complete inadequacy of a libertarian individualistic framework for analyzing the ethics of enhancement in sport. The pallid abstraction of "positional goods" points in the right direction, but it captures only weakly the stark reality athletes face when they compete against others who gain a palpable advantage from a drug or other technology. A drug that provides a five percent edge trumps a three percent advantage in talent and dedication. An individual's decision to use a performance-enhancing drug affects everyone he or she competes against. Given the realities of competitive sport, the libertarian view is in effect tyrannical: if some athletes are free to use drugs, just about all athletes who hope to compete at that level will be obliged to do the same or else consign themselves to frustration and failure. That, at least, is what I've learned from listening to athletes.

You will find much to agree with, much to disagree with (including from this Preface). But when you read this book you will gain abundant insights and your understanding of sport, technology, and human nature will surely be enhanced.

President Emeritus
The Hastings Center
Garrison, NY, USA

Thomas H. Murray, Ph.D.

Acknowledgements

New technologies with the potential to enhance athletic performance present scientists and practitioners with important moral choices and call upon ethicists to help identify and elucidate values like health, freedom and fairness and even more fundamental concerns like human dignity and human nature. Interdisciplinary collaboration and international networking is needed to deal with such broad and deep challenges. To assist that process, the newly created Interfaculty Research Group Sport and Ethics of the KU Leuven, together with the Bioethics Institute Ghent, organised the seminar *Enhancement and human nature in medicine and sport* (Leuven, 18 March 2011), the papers of which constitute the core of the volume at hand. In the following months several other authors submitted papers to complete and enrich the book *Athletic enhancement, human nature and ethics: Threats and opportunities of doping technologies*. Significantly, the interdisciplinary scope and the human nature focus seemed to arouse a lot of enthusiasm. No less than three international meetings created opportunities to meet colleagues, negotiate contributions and discuss the nature of the book. These were consecutively: *Philosophy of sport in Europe*, the first conference of the European Association for the Philosophy of Sport (EAPS) in Prague, 19–21 May 2011, *Human enhancement: medical, ethical and legal implications*, the Brocher-Hastings Center Summer Academy on the ethics of human enhancement in Hermance (Geneva), 4–8 July 2011 and the *Anti-doping. Rational policy or moral panic* conference at Aarhus University, 18–19 August 2011, organized by the International Network of Humanistic Doping Research (INHDR). The editors welcomed these opportunities and owe thanks to the organisers and to many colleagues among which Mike McNamee for his ever inspiring ideas. All these contacts and efforts resulted in the present compilation of original papers, presenting a sustained, multi-faceted analysis of the most foundational questions about human nature, humanity and natural talent as challenged by doping. The editors Jan Tolleneer, Pieter Bonte and Sigrid Sterckx and the members of the editorial committee, Andreas De Block and Paul Schotsmans, thank all the contributors and collaborators and hope that the book will play a significant role in the wider understanding of athletic enhancement and in the on-going process of interdisciplinary research and social debate with a view to ever more ethical and aesthetical sports. Last but not least, we express our deepest gratitude to Dr Julian Cockbain for his careful and thorough language editing of the entire manuscript.

Contents

Contributors

Pieter Bonte Faculty of Arts and Philosophy, Bioethics Institute Ghent (BIG), Ghent University, Ghent, Belgium

Barbara Broers Unit for Dependency in Primary Care, Department of Community Health and Primary Care, University Hospitals of Geneva, Geneva, Switzerland

Roger Brownsword The Dickson Poon School of Law, King's College London, University of Sheffield, London, UK

Andreas De Block Centre for Logic and Analytical Philosophy, KU Leuven University, Leuven, Belgium

Trijsje Franssen Department of Sociology and Philosophy, University of Exeter, Exeter, UK

Denis Hauw Institute of Sport Sciences, University of Lausanne, Lausanne, Switzerland

John Hoberman Department of Germanic Studies, University of Texas, Austin, TX, USA

M. Andrew Holowchak Department of Philosophy, Rider University, Lawrenceville, NJ, USA

Marianne Raakilde Jespersen Department of Public Health & Sport Science, Aarhus University, Aarhus C, Denmark

Eric T. Juengst UNC Centre for Bioethics, University of North Carolina, Chapel Hill, NC, USA

Bengt Kayser Institute of Movement Sciences and Sports Medicine, Faculty of Medicine, University of Geneva, Geneva, Switzerland

Christian Lenk Institute for History, Philosophy and Ethics of Medicine, Ulm University, Ulm, Germany

Tara Magdalinski UCD Centre for Sports Studies, University College Dublin, Dublin 4, Ireland

Michael J. McNamee Department of Philosophy, Humanities and Law in Healthcare, College of Human and Health Sciences, Swansea University, Swansea, UK

Darian Meacham Department of Arts, Faculty of Arts, Creative Industries and Education, University of the West of England, Bristol, UK

Paul Schotsmans Centre for Biomedical Ethics and Law, KU Leuven University, Leuven, Belgium

Sigrid Sterckx Faculty of Arts and Philosophy, Department of Philosophy and Moral Sciences, Ghent University, Ghent, Belgium

Jan Tolleneer Faculty of Kinesiology and Rehabilitation Sciences, KU Leuven University, Leuven, Belgium

Chapter 1
Introduction: Human Nature as a Promising Concept to Make Sense of the Spirit of Sport

Pieter Bonte, Jan Tolleneer, Paul Schotsmans, and Sigrid Sterckx

Should sport revolve around natural talent or should athletes be allowed to enhance their bodies with biotech? Throughout 2012, a series of events shook the world of sports, sparking heated but often sensationalist or simplistic debates on these issues. In fall, for instance, the US Anti-Doping Agency presented a formidable doping dossier against seven-fold Tour de France winner Lance Armstrong. Before long, the International Cycling Union UCI had repealed all his seven victories, and the words of UCI president Pat McQuaid "Lance Armstrong has no place in cycling" became the instant headline of every major global news outlet. In the public eye, arguably the world's greatest athlete of his time took an unprecedented fall from grace to join the ranks of the 'doping sinners'. A few months earlier, in an incomparably different moral climate, Oscar Pistorius – the South-African double amputee running on so-called Cheetah prostheses – could be seen sprinting in the 400m relay finals of the London Summer Olympics. As he ran – or did he 'blade'? – as a man made of muscle and blood and fiberglass, and above all of will and skill, Pistorius rose to the ranks from where Armstrong was to fall: the hall of famers of heroic

P. Bonte (✉)
Faculty of Arts and Philosophy, Bioethics Institute Ghent (BIG), Ghent University, Ghent, Belgium

J. Tolleneer
Faculty of Kinesiology and Rehabilitation Sciences, KU Leuven University, Leuven, Belgium

P. Schotsmans
Centre for Biomedical Ethics and Law, KU Leuven University, Leuven, Belgium

S. Sterckx
Faculty of Arts and Philosophy, Department of Philosophy and Moral Sciences, Ghent University, Ghent, Belgium

J. Tolleneer et al. (eds.), *Athletic Enhancement, Human Nature and Ethics*, International Library of Ethics, Law, and the New Medicine 52, DOI 10.1007/978-94-007-5101-9_1,

overcoming. In a sense, both athletes decided to draw biotechnologies inside their bodies, and to meddle with nature, in order to enhance their ability to perform. To many, however, the comparison stops right after it started, as each use of biotech signifies incomparably different things: the first causing mass desperation and gloom about the increasing intrusion of artifice in the human body, the second mass inspiration and hope about the coming marvels of biomedical technology. One of the many ethical conundrums this volume seeks to clarify concerns what might make Armstrong's endeavor so appalling and that of Pistorius so appealing.

As many commentators have noted, it seems that the contemporary doping debate is in urgent need of more in-depth investigations of such issues. Ultimately, the most vexing problems posed by enhancing 'doping' and enabling artifice nestling deep within the body do not seem to be about health or fair play, although doping clearly does pose momentous problems on those fronts too. But if some forms of doping would be made available in an adequately healthy and fair way, they would probably *still* cause much concern – concern about *doping itself* would persist, no matter how much its circumstances would be tidied up.

These intrinsic concerns require us to look at doping through a different set of lenses than that of health, fairness and rights. Issues of authentic agency and personal accountability, of the appeal of natural grace in performance, of sports as a testing ground for the capacities of the human species with which all members of the species can identify, of sport as a display of the special natural gifts given to the talented, of sport as a display of ideal exemplars of man's nature; or alternatively, of sport as a display of man's protean nature (his capacity to adapt and transform himself), of sport as a showcase of his promethean nature (his tendency to transgress naturally given constraints on his existence) – all these foundational issues come to the fore when we start looking at doping through the lens of 'human nature'.

These foundational issues make up the 'spirit of sport': they constitute the more profound existential, ethical and aesthetic reasons why sports is perceived as a very meaningful practice to so many people. These foundational issues concerning human nature are not only of great philosophical relevance, they are equally of great practical relevance. Today, a categorical anti-doping position has been successfully established as the near-universal official consensus. Yet, as many authors have argued and as can be derived from the World Anti-Doping Code, the legitimation of this categorical denunciation of doping relies heavily on the so-called 'spirit of sport'.

However, surprisingly few attempts have been made to provide sustained and focused articulations of what this foundational spirit consists of and how it could warrant a categorical denunciation of doping. To the extent that attempts have been made to back up a categorical anti-doping stance on the basis of concerns unrelated to health or fairness, none have been without controversy. At the same time, behind the official and popular consensus that doping is intrinsically wrong, a number of increasingly vocal scholars have been developing elaborate arguments for the permissibility and even the promotion of drastic athletic enhancement technologies, in direct opposition to the global anti-doping stance.

In recent years, anti-doping institutions have taken an impressive lift-off in terms of popular support, budget increase and regulative and policing powers. Nevertheless, their fundamental rationale is frequently perceived by both pro and contra-doping authors as a "black box". That is, the 'spirit of sport' is commonly acknowledged to contain the foundational justification that fuels the anti-doping project, but it seems difficult to find out just how to open the box and find out what the spirit is. Now that a number of contrary positions have been presented, the foundational anti-doping rationale must be restated if it wishes to trump those contestations. High time, therefore, for the anti-doping institutions to take a renewed interest in their *raison d'être*.

In preparing this volume, we have endeavored to provide a platform for this urgent foundational discussion. In an effort to push things forward we have, in a sense, 'cut to the chase' of the doping debate. Instead of *arriving at* conclusions that issues of human nature matter, we invited authors to *start from* the assumption that normative conceptions of human nature and the related issues matter in this debate. In doing so, we hope this volume can do more than *indicate* what the foundational issues are, and begin the work of *investigating* them. Besides doing the relatively harmless work of clarifying what the deeper questions are that athletic enhancement technologies pose, we wanted to encourage attempts to do the more controversial but urgently needed work of trying to provide deeper answers to those questions.

We have sought to unite a group of authors that is highly international, involving American, Belgian, British, Danish, Dutch, English, French, German and Swiss authors. We have equally endeavored to make this volume widely interdisciplinary, bringing together philosophers, sport scientists, historians, medical professionals, legal scholars and sociologists. Also, considering how both sports culture and attitudes towards biotechnologies are rapidly changing, new insights from well-established experts are interspersed with the work of a younger generation of researchers. Together, they tackle a wide variety of issues, interconnected by the quest to come to a clearer understanding and appreciation of the roles that human nature and enhancement technologies play – and should or should not play – in sports.

For any given contributor to this volume, several of his or her arguments and insights are instantly challenged or complemented by the writings of their colleagues. Therefore, we present the reader with more than a mere summary of the chapters here. Instead, we draw attention to what has struck us as the most intriguing points of controversy and consensus amongst the contributors. Also, for each author we foreground the way in which she fleshed out the concept of human nature.

1.1 Part I: Conceptual and Theoretical Framework

Jan Tolleneer and Paul Schotsmans open the conceptual and theoretical part of this book with Chap. 2, Self, Other, Play, Display and Humanity. Development of a Five-Level Model for the Analysis of Ethical Arguments in the Athletic Enhancement Debate. They introduce a schematic framework to better grasp and relate the plethora of ethical issues athletic enhancement technologies raise. They draw up five axes, relating the (doping) athlete to (1) the self, (2) the other (the fellow athlete), (3) play (the game of sports), (4) display (spectator sports) and, as an underlying level which might be significantly affected by doping, (5) human nature. This model shows that the choice of an athlete to engage in doping inextricably has relational repercussions. The authors use the athlete's moral perspective to study the respect that is shown or not shown on the five levels mentioned. On each level, Tolleneer and Schotsmans present clusters of arguments pro and contra. Contra, doping can be said to respectively incur health risks, lessen the opponent's chances and freedom, undermine sports' gaming spirit, create negative role models and desecrate human nature. Pro, doping can be said to respectively increase autonomy, equalize opportunity, align sport with other social spheres susceptible to enhancement interventions, stimulate awe-inspiring heroism, and realize human nature understood as a Promethean nature – an image that is analyzed at length in Trijsje Franssen's contribution. Tolleneer and Schotsmans have amassed these arguments from the scholarly literature on the ethics of doping and human enhancement which they enrich with their own study on the spontaneous opinions on doping's (im)permissibility with 18 year old students, where they note a trend towards increasing permissiveness. By contrasting conservative agents on the one hand and liberal agents on the other the five-level model helps in pinpointing the main ethical questions concerning athletic enhancement. Furthermore, a socio-historical excursion focuses on the role of the Olympic movement in including more and more social groups: female athletes, worker athletes, professional athletes etc. The authors argue that the Mexico 1968 games were a kind of turning point in this broad development because, based on the new doping regulations, one suddenly started to exclude a certain group: the enhanced athletes. The arguments discussed and categorised on the two opposite sites of the five-levels research model explain why this doping ban can be seen as (il)legitimate. Yet, the enhancement-case provided by the 400 m runner Oscar Pistorius, which will be studied in depth in Tara Magdalinski's chapter, entails a new interesting moment in history. This dis/superabled athlete is given access to the Olympic competitions and took part in the 2012 London Games. How will the Olympic family deal with this issue in the future? What if the prosthesis technologies and training methods improve and a new generations of blade runners put the non-disabled athletes at a disadvantage? Will 'designer legs' remain morally acceptable? The authors portray the history as a chain of negotiations between all those involved – recognizes the fact that policies are not, and should not, simply be developed and implemented top-down, but take shape trough continuous negotiations, an approach explicitly argued for in the contributions of Darian Meacham, Roger Brownsword and Pieter Bonte. Pro and contra arguments are constantly discussed and balanced against each other, and Tolleneer and Schotsmans advance genetic modification as the most troubling challenge for tomorrow. Given the tendency to increased inclusivity, might 'designer genes' ever become acceptable too? Or can anti-doping exclusions turn the tide? The five-level model and similar tools can help in managing both these technological developments as well as the negotiations between all social agents involved.

In Chap. 3, Is Human Enhancement Unnatural and Would This Be an Ethical Problem?, **Christian Lenk** distinguishes three possible meanings of 'naturalness'. First, he discusses 'natural' in the sense of something which comes or stems from nature.

Here he takes to task 'moral naturalism', an ethical position stating that what runs counter to a supposedly objective ethical standard of human nature – an example of which, some have argued argue, is homosexual intercourse – can then be seen as unnatural, perverse and immoral. Lenk counters such moral naturalism with what is perhaps the most formidable obstacle to deriving conclusions of ethical *value* out of considerations of natural *facts* – such as in the debate on athletic enhancements, deriving the conclusion that such enhancements are wrong because they run counter to (some aspects of) human nature. That obstacle is the 'naturalistic fallacy', i.e. the failure to take into account that the realm of values and the realm of facts are two distinct realms, and that one should not base a conclusion of value on premises of fact or vice versa. In Chap. 6, Trijsje Franssen exposes the ubiquity of such fallacious arguments in both the contra and pro enhancement literature. Mindful of this fallacy, Michael J. McNamee seeks to present a spirit of sport that retains a proper appreciation of our natural human constraints in his contribution, whereas Pieter Bonte argues for an existentialist spirit based on an appreciation of how the naturalistic fallacy leaves us with a nature-less human condition, in which we cannot find meaning in the way nature *de facto* constrains us. Second, Lenk discusses 'natural' in the sense of normal or usual. Here, he follows the lead of the biostatistical theory of C. Boorse, wherein a species' nature is seen as 'a functional design empirically shown typical of it.' Lenk points out that enhancement interventions might – but need not – cause a harmful disequilibrium of one's physical or mental constitution. To upset a given natural balance would not, however, be problematic *as such*. Third, Lenk analyses 'natural' in the sense of essential or essence. Discussing the 'right to morphological freedom' advocated by explicitly pro-enhancement transhumanist such as Nick Bostrom, Lenk again finds no fault with the alteration of biological essentials as such. Rather, he warns of such alterations being made without proper regard for autonomy. He does, however, tempers Bostrom's vision of radical self-alteration as such practices of self-enhancement are likely to build on the kind of being one already is. Raising an issue that will be investigated at length by Darian Meacham as well as Roger Brownsword, Lenk warns for the possible intolerance towards those humans who would transform themselves into beings that stray too far from the normal human fold. Lenk concludes: "in my interpretation, the "natural" has no secret metaphysical meaning, but is rather a reasonable and pragmatic benchmark for our living and the evaluation of the acceptability of changes in the human body." Anticipating the response that, as a result, he thereby gives no principled objection to enhancement, Lenk argues that further fairness concerns would provide adequate reason to continue the categorical ban. His conclusion that, ultimately, there are no deeper concerns than those of health and fairness, is something with which many other contributors would beg to differ.

Pieter Bonte, for instance, explicitly suspends the issues of health and fairness in Chap. 4, Dignified Doping: Truly Unthinkable? An Existentialist Critique of 'talentocracy' in Sports in order to find out what the problems peculiar to *doping itself* may be. Bonte develops a three-leveled framework, contrasting natural talent to an enhanced body. Within this framework, a proper athletic accomplishment should: (a) originate in proper origins; (b) take place via proper processes; and (c) result in proper end states. Common pro-talent, anti-doping intuitions would respectively require that the accomplishment originates in *natural talent*, takes place via *effortful agency*, and results in a *perfected but identifiably human* end state. However, at all three levels Bonte reaches conclusions that upset those intuitions. According to Bonte, the belief that innate natural talent is superior to artificially enabled embodiments should be significantly tempered, as it may be riddled with a 'talentocratic' bias that is at best based on unfortunate misconceptions, probably rooted in psychological atavisms that should be denied ethical validity (an issue that returns in Andreas De Block's contribution), and at worst related to latent social Darwinist or eugenicist adorations of genetic superiorities (also discussed in Eric Juengst's chapter). Wedding a demanding conception of sports based on virtue ethics to an existentialist understanding of the human condition, Bonte arrives at a conception of 'dignified doping' that may help to better distinguish the many deplorable doping practices from the few dignified ones, without having to commit to a blanket anti-enhancement position. Such 'dignified doping' would have to consist in a person first courageously assessing her existential predicament as one that is 'foundationlessly free and ruthlessly responsible' and then proceeding to affirm that burdensome responsibility by making a deep, willful intervention in her absurdly obtained nature. With the humanist dignity of, for example, Pico della Mirandola, the doping athlete thus literally incarnates the belief that man is the creator and carrier of his own values. Such an explicit abandonment of the belief that one was given a meaningful natural mold by nature or God may well be detrimental to one's comfort and superficial happiness, but from this perspective is to be understood as an obligation grounded in one's human(ist) dignity. To engage in such an act of self-imposed body-modification in the field of sports seems very apt, as sports is one of the most natural habitats of the *Homo ludens*, whose dignity and grace lies in playing 'useless' games of one's own devise, in 'useless' bodies of one's own devise. Although apparently antagonistic to the position defended by Michael J. McNamee in Chap. 9, both positions share a 'thick' virtue ethics approach as well as a staunch criticism of 'thin' libertarian and utilitarian pro-enhancement theories. Bonte, however, radically challenges the perfectionism and innatism implied in a spirit of sport understood as 'the virtuous perfection of natural talent', which results in an equally deep, but very different commitment to athletic play as 'the virtuous exploration of bodily virtuosity'.

1.2 Part II: Transgressing the Limits of Human Nature

In Chap. 5, Subhuman, Superhuman, and Inhuman: Human Nature and the Enhanced Athlete, **Eric T. Juengst** delves deeper into the concern that doping 'dehumanizes'. According to Juengst, the concept of 'human nature' is often given a negative or contrast-definition, that is: a definition in terms of what human nature is *not*. "Human nature is not, by definition, subhuman, superhuman, or inhuman. The subhuman is less than human, like a beast; the superhuman is more, like a god; and the inhuman something simply different, like a machine." This negative approach has the benefit of not having to commit to a positive account, which would face the problems of gaining sufficient support and avoiding the naturalistic fallacy, all the while retaining the commitment to protect human nature in practical decisions about possible transgressions, such as doping. Juengst analyzes the fear, that athletic enhancement may lead to degradation into the subhuman, as a fear of a disruption of the ontological boundaries between species, between races or between sexes. Runners modified to obtain a cheetah-like running speed or female swimmers with a masculinized constitution would on these ground be dismissed as unnatural freaks – a point of concern that returns in the contributions of Darian Meacham, Marianne Raakilde Jespersen and Tara Magdalinski. To Juengst, such lines of reasoning run the risk of sliding into racism, sexism or social class bias. The second transgression, that into the superhuman, is analyzed as a concern for *hubris* or the defiance of higher (divine or natural) powers, and a concern for the maintenance of a proper appreciation of life's 'giftedness', which in sport often takes the form of admiration and celebration of the talents of 'gifted' athletes. As do Trijsje Franssen and Pieter Bonte, Juengst quotes Michael Sandel's *The Case Against Perfection* in this regard. The third transgression, that into the inhuman or artificial, is set apart from the first two. Here, the concern is about the dilution of human agency by artifice, and here Juengst engages with John Hoberman who finds that doping brings forth a crisis of identity, wherein athletes would increasingly be driven by artifice instead of performing on the basis of their own agency. Juengst attempts to rebut this by arguing that there is no fundamental distinction between engaging in a particular diet and engaging in a performance enhancing intervention – both can be conscious decisions made by athletes, and both can be accepted as legitimate elements of their 'victory narrative'. Citing Ronald Cole-Turner, he adds that an enhancement would not do away with the need for struggle and effort, but rather *relocate* that struggle by heightening the baseline of what can be performed effortlessly. Moreover, if enhancement would serve to level out differences in natural endowments, it would increase the focus on effort. General human agency need perhaps not be undermined by doping, Juengst concludes, but perhaps a particular kind of *athletic* agency is. Juengst notes how, in the evaluation of athletic performance, a particular emphasis is put on both the presence of natural talent and on decisive hierarchical ranking. Juengst sees a fourth interpretation of "natural" at work here, which gives a special standing to natural talents, "i.e., inherited characteristics over which athletes have no control and which are ultimately traceable to particular combinations of ancestors and their genes". In his view, this seems to provide the fundamental legitimation of WADA's categorical doping prohibition: "Performance enhancement is wrong for athletes to pursue and sports medicine to provide, we can say, because it compromises the ability of athletes to segregate themselves into genetically defined classes and the ability of sport to elevate genetically advantaged athletes from their disadvantaged competitors." Juengst concludes that to the extent that principled concerns about how enhancement interventions transgress human nature would indeed rest on such hierarchical genetic essentialism, "sport may have more in common with these extreme human vices [of blood feuds, racism, and genocide] than we would like to think".

In Chap. 6, Prometheus on Dope. A Natural Aim for Improvement or a Hubristic Drive to Mastery?, **Trijsje Franssen** provides a revealing analysis of how enhancement enthusiasts can appeal to human nature as readily as enhancement critics. According to a pro-enhancement reading of the Prometheus myth, so often referenced in the enhancement debate, it is our fundamental human nature itself which calls on us to enhance and transform our bodily constitution. That is our Promethean nature: not to resign in our given predicament, but to "steal the fire of the gods" and to boldly explore and enhance our existence. Sports may provide an excellent showcase for this spirit. As Julian Savulescu and colleagues have argued: "Far from being against the spirit of sport, biological manipulation embodies the human spirit – the capacity to improve ourselves on the basis of reason and judgment." Some, such as activist author Simon Young, even proceed to root the call to become 'the New Prometheans' in the obedience we supposedly have to pay to an "innate Will to Evolve – an instinctive drive of a conscious entity to expand its abilities in pursuit of ever-increasing survivability and well-being". Such religious overtones are most commonly heard among categorical contra-enhancement authors, and they too commonly reference the Prometheus myth, although they tend to draw on Hesiod's instead of Aeschylus' account, stressing the Titan's hubristic usurpation of powers that he should better leave be. Although in this volume Michael J. McNamee builds on the myth of Odysseus, in earlier writing he has called upon sport ethicists to ensure that "sports do not become the vanguard of Hesiod's Promethean project". Franssen connects McNamee's Prometheus-inspired conceptions of hubris, humility and human nature to those of Michael Sandel and Leon Kass. She sets apart the many distinctions between all these commentators, and also notes how a nod is regularly given to the naturalistic fallacy. However, the frequent transgressions of the (supposedly) unbridgeable conceptual divide between factual descriptions of human nature and normative endorsements of that nature remain a sore point – not only for advocates of a principled ban on athletic enhancement, as is debated by Juengst, Lenk, De Block, Bonte and others in this volume, but also for certain *pro* enhancement lines of thought.

In Chap. 7, Outliers, Freaks, and Cheats: Constituting Normality in the Age of Enhancement, **Darian Meacham** points out that a major reason for sports' enormous and global popularity, is that "organised sport, and the Olympic movement in particular, function in part as a kind of testing ground for the physical limits of 'man'". Even though the world's finest athletes are extreme outliers when compared to the performance levels of average persons, every human being can nevertheless identify with those top athletes, and cheer them on, as members of their own species who remain '*normally* abnormal'. Such universal showcasing was one of the explicit motives of the founders of the modern Olympics. However, this project of universal identification with the most excellent specimens of the species cannot be maintained in an era when athletes begin to use increasingly effective enhancement technologies, allowing them to *go beyond* the outer boundaries of what can be performed by natural, normal humans. For Meacham, this "violation of the 'normal'" is key. Mindful of a legitimate suspicion of oppressive norms of normality, which also returns in the contributions of Raakilde Jespersen and Magdalinski, from a phenomenological perspective, Meacham argues for the inevitability, and fundamental necessity, of a conception of normality acknowledged by all who inhabit a life-world together. Ultimately, for Meacham the challenge of human enhancement is this: "whether or not these differently natured groups could share a 'world' in a phenomenological sense: communicate and engage in common projects", such as meaningful sporting competitions. The advent of athletic enhancement is a fork in the road of modern sports: the Olympics can become "something like a UNESCO heritage site" and continue approximating (and policing) the boundaries of the normal, natural human body; or it can adopt an ethos of open-ended pro- and transgression, and "break free from the Enlightenment idea of 'man'". This, a choice between approximation and transgression, as is also noted by Pieter Bonte, may be seen as a choice between two antagonistic projects. Meacham zooms in on two contemporary figures who appear to be 'abnormally abnormal': Oscar Pistorius, a runner without lower legs, and Castor Semenya, a woman with a partly masculine physique. Meacham finds himself "more able to empathise with Usain Bolt, who runs faster than Pistorius, because we share in a general style of the human body that is constitutive of a scope or horizon of normality." According to Meacham, the advent of substantially different 'general styles of the body' such as that of Pistorius could be viewed as "a beginning of a fragmenting of the 'species'" – and such fragmentations can begin to cause significant disruptions of the capacity of empathy even well before they reach the point of effectively creating fully distinctive biological populations. Castor Semenya, whose 'abnormal' bodily constitution was not brought about intentionally and technologically but is instead an outcome of natural processes, casts another revealing light on the norms of normality upheld in sports, *in casu* the normality of the male–female binary. Meacham notes that where Pistorius was put on billboards as a symbol of heroic overcoming, Semenya's abnormality on the other hand has mainly met with negative reactions. To Meacham, however, her case demonstrates the extent to which our horizon of normality is a social construction, and one that can be restructured in the face of new experiences. In his view, the confrontation with Semenya's embodiment should rather result in a reconfiguration of the conventional male–female binary than in a negative valuation of Semenya and others who do not fit that binary scheme of normality. Citing, as does Christian Lenk, anthropologist Ludgwig Siep's article on normative aspects of the human body, Meacham concludes that besides a healthy suspicion towards possibly oppressive norms of normality, we should also recognize the pivotal role played by a shared 'traditional shape' of the human body in the constitution and maintenance of a sense of empathy and community that is mankind-wide. Since the Olympic Games – and the modern Olympian spirit of sport in general – can be seen as a ceremonial glorification of this communal human nature, Meacham has located a profound, principled argument to resist athletic enhancement technologies in sports. This, however, invites a political and legal interrogation, which is readily provided in Roger Brownsword's contribution, whether such a universal conservationist project might consist in one contra-enhancement value-community denying the rights of other, pro-enhancement communities to engage in their own particular athletic culture.

1.3 Part III: The Normative Value of Human Nature

In Chap. 8, Doping Use As an Artistic Crime: On Natural Performances and Authentic Art, **Andreas De Block** argues that doping is not so much an offense against ethical values than it is a violation of *aesthetical* values. Of crucial importance in the appreciation of both an artistic and an athletic performance, is the presence of original and undiluted authorship. For De Block, doping is problematic in this regard because it shifts (or splits) the attribution of the performance from the athlete to the doping agent. The performance thus becomes a forgery, which the doping athlete fraudulently tries to pass as her own – a critique that both Eric Juengst and Pieter Bonte try to deflect in their contributions. De Block is clearly aware of the naturalistic fallacy which makes it impossible to cash out the intuition that doping is somehow intolerably 'unnatural' in terms of the goodness of human nature. Therefore, he proposes that anti-doping advocates cease to argue on ethical grounds but instead appeal to aesthetic arguments. For instance, what is aesthetically impressive about an athletic performance is that it was hard to do considering the physical constraints of which the athlete has had to make the most – a theme that is highlighted in Michael J. McNamee's contribution. In this light, using doping to enhance the physical restrictions of one's embodiment becomes a meaningless, misguided project. For De Block, the persistent demand that athletic performances be based on one's given human nature can also be made intelligible by referring to the psychology of disgust, which has an evolutionary rationale as a defense against the intake of toxic food. Later, the disgust response was coopted as a response to immorality – connecting the immoral to the 'impure' and 'alienated'. This is one among several reasons pointed out by De Block why the resistance to doping is often thought to be ethical but may in fact be aesthetical. In doing so, he provides an evolutionary psychological account of the common disgust for doping that shows interesting analogies with Darian Meacham's account, which is based on a radically different (phenomenological) theoretical framework. De Block concludes that although such reactions should be denied ethical validity, they can nevertheless remain legitimate as deeply held aesthetical preferences. As such, repugnance, *aesthetically* understood, can be a sufficient basis for a given community to demand that athletes who wish to partake in that community's sporting events, comply to a categorical doping ban.

In Chap. 9, Something from Nothing or Nothing from Something? Performance-Enhancing Drugs, Risk, and the Natures of Contest and of Humans, **M. Andrew Holowchak** looks at sports and doping from an ancient virtue ethics perspective, first treating the question of how doping may wrong the institution of sports, after which he investigates how doping may wrong the doping athlete herself. Mentioning his own use of doping agents in the past, Holowchak's conclusion is nonetheless double negative: doping is inconsistent both with the Aristotelian conception of human nature and with the nature of sports. According to Holowchak, central to the nature of sports is the idea of investing effort and skill in order to accomplish improved performance. Certain ergogenic aids, however, ranging from drag-minimizing swimming suits to performance enhancing drugs, deliver improved performance *without* added effort or skillful use: they offer a 'something for nothing'. Other such aids, however, may *increase* the need for effort and skill. One might argue that the aids that deliver a something for nothing harmlessly are good nor bad and could be permitted, and those that harmlessly increase effort and skill should even be encouraged. Yet for Holowchak this would leave a deeper concern unresolved. According to the hallmark definition provided by Bernard Suits, a game is a "voluntary attempt to overcome unnecessary obstacles." This implies that in playing a game, one must *not* use the most efficient means to achieve the game's goal. A competitive advantage must be achieved through improved skill or heightened expenditure of effort. Holowchak gives the example of novel golf clubs that would correct for a wide variety of swing errors. This would be of great benefit to golfers with a poor swinging technique yet provide little benefit to golfers who have mastered, through skill and effort, the art of golfing. According to Holowchak, all performance enhancing aids that would have such a diminishing effect on the role of effort or skill, can be legitimately banned. Performance enhancing drugs, however, need not have this handicapping effect on effort and skill. On the contrary, citing his own past use of steroids, Holowchak elucidates the way in which steroids operate on the body and on the internal practice of sports: by making training more rewarding and by reducing the need for recuperation between training sessions, it can actually jolt one's eagerness to train more and train harder. Moreover, if all competitors would use such a drug, it would probably provide equal advantage to all, not handicapping elite athletes. In this sense, therefore, performance enhancing drugs need not run counter to the nature of sports. When investigating the issue of human nature, Holowchak does find an intrinsic argument against doping. First he notes how one cannot shirk the issue that many doping agents may be very dangerous, even if that is 'only' a probabilistic and not an intrinsic feature of doping in general. He also criticizes liberal arguments that one should not paternalistically deny individuals the right to choose their own manner of living, even if they would choose high-risk lifestyles such as using doping – an argument that is also prominent in Roger Brownsword's contribution. He then goes on to present his own argument for the wrongness of doping. To Holowchak, athletic accomplishments are praiseworthy only when they promote virtue and social cohesion. Winning in itself does not count, it is the virtuous commitment to playing to win within the confines of the constituent rules that is key. If doping would be engaged in because of an obsession to gain an edge, the virtue-ethical foundation of sport would be lost, and the advantage gained would be morally vacuous. Also, if doping would be used to make one's superior athleticism stand out to the detriment of the communal dimension of sports, for Holowchak this too would be a 'nothing for something'.

In Chap. 10, Transhuman Athletes and Pathological Perfectionism: Recognising Limits in Sports and Human Nature, **Michael J. McNamee** also starts from the theoretical background of Aristotelian virtue ethics to conceive of sports as a perfectionist practice – not just in the superficial sense that a sports discipline often centers on the undistorted execution of certain movements, dynamic strategies etc., but also in a more fundamental sense: as a virtue-driven endeavor to realize, perfect and honor one's human nature. To McNamee, our given human nature, properly understood, can continue to serve as a moral compass even if the naturalistic fallacy is to be acknowledged. McNamee contrasts his virtue-ethical view of a life well lived with the dominant modern notion that subjective wellbeing ought to be the standard for judging accomplishment. In McNamee's view, however, "the idea that in seeking a life of perfection, what is being perfected (what excellences are being sought) is to be understood against an understanding of what it is to be human, or what nature is to be found in human being." Against this backdrop, he identifies as thoroughly misguided transhumanist notions of athletic and human perfection according to which the limits of human nature are imperfections that should be superseded. First, McNamee argues that transhumanism has a lopsided focus on 'technical' athletic perfection, fetishizing the error-free, perfect execution, such as the perfect swing that produces a hole-in-one. The transhumanist vision of perfection is also based on an overly restricted negative definition thereof: perfection as the mere lack of blemishing imperfections. However, such a perfect performance might still be very flawed, for instance if it was brought about by doping. Echoing Andreas De Block, McNamee identifies an essential aesthetic dimension to sports that consists in refraining from using the most efficient means to realize a goal: it is precisely performance *within constraints* that makes it meaningful and truly valuable. Second, the transhumanist vision of human perfection as the absence of all imperfections is said to lack a *telos* – a positively defined aim for perfectionist activity, given in an entity's nature. Here, Pieter Bonte's contribution could be read as a challenge to both McNamee's and transhumanist perfectionism: on an existentialist theory, there is no guiding human essence that preceeds human existence. Third, McNamee notes that transhumanism also seems to hold a misguided 'absolute' conception of human perfection, as the realization of all of one's potential *together*. An impossibility, as one cannot be all things at once, and because one's potentials do not have a fixed value that is independent of the specifics of one's developmental trajectory. To develop his positive, human nature-based account of perfection, McNamee builds on the myth of Odysseus, who declined a 'perfect' life of immortality and endless joy and instead willfully returned to his finite and fragile human life. For McNamee, taking up another form of embodiment, no matter how 'perfected' that might be, should be resisted on account of it being too alien and inauthentic for human beings.

1.4 Part IV: Socio-Cultural and Empirical Approaches

In Chap. 11, "Definitely Not for Women": An Online Community's Reflections on Woman's Use of Performance Enhancing Drugs in Recreational Sports, **Marianne Raakilde Jespersen** draws attention to the use of performance enhancing drugs by women. She takes a look inside the doping community itself, analyzing in detail how such female doping practices are received on the online bodybuilder community forum *Bodyhouse.dk* – a forum where people in doping subcultures discuss their doping behavior with little inhibition. On an anthropological rendering of human nature, the use of drugs of some form or other can be seen as an integral part of universal human behavior. Raakilde Jespersen cites Donald Brown's list of human universals, which includes "mood- and consciousness-altering techniques and/or substances". Wedded to an "imperative to perform" which Raakilde Jespersen, following John Hoberman, finds pervasive in twenty-first century western culture, a culture of performance enhancing drugs has taken firm root in large networks of both elite and amateur athletes. Interestingly, within these contemporary athletic drug cultures, the male use of them is regularly legitimized as a naturally manly act. The female use, however, will regularly be seen as an unnatural desecration of feminine nature. A machismo philosophy runs through the *Bodyhouse* forum, contrasting male will power and rational self-control – the required traits to dope like one should – to female whimpering and irrationality. Female use of performance enhancing drugs is regularly scorned for the possible masculinization it brings about, and for the risks it poses to women's reproductive function. These findings of Raakilde Jespersen's study of the *Bodyhouse* rhetoric thus seem to confirm Tara Magdalinski's analysis in *Sport, Technology and the Body:* "the use of technologies to take women beyond feminity is rejected." In an interesting addition to Darian Meacham's conclusion that human enhancement technologies negate vital conceptions of normal, natural embodiment, *within* certain doping practices such as those in bodybuilding milieus, conceptions of a normal, natural division between male and female body types can paradoxically be affirmed with a vengeance.

In Chap. 12, Toward a Situated and Dynamic Understanding of Doping Behaviors, **Denis Hauw** conceives of human nature as being profoundly social and dynamic. He leaves the analysis of the normative worth of our biological constitution and of athletic enhancement to others. Instead, in turning from 'human nature' to 'the nature of human activity', Hauw takes the wrongness of doping as a given and proceeds with an analysis of how circumstances can conspire to create the scene of an athlete taking dope. Similar to Bengt Kayser and Barbara Broers, Hauw concludes that current anti-doping programs are too narrowly focused on a principled denunciation of the 'doping sinner', to the neglect of practical policies that might remedy the failure to effectively bring down the number of doping athletes. Where Kayser and Broers present a pragmatic containment strategy, Hauw presents a pragmatic avoidance strategy, making use of six strands of psychosocial studies combined with specific empirical studies on doping behavior. For instance, several studies have analyzed how doping cultures can take root and proliferate in certain athletic environments, sometimes to the point of becoming subcultures complete with a (deviant) sense of pride and belonging – for example, in Marianne Raakilde Jespersen's contribution we can catch a glimpse of the inner life at the *Bodyhouse* online community. Other studies discussed by Hauw shed a light on the individual trajectories of doping use, from first experimentation to entrenched habit. Such methods help to identify what factors may trigger doping use, which in turns yields cues for effective prevention or rehabilitation. The myriad contextual issues identified by Hauw dispel the naive belief that doping use is essentially a rational choice made by an immoral athlete, a choice to 'sin' for which the athlete carries all responsibility as an isolated individual. Indeed, Hauw cites studies on doping athletes' own perception of doping, one of which revealed, for example, that the main thing on the interviewed doping cyclists' minds was not so much some excessive drive towards superhuman performance, but rather a concern to better cope with the daily stresses of elite athlete training, such as exhaustion.

In Chap. 13, Restoring or Enhancing Athletic Bodies: Oscar Pistorius and the Threat to Pure Performance, **Tara Magdalinski** recalls how modern sports was in part founded as a romantic project for the liberation of natural man from the alienating industrial world which suppresses his spontaneous self-expression. This project was conjoined with the cult of the natural world as a wild and pure realm unblemished by industrial civilisation. Engaging in sports was thus seen as an invigorating antidote to the ills of technological culture. This romanticism often resisted all manner of scientific and technological intervention in the natural sporting activity, such as advances in clothing, equipment or training facilities. Much of this resistance has crumbled, as contemporary sports are positively saturated with practices of scientific surveillance and technological assistance. It seems, however, that despite its many battles lost, the romantic insistence that sports honour 'the natural man' has focused its attention to defending the natural human body against the encroachment of invasive, human nature-altering biotechnologies. As such, sports can still be believed to display pure participants striving to establish the limits of the natural human's physical capacity – a project which other contributors connect – complementary or contradictory? – to the Enlightenment (Darian Meacham), Vitruvian perfectionism (Pieter Bonte) and a Vatican spirit of sport (John Hoberman). Magdalinski interrogates this contemporary insistence on natural sport via the controversial presence of Oscar Pistorius, a runner who competes in the Olympics despite having prostheses instead of natural lower legs. When Pistorius and analogous persons start demanding access to the 'able-bodied' leagues, they appeal to the inclusive universalism that is a central part of the Olympic spirit. At the same time, however, they deeply disrupt another central point of those competitions: the collective search for the natural contours of mankind. The conflict is confounded by the fact that the R&D of cutting-edge prostheses has advanced beyond the mere mimicry of the 'original model' of human physiology. If alternative (bio-)mechanisms yield better enabling results, such pathways will be pursued, even if they deviate from the ontology of *Homo sapiens* nature. As such, they reveal how impure and open human nature can be, how deeply natural artifice can come to be, and how many legitimate options for deviance from the natural norm this can open up. Here, Magdalinski sees suspect norms of 'natural normalcy' surface in an unwillingness to see deviant bodily constitutions as anything other than 'disabled' when compared to normal human nature – norms that are paternalistic, oppressive, and not only ethically but also factually mistaken. Magdalinski concludes with applying these insights to the struggles Oscar Pistorius has had to go through to be granted access to the 'able-bodied' competitions. For instance, IAAF's Elio Locatelli has protested that Pistorius' participation would disrupt the 'purity' of those competitions. Magdalinski strongly rejects such arguments: "efforts to label Pistorius' running blades as 'unfair' or 'advantaged' and to scientifically render his performance not 'proper' running, overlook the fact that regardless how efficient his prosthetics may be, Pistorius is still missing his two lower legs, and his technological aids can perhaps not be accurately classed as an 'unfair advantage' until athletes are willing to forgo their legs in favour of these running blades." Arguments presented by Meacham, Brownsword and Bonte may, however, share with Magdalinski the full recognition of Pistorius' athleticism but equally recognize the requirement that a competition, in order to remain meaningful, requires adequately similar bodily constitutions of all competitors.

1.5 Part V: Practices and Policies

In Chap. 14, Sports Physicians, Human Nature, and the Limits of Medical Enhancement, **John Hoberman** hones in on the role sport physicians play in enhancing athletes' performance levels. A traditional conception of medicine holds that the aim of the physician should be "the restoration of a physiologically deficient organism to its former, 'normal' state." The Vatican anthropology provides a good example of what this traditional conception amounts to. Here, the pivotal value of the restoration and conservation of one's given nature is grounded in the theological doctrine that considers each individual human being to be a unique creation of a Creator: this given form is inviolable, its boundaries are divinely sanctioned. On these grounds, Pope Pius XII denounced doping as early as 1955 as "the error of claiming the right to dispose unconditionally of his body". In a later critique, he chided the manipulation of human biology in general as a way of treating humans no different than animals – what Eric Juengst discusses as the concern for dehumanization as 'stooping to the subhuman'. At the same time, in the twentieth century the Vatican also lauded sport as a potentially pious practice of self-control, self-abnegation and purposeful suffering. Hoberman, however, is sceptical of this analogical *rapprochement* considering the transgressive dictum of the modern Olympic Games of *Citius, Altius, Fortius* (which, as Tara Magdalinski, Darian Meacham and Pieter Bonte note, may also be given an *approximating* reading). To Hoberman, the more convincing reading is that of a "perfect opposition" between the *Athleta Christi* and the modern Olympian, as they present "the sharpest possible contrast between competing conceptual models of 'human nature'." The anthropological model to contrast with that of the Vatican is that of the *Leistungsprinzip* or 'performance principle', according to which the human organism must adapt itself to requirements of efficiency and productivity. Projects of human enhancement would then be perfectly in line with this principle. Where Bengt Kayser and Barbara Broers provide a historical account of how the anti-doping culture and policy was created, Hoberman provides a historical account of how pro-doping subcultures were formed in medical milieus. Medico-scientific thinking on the limits to human athletic performance – and the possible transgression of them – is traced to the final decades of the nineteenth century, and Hoberman provides a remarkable overview of the changing perceptions of these issues throughout the twentieth century. Starting in the late

1970s, an increasing number of scientists argue that the Olympic project of *Faster, Higher, Stronger* was drawing to a close – that is, only medicine and biotechnology could procure (quite likely perilous) ways to breach the upper performance limits of the human body. According to Hoberman, medical professionals began such searches in earnest during the 1960s. The phenomenon of 'celebrity sports physicians' began to occur. These would at times be charismatic figures with ambiguous aims: on the one hand they were seeking ways to protect the athletes against increasingly high-risk levels of stress, while at the same time applying enhancement interventions to athletes to aid them in realizing epic performances. What is more, Hoberman shows how some physicians even went so far as to actively advocate the enhancement of human performance levels as a valuable and legitimate aim of their profession. Hoberman concludes that elite sport physicians risk picking up a troublesome sense of entitlement, which has lead and will continue to lead to attempts to transgress the traditional boundaries of both human nature and of the medical profession.

While Roger Brownsword challenges the prohibitionist approach by accepting the existence of a doping community as a legitimate *political* fact, in Chap. 15, Anti-doping Policies: Choosing Between Imperfections, **Bengt Kayser and Barbara Broers** challenge this approach by accepting the existence of a doping community as a *sociological* fact. Where for Brownsword a paternalistically imposed categorical prohibition is objectionable, for Kayser and Broers it is *impracticable*. Kayser and Broers conceptualize 'human nature' as the persistent, universal traits of humans. As Raakilde Jespersen also argues, drug use may be one of those universals of human nature. Thus, Kayser and Broers consider it utopian and ill-conceived to aim for a full prohibition and eradication of drug use. Their contribution is based on a comparison of WADA's 'war against doping' with the attempted and failed 'war against (psychotropic) drugs'. According to Kayser and Broers, the two campaigns have much in common as regards their intent, their organization, and their effects: policies based on repression and surveillance, large claims on public budgets, black markets controlled by organized crime, attempts to shape internationally harmonized legal frameworks, ideology and political convenience anchored in media-fuelled moral outrages, the stigmatization and marginalization of drug users, continued infractions on a large scale, and unsupervised, high-risk practices of drug intake. Kayser and Broers criticize various aspects of WADA's current anti-doping policy: the troublesome official definition of doping, its overly broad scope, the excessive severity of the 'whereabouts' and urine control surveillance regime, the dubious methodology of the 'biological passport' system, disproportionate punitive measures, the inability to realize significant behavioral change among athletes and their entourages – the list goes on. Their final verdict is damning. The proliferation of WADA-style zero-tolerance policy is said to have "worrying characteristics of a dystopia of Orwellian kind." As an alternative, Kayser and Broers propose a utilitarian, permissive *harm reduction* policy. Marshaling empirical studies that show the effectiveness of such policies in other domains, such an approach may in their view also result in a better control and containment of problematic doping practices in sports. This proposal differs significantly from the (cluster of) approach(es) proposed by Denis Hauw, who does not challenge the categorical anti-doping rule. It also differs significantly from a libertarian *laissez faire* approach. On the contrary, general testing would still take place, although based on a stricter health-related agenda. Kayser and Broers readily admit that their harm reduction approach would be messy and imperfect. Yet they emphasize the need for 'realism', which means understanding drug use as an irremediable part of human nature and doping use as a irremediable part of perfectionist sports.

In Chap. 16, A Simple Regulatory Principle for Performance-Enhancing Technologies: Too Good to Be True?, **Roger Brownsword** examines whether the following basic regulatory principle can be upheld: it is permissible for competent agents (such as Olympic competitors) to use enhancers unless either (i) use causes harm to others (who have not consented to this risk) or (ii) the user has freely agreed to act on a no-enhancement basis. Applying a paternalistic policy in sports to safeguard athletes' best interests (most noticeably their health), as WADA and its affiliates arguably do today, would fly in the face of the prevailing ethical doctrine applied in medical settings today, where paternalism has largely been abandoned in favor of a doctrine based on free and informed consent. This tension is also noted by John Hoberman in his contribution, although Hoberman draws the largely opposite conclusion that it is the modern 'consumerist' view of medicine, penetrating the sports world and fueling a permissive stance towards doping, that needs revising, and not the categorical prohibition of doping. Brownsword considers how an application of the principle of informed consent would indeed "revolutionise the rulebook for sports" by permitting enhancement to those who properly consent to it. However, at the same time, this principle would also allow for athletic associations in which the categorical enhancement prohibition remains in place. Brownsword proceeds to stress-test these hitherto highly abstracted notions of harm and consent. Echoing the relational issues mapped by Jan Tolleneer and Paul Schotsmans, competing interpretations of harm to competitors, to spectators and other parties – for example utilitarian versus communitarian interpretations – will yield very different conclusions about when an unacceptable risk of significant harms to relevant others is reached. In unpacking the concept of free and informed consent, Brownsword also highlights how much devil is in the detail. For instance, how is one to ensure that there has been no undue inducement to enhance – by one's sponsors, one's coach, one's physician, one's fan base, one's competitors? On the opposite side of the question, the possibility of undue inducement to abstain from enhancement, Brownsword intuits that this seems to raise less concern. However, both Pieter Bonte and Eric Juengst argue that upon closer inspection, this may pose a truly foundational problem: less talented athletes may be induced (and indeed, given the current regime of near-universal prohibition: forced) to forego enhancement in order to conserve the hegemony of the genetically superior 'talentocracy'. In sum, Brownsword is open to the possibility that the minefields of harm and consent may provide strong objections to athletic enhancement. But an altogether different objection concerns the categorical prohibitionism. Often rooting their argument in the hotly debated notion of human dignity, prohibitionists think it necessary to impose their prohibition universally. To Brownsword, such a "imperialist version of communitarianism" is unjustified, and he goes on to present a proposal for the regulation of enhancement via ongoing public deliberation within a community of rights. Within such a community of rights, it remains possible that substantially communitarian policies, such as a categorical doping ban, would be decided upon – with the caveat that such communitarian prescripts should not impinge upon basic generic rights of individual community members. Following Darian Meacham's argument, however, such a liberal position might indeed become too good to be true should enhancement practices begin to fracture the sense of membership in a single 'species' community. Such alienation may undercut the willingness to grant the all-too-alienated others equal moral standing and equal rights.

Part I
Conceptual and Theoretical Framework

Chapter 2
Self, Other, Play, Display and Humanity: Development of a Five-Level Model for the Analysis of Ethical Arguments in the Athletic Enhancement Debate

Jan Tolleneer and Paul Schotsmans

2.1 Introduction

The plethora of reactions to the release on 10 October 2012 by the United States Anti-Doping Agency of abundant evidence of doping use by the seven-time Tour de France champion Lance Armstrong (Mackey 2012) confirmed what many already knew: commenting on the moral dimension of doping use and the doping ban is a complex challenge, which can shoot off in numerous directions. And yet, or perhaps precisely because of this, it is a subject that is on everyone's lips. Few people miss an opportunity to voice their own personal point of view in the wake of such a media storm. Whether you are passionate about competitive sport or not, whether you are well informed about the benefit and risks of performance enhancement or not, almost everyone is seduced into joining this great game of social debate. The play essence within sport seems to be extending to the sporting ethics debate: sport as 'die wichtigste Nebensache der Welt' (the most important of the trivial subjects in the world) is broadening into the sporting debate as 'the most important of the trivial debates in the world'. This debate is being played out in the margins, but paradoxically it provides an opportunity for the articulation of concerns and values, such as respect, integrity and justice, values that are at the heart of ethical consciousness and social sensibilities. The aim of this contribution is not to look for an anthropological explanation for this remarkable mechanism, nor to produce a meticulous sociogram of individuals and institutions participating in the debate, however important these endeavours may be. Instead, this chapter focuses on the content of the discussion

J. Tolleneer (✉)
Faculty of Kinesiology and Rehabilitation Sciences, KU Leuven University,
Tervuursevest 101 – bus 1500, 3001 Leuven, Belgium
e-mail: Jan.Tolleneer@faber.kuleuven.be

P. Schotsmans
Centre for Biomedical Ethics and Law, KU Leuven University,
Leuven, Belgium

J. Tolleneer et al. (eds.), *Athletic Enhancement, Human Nature and Ethics*, International Library of Ethics, Law, and the New Medicine 52, DOI 10.1007/978-94-007-5101-9_2,

itself and the nature of the arguments used in it. On what ethical grounds do some people oppose the use of performance-enhancing drugs? On what ethical grounds do others think that opinions and regulations should be revised?

The main purpose of this contribution is to outline the most important ethical arguments deployed by what could be called the conservative and the liberal camp. More specifically, we hazard a go at developing a logical and all-embracing model that offers a place for the many arguments for and against. We will expound upon the development and the meaning of the model, and test out the practicability of the various levels. Particular attention will be paid to the most fundamental level, which we refer to using the concept *humanity* or *human nature*. By way of conclusion, the ethical boundaries that are guarded or moved at that level will lead us to a short historical exercise, whereby the debate on the acceptability of performance enhancement is placed in a broader time-frame. However, first let us clarify what we mean (do not mean) by performance enhancement on the one hand and sport on the other.

Performance enhancement is not a recent phenomenon in sport; on the contrary, it is an essential and constituent part of it. The will to be better than fellow contestants here and now (competition) or than competitors anywhere and at any time (record) explains the training efforts that athletes put themselves through. Training as a means of improving performance is morally highly valued (we will come back to this). Winning on the basis of planned preparation and self-developed body capital became the essence which made sport a symbol of an achievement- and advancement-oriented society. Modernisation and scientification have opened the door (irrevocably?) to new technologies. The clap skate that was introduced in 1997 is a much-cited example of an aid that increases efficiency. Much like tennis rackets and racing bikes, clap skates are forms of what Butryn calls 'implement technologies', instruments that athletes use to realise their sporting goals. These innovations result in a sort of performance enhancement that usually ends up by being accepted and appreciated, sometimes after an initial phase of controversy. Obviously this contribution does not focus on improvement resulting from these kinds of aids but on the improvement of the athlete himself. Butryn (2002: 112) speaks here about self 'technologies': "Self technologies consist of innovations which alter the physical and/or psychological make-up of athletes who employ them, and which therefore ultimately confound traditional definitions of human performance". The improvements made to the athlete's body are called 'biotechnology'. The kinds of biotechnology distinguished here are: the use of products and methods referred to as doping, the insertion of prostheses, and surgical operations. The modification of genetic material is also a form of technology that requires special attention.

Today, in contrast to the phenomena of training and external technology, the use of these 'self technologies' to improve performances is heavily debated in society or at least in the social phenomenon called sport. What is so special about this phenomenon? Historian Allen Guttmann (2004) defines sport as an 'autotelic physical contest', where 'contest' refers to competition and falls entirely within the concept of 'play', a semantic field characterised by intrinsic motivations and also principally by the absence of social pressure and consequences. More specifically it falls within the kind of 'play' for which there is a separate word 'game' – at least in English – alluding to the 'regulated' and 'rule-bound' nature of the activity. It is, incidentally, no surprise that both historians and philosophers and ethicists like to build on Johan Huizinga's work *Homo ludens*. This inter-war Dutch cultural historian revealed the quintessence of sport: the participation in a game that exists outside of reality and in which unnecessary obstacles need to be overcome according to well-defined rules. Suits (1988: 11) defines it as an attempt to achieve a specific state of affairs (pre-lusory goal), using only permitted means (lusory means), where the rules prohibit the use of more efficient means (constitutive rules) and where such rules are accepted precisely because they make possible such activity (lusory attitude). The adherence to 'constitutive rules' and the acceptance of a 'lusory attitude' is often referred to in the literature using the double concept of formal fair play and informal fair play. Both the enforcement of and the adherence to equal starting opportunities is not accidental but rather something upon which the foundations of sport are built. If sport is defined as the ranking of 'inequality' of performances, this can only happen if the 'equality' of participants at the outset is guaranteed. The sporting competition is therefore an instrument for measuring well-prescribed skills and ranking athletes according to their lusory qualities, an instrument whose validity is determined by the equality of non-lusory elements (De Wachter 2000). It is important to bear in mind these properties of sport when we study the use and abuse of biotechnology within this unique sphere, and when we develop a model for articulating the ethical arguments and placing them in a logical context.

2.2 Creation of an Ethical Research Model

When ethics is essentially about right and fair behaviour and its counterpart wrong and unfair behaviour, then in the context of this contribution we equate this, somewhat pragmatically, to showing or not showing respect, respectively. When we examine the ethical dimension of attitudes, dispositions and actions, we have to specify which moral agents are involved. In establishing our model we have deliberately chosen the standpoint of the athlete. We are interested in the athletes' deeds and words and the ethical dimension thereof as discussed by commentators. Respect that is shown or not shown by the individual athlete is the focus of the exercise we are conducting here. But what sort of respect are we talking about, in other words what or who must be respected? Who or what is under discussion, what is at stake when this athlete decides (not) to use doping? This question lies at the basis of the five levels of our model, which was inspired by the relation-axes approach (Fig. 2.1) as used in comparative pedagogy (Wielemans and Chan 1992: 20) and comparative physical education (Tolleneer 2011).

First, the athlete maintains a relationship with himself and has a responsibility for himself. Here, the ethical arguments for and against the permissibility of doping refer therefore to *the self*. Secondly, taking part in doping or complying with the doping ban must be tested against the respect shown to *the other*. The 'other' refers here not

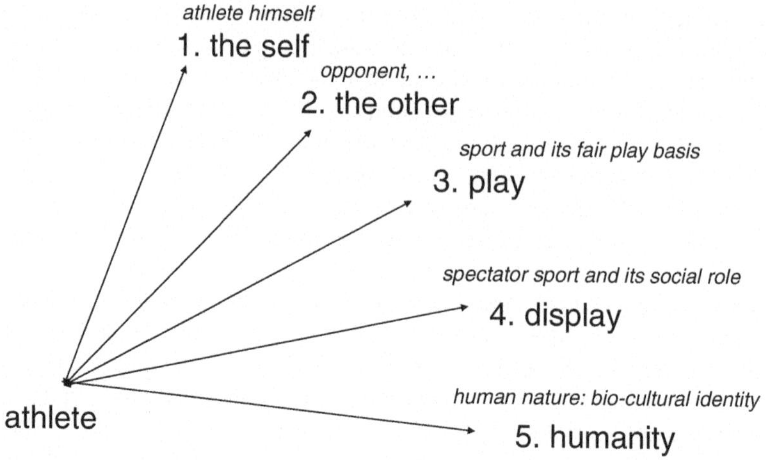

Fig. 2.1 Relation axes as a basis for the five-level model for the analysis of ethical arguments in the athletic enhancement debate

only to sporting opponents, but all those involved in a sporting activity such as team mates and coaches are under scrutiny. The third relationship that we are studying is the relationship between the athlete and the *sport*, the cultural phenomenon resulting from the historical process and characterised by the above-mentioned essence of play and by the accompanying formal and informal fair play. This game can be played – and its moral dimension evaluated – in the simple constellation of pure play, during a training competition or in the intimacy of a village playing field without spectators. However, in high-performance sport as we know it, there is another dimension: the fourth relationship concerns play that becomes 'display', the sport that is raised to the square to be become *spectator sport.* A social dimension is added here, which in our opinion is very significant as such and also because it forms a bridge to the most fundamental level, namely the relationship with *human nature* itself.

The various ethical arguments that make the current-day doping debate so colourful can be linked to the above-described relation-axes. This is precisely what we did in an earlier study when we tried to categorise the opinions of commentators (Vangrunderbeek 2007; Vangrunderbeek and Tolleneer 2011). It concerned the relation between the athlete on the one hand and himself, the other, the sport and the spectator sport on the other (the human nature category was included in the model only later). The study involved, more specifically, 555 eighteen-year-old physical education and kinesiology students who were writing a paper about doping, in which they also had to proffer their own personal opinions. Our discourse analysis showed that, between 2003 and 2005, the number of students arguing for zero tolerance had fallen from 85 to 53%, the number opting for tolerance had risen from 5 to 26%, and the number who had no outspoken opinion rose from 10 to 21%. More significant in this context, however, are the precise arguments they used in support of their position: in that period 32% of the arguments of the zero tolerance supporters were concerned with the moral responsibility of the athlete to *himself* (in particular, in connection with health) and 50% of the arguments

Athletic enhancement? NO because it e.g.:	What is at stake?	Athletic enhancement? YES because it e.g.:
threatens my health	1. The self	helps in fulfilling my aspirations
reduces the opponent's chances	2. The other	guarantees equal chances
undermines the spirit of sport	3. Play *Sport*	aligns sport with other cultural phenomena
creates negative role models	4. Display *Spectator sport*	reinforces sport's heroic character
desecrates human nature	5. Humanity *Human nature*	fulfils the mission to push back frontiers

Fig. 2.2 Five-level model for the analysis of ethical arguments in the athletic enhancement debate

were concerned with the moral responsibility to *the sport* (in particular, in connection with the fair play element). The two other categories made up the remaining 18%: *the other* (in particular, taking away the opportunities of the opponent) and *the spectator sport* (in particular, creating negative role models).

The model that emerged in the context of the above-mentioned study seemed to be a useful means of categorising ethical arguments. As demonstrated in Fig. 2.2, the different levels create a framework for the logical positioning not just of the arguments of those that argue for zero tolerance, but also of those who advocate a relaxation.

We will now elaborate on the right-hand side of the model using some short examples. One can approve of performance enhancement because on the *first level* it is regarded as fulfilling one's own sporting aspirations and therefore improving well-being. Arguing for tolerance can be done, on the *second level*, by emphasising the creation of equal opportunities for all. The *third level* accommodates arguments of those who claim that sport shouldn't be any different from e.g. music and studying, where no control is organized. Those who think that enhanced athletes should not be punished, can substantiate this on the *fourth level* by stressing their role as celebrity or figurehead: their heroic deeds, displayed in sporting spectacles and the media, reinforce the image of sport as a general source of inspiration. Most people are far more familiar with the arguments on the left-hand side of the model: the conviction that doping is not ethically responsible because the athlete, firstly, damages his health, secondly disadvantages the other, thirdly undermines the aim of the sport and fourthly presents a negative role model. The important fifth level will be dealt with in Sect. 2.4. Anyway, the familiarity with the arguments demonstrates on the left-hand side once again how much social backing there is for the conservative opinions and regulations of the International Olympic Committee (IOC) and the World Anti-Doping Agency

(WADA). The model we present here provides genuinely coherent arguments and counter-arguments on every level. In this sense, this tool can hopefully help provide an understanding of the ethical arguments of both camps in their coherence and question the grounds on which they are based. Starting with the first four levels of the self, the other, play and display, let us more accurately pinpoint and illustrate what the left-hand column of the conservative camp and the right-hand column of the liberal camp are composed of.

2.3 What Is at Stake?

Respect for the physical and psychological integrity of the person of the athlete is on top of the list of moral elements. It will come as no surprise that health heads up the list as an important argument in the study mentioned. One of the students stated: "…in my opinion these athletes could die any minute. I don't understand why they go so far and endanger their own health and life. What are the limits?" Vorstenbosch (1998) is one of the numerous philosophers who use the argument of respecting health as a justification for the doping ban. Although he places it in what he calls an 'external argumentation context', he sees *non-harm* as an important element. But it is precisely this respect for the person of the athlete that is invoked by those who question that zero tolerance. "I am in favour of controlled doping use. In my opinion, it would solve a number of problems created by the current anti-doping policy" says another student (see Vangrunderbeek 2007). Legalising doping would facilitate supervision and it would limit the risks. Philosophers have been quick to say that elite sports is *in itself* problematic for the health of the athlete. Those who see no problems there, but who do have a problem with doping use, seem to be inconsistent, according to Brown (1997) and Lavin (1987). These same authors write that the health argument is also frequently invoked too simplistically: after all, not all forms of doping are (equally) harmful. Finally, on the first, *self* level of doping tolerance the autonomy argument comes primarily to the fore. Those who respect the athlete grant him the right to take decisions for himself, whether or not that decision is for doping, whether or not the decision is to risk his health. The paternalist-liberalist discussion unfolding in this context is a crucial element in understanding the doping debate and also in testing the research model, in the context of this chapter, on its practicability. We will therefore come back to this issue in the discussion section of this chapter.

Ethical behaviour also involves allowing others to reach their potential or at least ensuring that they are not illegitimately disadvantaged. On the second level of our model, that *other* is under discussion. An athlete who takes drugs might lessen the other's chances of sporting and financial success and thus causes – or threatens to cause – him substantial and avoidable harm, which is unethical. An extension of this argument is *coercion*. The students also claim or underscore that if you choose doping "…you also oblige others to use doping. And then you indirectly put the health of others in danger". Thus, the 'doping offenders' are depriving the others of that very freedom-to-decide which they claim for themselves. If the others want to hold on to the same opportunities to win, they are challenged to take drugs as well (Houlihan 1999), creating a downward spiral. Yet, nobody is forced to do these things, according to the commentators on the right-hand column of the model. Those not wishing to play this game, can stand on the sidelines. And if doping is immoral for these reasons, then according to Parry (1999) legal methods such as altitude training should also be forbidden, and even training as such: those who train hard curtail the freedom of others and oblige them to train hard or even harder. A number of commentators, including students, argue for partial legalisation of doping and for the relaxation of doping rules because then the rights of both the *the self* and the *other* are protected: "Maybe WADA should allow doping up to a certain point, because if everyone can use the same substance, then nobody has an advantage". But is it not precisely this change in the game and its rules that the authorities are obstinately trying to prevent? This is where sport itself, on the third level, comes under debate as a social phenomenon.

Athletes who violate the rules of the game are cheating and therefore betraying the sport as such. There are many, indeed, who include doping rules under the rules of the game, and associate their violation with unethical behaviour. "…it's quite simple, there are rules and they are there to be respected. It doesn't matter whether you agree with those rules or not. As an athlete the only thing that counts is not breaking the rules", writes one of the students. Anyone using anything other than *lusory means* is breaching the fair play element of the sport and is effectively placing himself outside the game. Frans De Wachter (2000) maintains that doping introduces a crucially non-lusory element to sport which sweeps away the equality of opportunity that is so fundamental. In that respect, the decisive justification of the doping ban is the guarantee of equal opportunities. Sport itself is damaged as a cultural phenomenon and as a social practice. And anyone bringing the integrity of the game into disrepute is socially and morally in the wrong. Many, therefore, will ascribe an important role to the third level of our model. Doping harms the game and, according to Vorstenbosch (1998), creates another social practice: the athlete turns his back on the game and creates a will to get results and win, without any play moment. But opponents raise the point that other methods, which equally undermine fair play, are not penalised. Moreover, the unequal access to medical and scientific facilities such as hypoxia chambers and what was described above as *implement* technologies, and even the unequal availability of genetic material, can call for a critical reappraisal of the revered equal opportunities principle (Mehlman 2009). So why should doping be singled out for a ban? Authors such as Morgan et al. (2001) confirm that, under the current regulations, doping users must be regarded as offenders, but at the same time they pose the deeper question about the disproportionality of the punishment and the justification and reason for existence itself of the regulations in question. And why do these regulations and the accompanying punitive measures exist in sport and not in other areas of society? Why are archers checked for beta-blockers while violinists are not, even when they are taking them in a high-level competition? And why should a fitness addict working on his health and appearance in the privacy of a gym run the risk of being tested, while a student using stimulants does not? Might there be other reasons why sport is targeted, reasons on another level? That brings us to the fourth level of the model. When the fitness devotee steps outside the privacy of his gym and climbs onto a podium, for example, to take part in a public bodybuilding show or competition, there is an added social and moral dimension, all the more crucial if there are television cameras trained on him.

Our discourse analysis seems to show that many of the statements made by the 18-year-old students were written down intuitively and impulsively, and yet, or perhaps therefore, they are instructive: "How can a child of 12 look up to his favourite basketball player if he knows he has taken banned substances to enhance his performance? When he turns 16, this boy will imitate his idols, and start injecting and swallowing banned substances". The importance of the athlete as a role model was famously and expressly highlighted by the judge who sent the sprinter Marion Jones to jail in 2008 and degraded her from a charismatic sprinter to a vulgar doping offender: "Athletes in society have an elevated status.... They entertain. They inspire. And perhaps most importantly, [they serve] as role models for children around the world." Lying to children, instead of setting them an example of 'hard work and sportsmanship' and of 'how to win and lose within the rules', came at a serious cost to Marion Jones (Esposito 2008). Doping must not only be punished, it must be *seen* to be punished because it represents a danger to society. The legalisation of doping in sport would, in the eyes of many commentators, give the green light to drug use in the wider community. Allowing doping users to go unpunished would also be seen as disastrous because it would send a signal to young people that rules do not really have to be respected (Vangrunderbeek 2007). As previously mentioned, our model uses the perspective of the individual athlete and of the (im)moral dimension of his actions. The episode involving Marion Jones illustrates that the moral responsibility of the athlete is no less important when it involves a media figure, when sport becomes a spectacle – quite the opposite. And although authors like Mumford (2008) substantiate the opinion that we should not saddle stars with excessive ethical accountability – which they haven't expressly chosen – many of them are convinced that this accountability is now part of the ethics or the deontology of top athletes. After all, they are richly rewarded in terms of media attention, prestige and/or financial gain. But in this respect again, there is a flipside to the medal.

On this front, doping proponents can argue that if doping products improve performances, sharpen records and enhance the spectator experience then the public has everything to gain from them. Audiences and media users have an appetite for heroic deeds. What's wrong with medically supervised EPO treatment if it helps cyclists to conquer one Alpine col after another and excel for 3 weeks at a stretch? The concept of excelling is significant in their argumentation. People are entitled to and need role models that do their utmost to achieve their goals and make a profit for themselves ... or for the leader of the cycling team. In a certain sense, the role model argument is thus reversed in this case. The logic of spectator sport is extended and the doped athlete is not vilified but revered as a source of inspiration for us all. In that sense, EPO is useful in that it helps athletes to act as god-like icons and helps spectator sport to fulfil its role as a supplier of modern myths. How can something like that be unethical, is the question put by proponents of doping tolerance. And here is the sticking point for the proponents of zero tolerance: there is something fundamentally wrong when EPO gives an athlete the wings of a god-like being, when steroid use degrades him to an animal-like monster, when implanting artificial ligaments reduces him to a machine. A fifth level was needed to provide a place for these fundamental arguments in our research model. This level concerns the deeper dimension of humanity and of human nature.

2.4 And What if Humanity Is at Stake?

The creation of any research model is a work-in-progress. The authors of this contribution have been able to test the practical as well as the philosophical utility of this model on colleagues from the multidisciplinary circles of the Interfaculty Research Group Sport and Ethics at the KU Leuven, the international research network Ethics of Enhancement in Medicine and Sport (EEMS) and the European Association for the Philosophy of Sport (EAPS). On that basis they have refined the model and also added a fifth, more fundamental level. After all, certain arguments transcend individual athletes and their opponents and go beyond the specific institutions of sport and spectator sport. They touch the very essence of human existence. Returning to the previously discussed relation-axes (Fig. 2.1), we posit that the athlete, by renouncing or embracing performance enhancement, also assumes an ethical position vis-à-vis human nature as such. This translates into the fifth, more elementary and more fundamental level of Fig. 2.2, and in the examples presented there of arguments for and against. An athlete who reaches for artificial biotechnological aids is regarded by many as short-changing humanity and defiling human nature. Others will support the inversion of this argument: for them, humanity is all about seeking out and moving the boundaries.

In the midst of the whole spectrum of arguments the ‘human nature’ argument stands out in all respects as the most fundamental: those pronouncing against performance-enhancing substances do so out of a deep unease, an intense indignation. This repugnance has seldom manifested itself so acutely as when Ben Johnson was caught using anabolic steroids in 1988. Not only did he have to give back his Olympic 100 m medal, he was also written off as a non-human, so to speak, by public opinion. The deep-rooted conviction that one should not tamper with human nature forms the foundation of the World Anti-Doping Agency (WADA) which was set up in the aftermath of the Festina scandal during the Tour de France a decade later, in 1998. The sporting world is crying out for the limits of human nature to be respected.

Curiously enough, the quintessence of sport, summed up in the Olympic motto *Citius, altius, fortius*, also demands that the boundaries of human ability are pushed still further. By the end of this contribution we hope to have moved a step nearer to a solution for this remarkable paradox, but first let us place our specific hypothesis within the wider problem of the relentlessly advancing biotechnology, the bioethics of general enhancement science and the conservative and progressive standpoints in relation to this.

The study of the moral dimension of ideas, behaviours and policy implementation is centuries old, but bioethics as such only developed after the Second World War, in the wake of revolutionary scientific innovations (Reich 1995; Bryant et al. 2006; see also Belmans 2009). Traditional ethical principles often need to be adjusted before we are able to pass judgement on, for example, organ transplants and in vitro fertilisation – procedures that may risk causing as many problems as they resolve (Shannon 1997). Should man interfere with his own nature? Should all that is possible be allowed? These questions are becoming weightier and more urgent with the development of genetic modification, transgenic technology, bionic implants and brain/computer interfaces (Smith 2005: 1). In *Our posthuman future: consequences of the biotechnology revolution* Fukuyama (2002) makes a plea that ethicists be attached to every scientist and for biotech innovations not only to be ethically supervised but also regulated and controlled on the international level. Genetic manipulation is a major problem according to bioconservatives such as Francis Fukuyama and Leon Kass. On the other side of the fence are the bioliberals or techno-progressives such as Nick Bostrom and James Hughes who support liberation, and who see great opportunities instead of threats for man in the advent of such technologies (Agar 2007). Some of them, the 'transhumanists', even predict that there will be new beings in the future, signifying the next step in evolution. With the development of science, we are not only able to manipulate our environment, we are also able to adapt the biological structure with which we manipulate that environment. Just as we created hunting spears and GPS devices to make life easier, so can we now improve the *tool* that is our body (Miah 2007:13). Life is taking control of itself and will determine its own fate. The lines of separation between machine and human being and between tool and tool-user are blurring, creating a 'cyborg' or 'hybrid' (Simon 2003).

The transhumanist movement does not regard itself as a break with history but as a continuation thereof, an extension of the legacy of the ideas of humanism: just as educational practices were the instruments of humanism, so are biomedical techniques the instrument of transhumanism. The 'transhuman' concept refers to the transition from the current (unimproved) human to a 'posthuman' being that is completely developed, whose 'hardware' and 'software' will be upgraded to such an extent that we simply won't be able to relate it to our human beginnings, just as a chimpanzee cannot imagine what it is like to be human (Bostrom 2005). Sport can help to stir up that power of imagination. Where else would the speed of a leopard and the power or a gorilla be more useful than in athletics championships which are about being faster, higher and stronger? Is it possible that athletes are becoming the pioneers of posthumanism, the first real post-humans? Pronger (1998) is already a devotee of this line of thought: seduced by the miracle drugs of science, athletes can be the first to technologize their bodies and remove the boundaries between animal, human and machine. Back in 1992 John Hoberman wrote: *"It is genetic engineering that promises to bring about the most profound biological transformations of the human being, and it is likely that this technology will be used to develop athletes before it is applied to the creation of other kinds of human performers"* (Hoberman 1992: 286). Although a major revolution in sport did not materialise, 20 years later in 2012, advances in genetic research are proving to be a real upcoming challenge. A growing number of sport ethics discussions focus on this subject.

2.5 Doctoring Genes: Threats and Opportunities

The International Olympic Committee Medical Commission volume *Genetic and molecular aspects of sport performance* (Bouchard and Hoffman 2011), a state-of-the-art reference book, clearly shows the complexity of the issue: *the human gene map for performance and health-related fitness phenotypes* (Bray et al. 2009) features more and more candidate genes, rendering applications in the area of sport ever more realistic. The abuse of somatic cell-modification is at this moment realistic enough to be taken seriously (Van Hilvoorde et al. 2007). One of the most recognizable candidates is the EPO (erythropoietin) gene. A therapeutic study with anemia patients has already demonstrated how the introduction of the EPO-gene can enhance oxygen supply and thus improve athletic endurance (Sebestyen et al. 2007). An inquiry about the EPO gene therapy construct Repoxygen on the black market was, according to Baoutina (2011), the first indication of the imminent reality of gene doping. Other candidate genes like Insulin-like-growth-factor 1 (IGF-1), PPARδ and PEPCK are proven to enhance respectively muscle mass, endurance and metabolism in mice, giving rise to the 'Schwarzenegger-mouse' (Azzazy et al. 2009). Above all, it is important to be aware of the complexity of the relationship between genotype and phenotype (Sheridan et al. 2006). The difference between improving muscle mass in mice and making a perfect athlete remains enormous. Every contest will have parameters where no genetic intervention will help. Allison (2005) for instance even contends that *"there are no drugs for tactics, judgment or even ball-control"*. In any case, not all (aspects of) sports are equally at risk. One has to distinguish between so-called non-vulnerable sports (e.g. football and tennis) and vulnerable sports (e.g. cycling and weightlifting).

Even though genetic modification of athletes is likely to remain undetectable, WADA has nevertheless issued a principled prohibition in accordance with its Declaration of St. Petersburg (11th June 2008): *"The transfer of cells or genetic elements or the use of cells, genetic elements or pharmacological agents to modulating expression of endogenous genes having the capacity to enhance athletic performance, is prohibited"* (WADA 2009). Yet, questions arise concerning the ethical foundations of WADA's policy. For each concern advocating zero-tolerance, there are more liberal thinkers presenting counter-arguments that justify a *medically supervised doping policy*. As scientists predict the arrival of genetically modified athletes in the near future (Schneider and Rupert 2009), there is indeed a growing interest among philosophers of sport in the corresponding ethical questions.

The genetic modification of athletes seems to be eroding human nature itself. The question thus significantly transcends the limits of sport. Enthusiasts predict the arrival of the very first cyborg athletes (Miah 2003), philosophers speak of potential threats to our dignity, social equality and *humanity* (Smith 2005; Culbertson 2007). Here we focus on the concept of humanity which in the context of this contribution corresponds to the concept of human nature (Belmans 2009; McNamee 2007; Belmans and Tolleneer 2009). In this we understand 'nature' to be not so much natural (as opposed to artificial) or normal (as opposed to abnormal), but: that which characterises humans, the 'identity' that makes humans human, and that Christian Lenk calls the essence or *decisive quality* in his philosophical-conceptual exercise later on in this volume. Most authors in this volume are implicitly or explicitly concerned with the question of whether athletic enhancement does or does not erode this human nature, and with the question as to whether this is or is not an ethical problem. Does athletic enhancement change man into something else? Into an animal, a god or a machine, which would make him in the words of Eric Juengst in Chap. 5, a subhuman, a superhuman or inhuman? The direction in which and the intensity with which the identity is reshaped must also therefore be subject to ethical research, and that is precisely the challenge of this entire book. Translated into the fifth level of our research model: on the left-hand side are the arguments that conservatives use to protect human identity in the context of (Olympic) sport and, in the words of Darian Meacham further in this book, to celebrate and preserve "the heritage of the 'traditional shape' of the human body as normative". On the right-hand side is a list of the arguments used by the protagonists of change to allow the (post)human athlete to reach his full potential. This athlete actually has the right and the task to push the boundaries of athletic ability and human identity. At least, that is the opinion of authors such as Stock, Young and Savulescu (discussed by Trijsje Franssen in Chap. 6 of this volume). The question as to how humanity comes into play with the enhancement and genetic modification of athletes is more fundamental than the questions posed by the four other levels of our model.

Several authors, incidentally, stress that the ethical arguments on the fifth level are 'more intrinsic' in nature than the others, more intrinsic for example than health, fair play and equal opportunities. In our view, the concept of intrinsic requires some qualification here, and we will return to it briefly later in the discussion. But, first we would like to examine the ethical impact of the latest and most radical technologies. If the fifth level discusses the possible redefinition of humanity itself, how does this reflect on the other levels? How does it affect the arguments put forth by both camps on the levels of spectator sport, sport, the other and the self? In the process of answering this question, we would also like to make an attempt to further substantiate the term respect, which has been used somewhat generically up till now. For each level we will give a short example of a core value at issue now that genetic and other extreme forms of enhancement are on the horizon: integrity, fairness, freedom and health.

2.6 Integrity, Fairness, Freedom and Health

Let us try to give more substance to the – both challenging and abstract – humanity notion and use it as a tool to bring the four other levels (Fig. 2.2) into perspective, while moving back upwards in our model.

Top athletes display excellence, be it inflated or not, in stadiums and on television and smartphone screens. It is their privilege or fate to function as public figures in this way. Spectators can oppose or identify with this and they can project their fears or dreams onto them. There's no question that in this regard the latest enhancement techniques render the problem more significant than traditional steroids or EPO. Gene doping threatens to undermine the concept of excellence itself and to nullify the authenticity of the performance, which is the greatest concern of the conservative camp. In this respect, the 'integrity' of the elite athlete, as a role model, is put to the test. We expect our icons to be real humans of flesh and blood, and not robots. By punishing offenders one wants to safeguard the belief in pure sports and in athletes with integrity as well as protect sport and media consumers from what has been called expectation disappointment. In this respect, WADA's explicit gene doping ban is a sharp reflection of the *unease* which most commentators feel about the genetic modification of athletes in particular and of humanity in general.

What is the point, ask liberal commentators. As illustrated by Andrew Holowchak later on in this volume, enhanced training methods and even steroid use in American baseball has never been a problem for the spectators, they simply increase the entertainment value. Of course, the question remains as to whether this argumentation can withstand the potentially dramatic increase in the impact of technology. At the moment, the answer to this question seems to be a positive one if we look at the case of Oscar Pistorius. Athletes like him, and like Aimée Mullins, become a new sort of role model, whose integrity is not questioned. They become a source of inspiration for disabled people and for the whole of society. Pistorius seems to use his equipment to show that he can be a complete human athlete despite the fact that he is missing his two legs, even evolving in the direction of a superhuman, not a disabled athlete anymore, but a superabled athlete. The Oscar Pistorius case is informative when it comes to discussing the blurred boundaries between disability, ability and superability and between therapy and enhancement. Some people call him the first posthuman. We don't entirely agree with that, for Pistorius is only superabled when he wears his Cheetah prostheses. When he removes his 'designer legs' he returns to his human proportions. So we prefer to talk about a posthuman performance rather than a posthuman athlete. How different is the picture when it comes to genetic modification? Most people think the ethical consequences will be much more radical: you can remove your 'designer legs' but you cannot remove your 'designer genes' (Belmans 2009; Belmans and Tolleneer 2009). Doping tolerance, others argue, would on the contrary create a sports world free from hypocrisy and would benefit the quality of the spectacle and record-hunting. In this respect gene doping would not destroy sports but, quite the reverse, make it more attractive for the audience, the sponsors and the media (Fillip 2007). An athlete who undergoes genetic modification cannot in this respect be accused of ethical misbehaviour, or of desecrating humanity or 'dismantling' its image. Rather, by using the podium of sport, he shows the path that humanity can take. Spectator sport is a powerful metaphor for how one fulfils the human mission to push back frontiers. But is it the job of sport to become a public laboratory for experiments about exceeding the limits in this way?

Couldn't or shouldn't sport continue to be reduced to its essence, i.e. a physical contest that doesn't require extrinsic goals or spectators to be a game? If we leave aside spectator sport and zoom in on sport as such (extract the square root of spectator sport), and look at its essence on the third level, is there room in this pure game for the latest technologies? Absolutely not, is the response of doping opponents, because then the game is destroyed: in essence sport requires equality of opportunities.

Winning or losing must be the outcome of the physical and moral excellence of the competitors and not of other elements (Lenk 2007). Doping, and *a fortiori* gene doping, introduces external, non-ludic elements which threaten the heart of sport and spoil the so-called 'level playing field'. As Tara Magdalinski mentions, further on in this book, for many people the latter is less endangered when it concerns technologies applied to the body's surface (in a certain sense Pistorius' prostheses are not considered unnatural). The 'level playing field' is judged to be definitely more endangered when it comes to hidden and internal technologies of which gene technology is, by its very nature, certainly the most radical example. It harms the sport and the ethos on which it is built, it corrupts, in the true sense of the word, the heart of the game, which is a testing procedure for lusory competences, as the result of a long and delicate historical process.

If sport means the testing of excellence, the defenders of the doping ban stress that it is all about 'human' excellence and not about something like 'superhuman' excellence. Genetic modification of somatic cells, and certainly the future modification of germ line cells lead to another kind of humanity. The participation of genetically engineered athletes would radically change the nature of the game and would require other types of sport. Yet, transhumanists are not convinced that it would have these problematic consequences. Gene doping would increase the athlete's accomplishment and would add humanity to his activity, similar to the way in which Jimi Hendrix would have added humanity to the composing and performing of a song like Hey Joe or Little Wing by taking, for instance, LSD. As they see it, the use of doping enables the removal of unfair genetic inequalities between athletes. Were equal opportunities not the essence of the game that we call sport? The more equal the athletes at the start, the better they can be selected on the basis of inequalities in their performance. According to this camp, athletes who (in the future) make use of gene doping should not be accused of dealing unfairly or disrespectfully with the phenomenon of sport.

In all ethical conceptual frameworks a significant place is dedicated to respect for the other. The second level of our own sport-related model questions the way in which the other person is at stake here. If technology, with the development of genetic engineering and other methods, becomes more and more radical, what are the implications for the moral responsibility of the athlete? Does he disadvantage his fellow contestants if he has his body cells and later his germ cells modified? Is gene doping merely an extension of regular doping, or is it a violation, in that the ethical questions raised suddenly become much more serious? We think the latter is the case. The manipulation of somatic cells and later germ cells incurs so many (health) risks that the above-mentioned mechanism of coercion is seen in a totally different light. This sort of problem requires more ethical-philosophical and transdisciplinary research but many authors agree that there is a risk of a disastrous downward spiral. If the new technologies are further perfected, doped athletes will acquire increasing advantages and pure athletes will have to choose between suppressing their aspirations and abandoning the competition on the one hand, or entering the competition and using the same means on the other. Not only does this fundamentally restrict their freedom, they are also led down the same path of exposing themselves to serious health risks. The whole freedom-coercion problem is further intensified by the new technologies. The defenders of the doping ban reinforce their stance and push further for a refinement of tracing systems and an intensification of controls. Conversely, transhumanists and other doping defenders, maintain their stance that there is no injustice caused to fellow contestants because everyone has the freedom to dope or not to dope, to compete or not to compete. For them, the freedom argument is paramount. They believe that tolerating gene doping and guaranteeing access to everyone would more effectively satisfy the equality of opportunities principle than the doping ban. But what about the health argument, can they just ignore this?

It is precisely on the very first level of the model that the ethical problem becomes more significant as science and technology make dramatic advances and the limits of what is human are approached or extended. The physical and psychological wellbeing of the individual is a key element of Western civilisation and of universal human rights. Performance enhancement is not an abstract concept if we look at the individual athlete looking for or steering clear of doping. It is about the alteration and technologization of a tangible body here and now, and those who dope are seen to be dealing immorally with their own health and person. In her contribution, Marianne Raakilde Jespersen provides a remarkable real-life example of how, among a number of fitness adepts, this 'rashness' is accompanied paradoxically by a discussion about 'respect' for the body and for human nature. In any event enhancement technologies raise stark questions, in our intensely medicalized (sporting) world, about the health and no-harm principle. It comes as no surprise then that enhancement critics use this argument explicitly when discussing gene doping. On the therapeutic level there is still a long way to go, so imagine what this means for risk assessment and management in the grey area of non-therapeutic applications? Pitsiladis and Scott issued the following warning in the groundbreaking theme issue of *The Lancet* in 2005 (S17)*: "the long-term effects of gene doping will inevitably be discovered to the misfortune of those who experiment."* But doping enthusiasts stand by their opinion that any concern around gene doping continues to be at odds with the concern about the inherent dangers of practising (some) sports themselves. A number of sports philosophers take a positive line on this as well: why should gene doping be banned if it can help individuals realise their life's ambitions and dreams? If gene doping produces talent which the athlete doesn't yet possess, and if as a result it increases his life's chances, what is the objection? From among these sports philosophers, Claudio Tamburrini (2005) even argues for the use of germline cell modification, and furthermore, seems to see no real danger there for personal autonomy. Biomedical scientists, including Eric Legius (2010) are inclined to make us think more realistically. They maintain that all the talk about somatic and in particular about germ-cell gene doping is premature in this respect, and that there is still a long way to go in genetics, and more especially in research into the link between genotype and phenotype. The actual role that (original and manipulated) genetic material plays in the practice of sport will not be easily uncovered. This goes for doping use in 'susceptible' sports (such as long-distance running and weightlifting), and to a much greater extent in 'non-susceptible' sports (such as football and polo). Not that any of this will prevent the transhumanists from outlining and substantiating their vision of the future. Is it feasible that these ideas will gain ground in future and that in the long term they will undermine social support for the (gene)doping ban? Will doping practices result over time in changes to the disciplinary and general regulations? And would legalisation remove the problematic tension between therapy and enhancement and put an end to the battle between doping and anti-doping technology (Kayser et al. 2007)? A brief look at the wider picture of sporting history in the twentieth century may help provide an answer to such questions.

2.7 Inclusion and Exclusion of Athletes

There is always a risk when summarising the arguments for and against doping of not doing justice to the rich details of reality. Those who design a research model for logically categorising these arguments even consciously embrace this risk. The reality of behaviours and ethical judgments and pronouncements is obviously very diverse. With their exercise in abstraction some philosophical speculations also tend to fall short of the mark. Texts that give full vent to the rich nuances, such as the empirically based study of Denis Hauw, therefore earn their place in this volume. They make clear that *the* doping user and *the* use of doping do not exist. There are only individuals and systems which, for example, are very much a victim of their time. We think that the ethics of doping and anti-doping would benefit from an even broader time perspective and a social-historical retrospective. This section is not specifically concerned with the inclusion or exclusion of doping athletes, but looks at the position that the dominant sporting culture has adopted in modern history in relation to groups of athletes in general. We may be able to draw conclusions from that and formulate tentative hypotheses about what the future may mean for doping and anti-doping.

Figure 2.3 gives an overview of the specific groups that have been embraced or rejected by the sporting world. What do we mean by sporting world? Here we mean the Olympic movement and by extension the global subculture of high-performance sport, which, as John Hoberman says in his contribution to this volume, is based on the *Citius Altius Fortius* ideal and the ensuing ideas and practices. Modern sport, as it emerged in England and took shape in the nineteenth century, was largely an occupation of the elite, of the so-called *leisure class*. More specifically, it was about 'untrained, male, white, upper class, amateur and abled athletes'. In the context of general social developments that were undoubtedly related to the emancipation of minorities, other groups gradually joined this world. We will run through the groups that were included consecutively in the family (+), forcing us once again … to schematise somewhat. Four Olympic Games (OG) appear to be milestones in this respect: Amsterdam 1928, Mexico 1968, Barcelona 1992 and London 2012.

Initially, out of a deeply ingrained sense of fair play, the gentleman amateurs rejected any form of specialisation and (intensive) training, which they associated with the working culture of the lower classes (Renson 2009: 11). They were

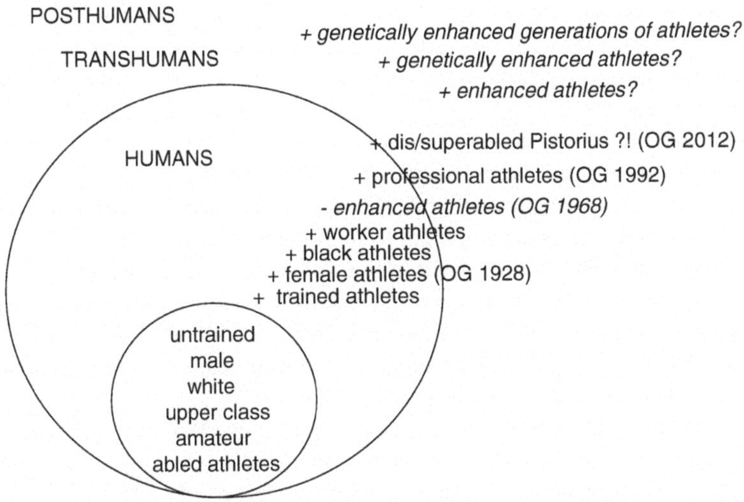

Fig. 2.3 Expansion (+) and reduction (–) of the 'Olympic' family as a result of socio-cultural change and ongoing negotiations between the agents (OG = Olympic games)

convinced that real sport was more of an *avocation* than a *vocation* (Guttmann 2004: 98). The emphasis had to lie on the innate talent of the individual and the elite nature of their group. We can therefore assert, in the context of this contribution, that they were against enhancement on the basis of training and a planned and specialist approach. This was unfair, and 'undignified', to use a concept that Pieter Bonte uses in his chapter. At the same time as the general rift between social classes diminished, so the aversion to such sporting practices weakened with time, and was even turned on its head in part: during the inter-war period a clear appreciation for training and systematic preparation emerged. Training was even praised as the main source of achievement, more important even than innate talent. Trained athletes were welcomed in the family .

In the same period, more specifically at the occasion of the 1928 Games in Amsterdam, the International Olympic Committee finally opened the door of (athletic) sports to women, women who had up till then been forced to arrange their own International Women's Games. The sports world also gradually began to include black athletes (though not until 1947 in American baseball), and working-class athletes (in 1937 the international socialist workers' movement organised its own third Olympiad in Antwerp). From a sporting perspective, the inclusion of professional athletes is more revolutionary, and also more of a discussion point among philosophers. Although this also happened/is happening gradually, we use as an example here the Olympic Games of 1992 in Barcelona, where the American 'dream team' basketball players made a noticeable appearance. Now, therefore, 'enhancement on the basis of professional training' is also tolerated and legitimate. These developments run parallel with the rise in scientification and technologization. The latter also ensured that an athlete like Oscar Pistorius, albeit not without extensive negotiations, was allowed to upgrade from the Paralympics movement (an entity born after the Second World War; see also Van de Vliet 2012) to the Olympic movement. In this respect the London 2012 games can be seen as a landmark in history: a disabled athlete is given access to the family.

The acceptance of all these groups occurred in the wider context of a society that was exchanging exclusion and elitism for inclusion, egalitarianism and democratisation. Can this development in sport be extrapolated and can we assume that the transhumanists, as suggested in Fig. 2.3, will get their own way after long negotiations between the social agencies, and will be allowed within the (un)foreseeable future to participate in tournaments like the Olympic Games? Moreover, will that open the floodgates for posthuman and genetically modified athletes and the enhanced generations who will succeed them?

How will the great game of the social debate develop? Will the arguments of the doping enthusiasts on the right-hand of our five-level model (Fig. 2.2) be able to stand up to the arguments of their opponents on the left-hand side? On the one hand there are undoubtedly reasons for answering this question in the positive. The history of (Olympic) sport is the story of a community which is expanding with ever broader concentric circles, so to speak. In that sense, it isn't too much of a leap of faith to extrapolate the developments and predict that, over time, the minority of doping enthusiasts will be embraced. On the other hand, we shouldn't forget the resolute step taken with the 1968 Olympic Games. In Mexico the first athlete was disqualified for doping use. This is indicated as a symbolic moment (−) in our Fig. 2.3. Although it can be qualified in various respects, it seems that for the first time the IOC is vigorously and demonstrably excluding certain athletes: enhanced athletes. The obstinacy of this institute has increased dramatically since then. We all know the history; in the run-up to the Games in London, it was concisely summed up in the 3 March edition of *The Economist* (2012). When we see the systematic increase in efforts to ensure a doping-free sport, with the establishment of WADA and the vigorous policy of IOC chairman Jacques Rogge, then it seems improbable that this spiral will be reversed anytime soon. And yet, a growing number of commentators see not only the need for a relaxation, but also a gradual change in the tide of public opinion. In their opinion, the grey zone of illegality is too dangerous and the costs of fighting it too high. As far as fair play and a level playing field are concerned, a comparison with the ultimate abolition of amateurism and the acceptance of professionals is not out of order here. After all, hasn't the amateur rule long been undermined from the inside by the efforts of so-called amateurs, state-sponsored amateurs? Is this not comparable with the continuous undermining of the ideal of a doping-free sport, which will no doubt ultimately lead to the abolition of the current doping rules? Just as state-sponsored amateurs could not exist without the interest and alacrity of states, so athletes cannot dope without the lust for gain and the creativity of companies or circuits. Whether or not such analogies are useful, an important point is being made here, also touched on in the above-mentioned text by Denis Hauw: each case of doping is framed in a specific socio-cultural context. Is this not one of the areas in which our five-level model, by taking the perspective of the individual athlete, falls short?

2.8 Discussion and Conclusion

Before we list the shortcomings that are inevitable in a research model like ours, and before we formulate some more general conclusions in relation to the truth that the model tries to uncover, we will briefly dwell upon the possibilities of our analysis tool. The previously conducted empirical research and the illustrations we have provided in this chapter appear to demonstrate that the model with its five layers and two columns is indeed useful for identifying, positioning and inter-relating the most diverse arguments emanating from the social debate on doping. The simplicity of the model can be seen as a strength, starting as it does from (a) a well-described agent or *subject*, namely the individual athlete, (b) an all-embracing ethical orientation, namely respect, (c) five recognizable *objects* at which this respect is directed, namely the persons of the self and the other, the institutes of sport and spectator sport and humanity as such, and (d) two diametrically opposed camps, namely the doping critics and the doping advocates. For example: the first level of analysis helps the discussion to elucidate the physical and psychological well-being of athletes, which, judging by research into fitness-adepts, are given extremely multi-dimensional meanings, and also the debate about another significant point of interest, which sadly cannot be elaborated here, i.e. privacy. The central, third level, can be regarded as a key level in the sense that it deals with the play element of sport and is thus a defining factor in the unique, sport-specific nature of the model. The spotlight is turned on public events and mass media, with the focus on spectator sport in the fourth level. By acting as a role model, the moral engagement of the individual is entwined with that of media producers, and thus embedded in a broader, socio-cultural whole. In that respect, the model does not separate the moral position of individual from the wider context.

Nevertheless, it is precisely in this area that the limitations of our tool are revealed. If we start with the individual athlete, we often fail to give sufficient credit to the specificity of historical-cultural and socio-political circumstances. We also tend to assume that the individual is responsible for his own fate in all circumstances and can act with an autonomous will. This is not the case, for instance, if he is enslaved by the uses within one or other authoritarian subculture or – taking as an example the doping misdemeanours in the former DDR – by the sporting regime of a totalitarian state. At that point it is not (completely) the athlete's moral responsibility which is at issue, but rather that of other individuals, subcultures or political systems. Obviously, our model also falls short when we want to research doping in youth sport. We cannot expect children, as well as for instance persons with mental impairment, to take independent and informed decisions in relation to athletic enhancement. Few people can refute that, in such a case as this, forms of paternalism are recommended – in his essay *Beyond consent? Paternalism and pediatric doping* Michael J. McNamee (2009) argues for a 'soft paternalism'.

This brings us to one of the key points of contention in relation to the enhancement of (adult) athletes: the debate on paternalism versus liberalism. In the conservative camp, doping proponents lean towards the first, in the progressive camp the doping defenders lean towards the second. But by using such broad categories, do we not risk detracting from the reality and the many-hued ways of making ethical pronouncements? The trouble with talking about *the* bioconservatives and *the* bioliberal camp in the enhancement debate became all too clear in the discussions during the Brocher-Hastings Center Summer Academy *Human enhancement: medical, ethical and legal implications*, which took place in July 2011 in Geneva chaired by Thomas Murray. Extra attention must therefore be given to this aspect when using and further refining this research tool. We can also learn from contributions to this book, such as that from Andrew Holowchack who points out that it is quite difficult to distinguish harm to sport (our level 3) from harm to individuals (our levels 1 and 2). Other texts suggest, at least implicitly, that the division between play (level 3) and display (level 4) is also not so clear-cut. With respect to *human nature*, approached here as the bio-cultural identity of man, and in that sense corresponding to *humanity* (level 5), the question is whether this is on another, much more essential dimension than the four other levels. This observation is justified to a large extent and this insight is the real reason why, in Sect. 2.6, we have looked at how the notion of 'humanity', together with the approaching ethical dangers, is changing the perspective on the other four levels.

In our opinion, some philosophical and other reflections on doping pay insufficient attention to the fact that sport, as an autotelic physical contest, is essentially different from other phenomena in society. The formal and informal forms of fair play, which we mentioned and which form the essence of the third level of our model, are not inessential but rather essential for gaining an understanding of the phenomenon and for researching the ethical standpoints. They also help to provide a better understanding of the paradox we referred to in the discussion about the *humanity* level. On the one hand, the sporting world is calling for the boundaries of humanity to be more closely guarded and human nature not to be desecrated. On the other hand the athlete is increasingly confronted with the *Citius, Altius, Fortius* motto and challenged to go beyond the limits of human capacities. There is no inherent contradiction in this, as long as it can be done safely within the context of sport – where athletes are rated and ranked very obstinately on the basis of strictly-lusory qualities – and as long as, in Suit's terms, the pre-lusory goals and the lusory attitude are preserved. Should the boundaries of human performance capacity be pushed? Yes, but only within the limits of the constitutive rules of the game. The faster-higher-stronger element is essential, but it is bound by the principle of equal opportunities. Viewed from this perspective, we believe that fair play in the doping debate can be interpreted not as an inessential and *external argument*, as asserted by several authors, but as an essential and *internal argument*.

These elements very definitely make sport a unique and isolated institution inhabited by people who like to stick to their own rules. We think that sport is one of the modern ways of preserving deep-rooted myths, not least the myth of objective quantifiability. Around 1,900 gentleman-athletes imagined to have developed tools for quantifying innate talent, and they didn't want to see these tools derailed by training. The historical chain of negotations between all those involved has ensured that today there is a measuring tool for the combination of talent, training methodology and implement technologies, and nobody wants to see it derailed by self-technologies. Nonetheless, transhumanists dream of logically extending athletic enhancement, legalising doping, and celebrating the athlete as an ambassador of their movement. Although the mystery behind why sport is such a particular focus of the enhancement debate is still largely unresolved, it undoubtedly has something to do with the very visual, extremely physical and very measurable nature of the activity. But can sport also be regarded as a mirror with which to examine the threats and opportunities posed by general technological evolutions and by transhumanist tendencies? Our answer would be yes and no. On the one hand no, because sport is an atypical social phenomenon, situated outside society. It remains an "unproductive activity which is mostly exercised because of its inherent value and not to achieve other external goals" (Lenk 2007). On the other hand, yes, it can be considered as a mirror to study dangers and opportunities in broader society. Due to commercialization, scientification and mediatization it has become an integral part of society. Sport is so physical, concrete and spectacular that it can act as a magnifying glass through which to see and understand problems and possibilities. Hopefully, the analysis of the arguments for and against doping within sport, to which we have attempted to contribute with our five-level model, with this chapter and with the entire book, will in this sense be useful for the wider debate around transhumanism and all the hypes and hopes that surround it. Many authors appeal for differentiation and a certain pragmatism in the doping field that is so full of grey areas, and for "distinguishing in practice between acceptable and non-acceptable performance-enhancing methods" (Loland 2011: 330). All commentators involved are negotiating and playing along. Referring to Holm and McNamee (2011) we state that there are not yet "knockdown ethical arguments for or against physical enhancement .... Some arguments are persuasive against some enhancements in some contexts". Until then we will all continue to play an intensive part in the great game of social debate. In this area too, we acknowledge the *Homo ludens* hypothesis of Johan Huizinga: the history of society is unfolding like a game.

References

Agar, N. 2007. Whereto transhumanism: The literature reaches a critical mass. *The Hastings Center Report* 37(3): 12–17.

Allison, L. 2005. Citius, altius, fortius ad absurdum: Biology, performance and sportsmanship in the twenty-first century. In *Genetic technology and sport: Ethical questions*, ed. C. Tamburrini and T. Tännsjö, 149–157. London: Routledge.

Azzazy, H.M.E., et al. 2009. Gene doping: Of mice and men. *Clinical Biochemistry* 36: 435–441.

Baoutina, A. 2011. Performance enhancement by gene doping. In *Genetic and molecular aspects of sport performance*, Encyclopaedia of sports medicine 18. An IOC Medical Commission Publication, ed. C. Bouchard and E.P. Hoffman, 376. Chichester: Wiley-Blackwell.

Belmans, J. 2009. *Sport, posthumaniteit en bio-ethiek: Een terreinverkennende studie*. KU Leuven: Faculty of Kinesiology and Rehabilitation Sciences.

Belmans, J., and J. Tolleneer. 2009. Oscar Pistorius: een postmens? *Sporta Magazine* 4: 34–35.

Bostrom, N. 2005. Transhumanist values. In *Ethical issues for the 21st century*, ed. F. Adams, 3–14. Charlottesville: Philosophical Documentation Center Press.
Bouchard, C., and E.P. Hoffman (eds.). 2011. *Genetic and molecular aspects of sport performance*, Encyclopaedia of sports medicine 18. An IOC Medical Commission Publication. Chichester: Wiley-Blackwell.
Bray, M., et al. 2009. The human gene map for performance and health-related fitness phenotypes: The 2006–2007 update. *Medicine and Science in Sports and Exercise* 41(1): 35–73.
Brown, W.M. 1997. As American as gatorade and apple pie: Performance drugs and sports. In *Ethics in sport*, ed. J.W. Morgan, K.V. Meier, and A.J. Schneider, 142–168. Champaign: Human Kinetics.
Bryant, J., L. Bagotte la Velle, and J. Searle. 2006. *Introduction to bioethics*. Chichester: Wiley.
Butryn, T. 2002. Cyborg horizons: Sport and the ethics of self-technologization. In *Sport technology: History, philosophy and policy*, ed. A. Miah and B. Simon, 111–133. Amsterdam: JAI.
Culbertson, L. 2007. Human-ness, dehumanisation and performance enhancement. *Sport, Ethics and Philosophy* 2: 195–217.
De Wachter, F. 2000. De sportdroom bedreigd. Doping als filosofische kwestie. *Ethische perspectieven* 10(3): 156–162.
Esposito, R. 2008. Marion Jones gets six months in prison. http://abcnews.go.com/TheLaw/story?id=4120748&page=1. 11 Jan 2008 (Accessed 28 Mar 2012).
Fillip, F. 2007. Is science killing sport: Gene therapy and its possible abuse in doping. *EMBO Reports* 5: 433–435.
Fukuyama, F. 2002. *Our posthuman future: Consequences of the biotechnology revolution*. New York: Farar, Straus and Giroux.
Guttmann, A. 2004. *Sports: The five first millennia*. Amherst: University of Massachusetts Press.
Hoberman, J. 1992. *Mortal engines: The science of performance and the dehumanization of sport*. New York: Free Press.
Holm, S., and M. McNamee. 2011. Physical enhancement: What baseline, whose judgment? In *Enhancing human capacities*, ed. J. Savulescu, R. ter Meulen, and G. Kahane, 291–303. Chichester: Wiley-Blackwell.
Houlihan, B. 1999. *Dying to win: Doping in sport and the development of anti-doping*. Strasbourg: Council of Europe Publishing.
Kayser, B., A. Mauron, and A. Miah. 2007. Current anti-doping policy: A critical appraisal. *Medical Ethics* 8: 2.
Lavin, M. 1987. Sport and drugs: Are the current bans justified? In *Ethics in sport*, ed. J.W. Morgan, K.V. Meier, and A.J. Schneider AJ, 169–180. Champaign: Human Kinetics.
Legius, E. 2010. *Gene doping: Genetic aspects*. Paper Colloquium Genetic enhancement in medicine and sport, Science and Ethics, Leuven, KU Leuven, 19 Mar 2010.
Lenk, C. 2007. Is enhancement in sport really unfair? Arguments on the concept of competition and equality of opportunities. *Sport, Ethics and Philosophy* 1: 218–228.
Loland, S. 2011. Can a ban on doping in sport be morally justified? In *Enhancing human capacities*, ed. J. Savulescu, R. ter Meulen, and G. Kahane, 326–331. Chichester: Wiley-Blackwell.
Mackey, R. 2012. Can a race among doped cyclists be fair? One former armstrong teammate says no. *New York Times*. http://thelede.blogs.nytimes.com/2012/10/12/can-a-race-among-doped-cyclists-be-fair-one-former-armstrong-teammate-says-no/?ref=lancearmstrong. 12 Oct (Accessed 15 Oct 2012).
McNamee, M. 2007. Whose prometheus? Transhumanism, biotechnology and the moral topography of sport medicine. *Sport, Ethics & Philosophy* 2: 181–194.
McNamee, M. 2009. Beyond consent? Paternalism and pediatric doping (The Warren P. Fraleigh Distinguished Scholar Lecture). *Journal of the Philosophy of Sport* 36: 111–126.
Mehlman, M. 2009. *The price of perfection. Individualism and society in the era of biomedical enhancement*. Baltimore: Johns Hopkins University Press.
Miah, A. 2003. Be very afraid: Cyborg athletes, transhuman ideals & posthumanity. *Journal of evolution technology,* 13: 2. http://www.jetpress.org/volume13/miah.html. Accessed 9 Mar 2011.
Miah, A. 2007. Genetics, bioethics and sport. *Sport, Ethics and Philosophy* 1: 146–158.

Morgan, J.W., K.V. Meier, and A.J. Schneider. 2001. *Ethics in sport*. Champaign: Human Kinetics.

Mumford, 2008. *Why our athletes should not be our role models*. Paper annual IAPS conference International Association for the Philosophy of Sport, Tokyo, 10–15 Sept 2008.

Parry, J. 1999. *Ethics and doping*. Proceedings of the conference: 'The Limits of Sport', Institute of Catalonian Studies, Barcelona, June 1999.

Pitsalidis, Y.P., and R. Scott. 2005. The makings of the perfect athlete. *Lancet* 366(supplement 1): S16–S17.

Pronger, B. 1998. Post-sport: Transgressing boundaries in physical culture. In *Sport and postmodern times*, ed. G. Rail, 277–298. New York: State University of NY Press.

Reich, W. 1995. *Encyclopedia of bioethics*. Hemel Hempstead: Prentice Hall/Macmillan.

Renson, R. 2009. Fair play: Its origins and meanings in sport and society. *Kinesiology* 41: 5–18.

Schneider, A.J., and J. Rupert. 2009. Constructing winners: The science and ethics of genetically manipulating athletes. *Journal of the Philosophy of Sport* 36: 182–206.

Sebestyen, M.G., et al. 2007. Progress toward a nonviral gene therapy protocol for the treatment of anemia. *Human Gene Therapy* 18: 269–285.

Shannon, T. 1997. *An introduction to bioethics*. New York: Paulist.

Sheridan, H., B. Pasveer, and I. Van Hilvoorde. 2006. Gene-talk and sport-talk: A view from the radical middle ground. *European Journal of Sport Science* 4: 223–230.

Simon, B. 2003. *Posthumanism*. Minneapolis: University of Minnesota.

Smith, K.M. 2005. Saving humanity? Counter-arguing posthuman enhancement. *Journal of Evolution & Technology* 1: 43–53.

Suits, B. 1988. The elements of sport. In *Philosophic inquiry in sport*, ed. W. Morgan and K. Meier, 8–15. Champaign: Human Kinetics.

Tamburrini, C.M. 2005. Educational or genetic blueprints, what's the difference. In *Genetic technology and sport*, ed. C.M. Tamburrini and T. Tännsjö, 82–91. London: Routledge.

The Economist. 2012. Can the scientists keep up? Drugs and sport: The twists and turns of the long-running race between drug-taking athletes and boffins trying to catch them. *The Economist Technology Quarterly*, 2 Mar 2012: 15–17.

Tolleneer, J. 2011. *Comparative physical education and sport*. KU Leuven: Faculty of Kinesiology and Rehabilitation Sciences.

Van de Vliet, P. 2012. Anti-doping in Paralympic sport. *Clinical Journal of Sport Medicine* 22: 21–25.

Van Hilvoorde, I., R. Vos, and G. De Wert. 2007. Flopping, klapping and gene doping: Dichotomies between 'natural' and 'artificial' in elite sport. *Social Studies of Science* 37(2): 173–200.

Vangrunderbeek, H. 2007. *Sport, doping en ethiek: Discoursanalyse van geschriften van studenten en journalisten (1998–2006)*. KU Leuven: Faculty of Kinesiology and Rehabilitation Sciences.

Vangrunderbeek, H., and J. Tolleneer. 2011. Student attitudes towards doping in sport: Shifting from repression to tolerance? *International Review for the Sociology of Sport* 46: 346–357.

Vorstenbosch, J.M.G. 1998. Drogredenen over doping. Een pleidooi tegen legalisatie van dopinggebruik in de sport. *Filosofie en praktijk* 19: 169–183.

WADA. 2009. The World Anti-Doping Code. The 2009 prohibited list. International Standard 6. http://www.wada-ama.org/rtecontent/document/2009_Prohibited_List_ENG_Final_20_Sept_08.pdf. Accessed 19 Mar 2011.

Wielemans, W., and P. Chan (eds.). 1992. *Education and culture in industrialising Asia*. Leuven: Leuven University Press.

Chapter 3
Is Human Enhancement Unnatural and Would This Be an Ethical Problem?

Christian Lenk

3.1 Introduction

With the advent of human gene therapy at the start of the 1980s the enhancement of the human body and mind became an issue for medical ethics. We can say today that, after three decades of this discussion, we have extensive knowledge on many kinds of enhancement interventions and also on the connected ethical problems. Right from the start, the discussion on the improvement of human functions and traits raised some concerns regarding a possible violation of the limits and boundaries of human nature. However, despite this clear moral intuition, there were no explicit ethical or legal principles known which would exclude interventions with the purpose of enhancing or improving human nature. The existing principles of medical ethics focused instead on the respect for the patient's autonomy, which was sufficient to protect the patients against malpractice in the medical setting. On the other hand, the improvement of human nature with the aim of individual and collective health (i.e., therapy) has always been seen as a legitimate goal of medicine. Therefore, it was not clear, whether the perceived *unnaturalness* of enhancement interventions was problematic from the ethical point of view.

One can also add that from the side of philosophy – as the classical mother discipline of ethics in general and medical ethics in particular – there is and always was an attitude critical of ethical values which flow directly out of nature (like it is the case in the natural law tradition). Ethical values have to be discussed and justified, and the natural as such – be it positive or negative – cannot give us ethical

C. Lenk (✉)
Institute for History, Philosophy and Ethics of Medicine, Ulm University, Frauensteige 6, D-89075 Ulm, Germany
e-mail: christian.lenk@uni-ulm.de

J. Tolleneer et al. (eds.), *Athletic Enhancement, Human Nature and Ethics*, International Library of Ethics, Law, and the New Medicine 52, DOI 10.1007/978-94-007-5101-9_3,

orientation for human actions. However, such a statement is maybe not entirely uncontested, where many citizens see human action in the technical and industrial sphere as potentially dangerous and want to preserve nature and keep the environment as natural as possible. But what does this mean for human nature, for our body and our mind? For example, aesthetic surgery is unnatural – but is it therefore also unethical? Why should we force somebody to show his or her wrinkles as signs of biological age when the person could avoid this by a medical intervention? What is the ethical problem in such a case? Human nature has many positive qualities, but it also has negative qualities – it can be empathetic and intellectual, but also brutal and nasty. We can conclude from these thoughts, that it is indeed not fully self-evident to preserve human nature as it is. The examination of this question is therefore the aim of the present article.

3.2 Some Meanings of the Natural

The question regarding the 'naturalness' of an entity can be posed in several ways. I will not mention all possible meanings of 'natural' in the present article, but only three different meanings which are of relevance for the following text:

- *Natural* in the sense of 'something which comes or stems from nature': I make the supposition that this is the meaning most often used. The opposite meaning is then 'artificial'. Animals and plants are natural, human products are artificial.
- *Natural* in the sense of 'normal' or 'usual'. In English and German, we often use the word 'natural' or 'natürlich' as synonymous to 'normal'. This meaning of the word is especially interesting regarding our perception of the human body, because things which are (in the statistical sense) 'abnormal' are also often perceived as 'unnatural' (although it is quite normal for nature itself that there are deviations of all kind).
- *Natural* in the sense of 'essential' or 'essence'. We regularly talk of the 'nature' of something in the sense of a decisive quality. When a human being would lose this decisive quality that makes him human (whatever it may be), she would no longer count as a human person.

In the following, I will refer to these three different meanings of the natural. I will not analyse my question systematically for the three meanings, but come back to these definitions at different times as a starting point for my ethical considerations.

3.3 The Natural As According to Nature

I will start my analysis with the first described meaning of the natural: something which comes or stems from nature. This meaning normally produces no major problems with regard to entities of external nature. However, the distinction between the natural and the unnatural in this sense is difficult in the case of human beings, because humans are in general the source of the unnatural. For example, it is natural for *Homo sapiens* to produce tools for daily work, but paradoxically, these tools are then labelled as artificial or 'unnatural'. Therefore, it is certainly not natural that an elephant cycles to the supermarket, but it is difficult to say whether the distinction between the natural and the unnatural also makes sense in the case of humans. As a biological species, *Homo sapiens* should be categorized like his fellow animals, but on the other hand, the sphere of human culture and technology creates a massive separation of *Homo sapiens* from the rest of the fauna which makes it difficult to define the natural appearance of this species in the biological sense. Maybe it was not natural for the Stone Age Man to cycle to the supermarket, but nowadays it is at least normal, and we would therefore not call this an unnatural behaviour. However, some of the basic (or natural) needs of *Homo sapiens* (on a more abstract level) like the need for shelter, food, human society and communication have remained very similar (although the Stone Age Man did not have the IPhone at his disposal and in a sense suffered therefore from relative technical underdevelopment in comparison to our contemporaries). As it is often described in philosophical anthropologies, the human species appears as a kind of double-sided construction which has animalistic as well as more intellectual and cultural traits and abilities.

In his article on the normativity of human nature, Kurt Bayertz (2003: 135f.) examines the question, whether a human being can count as a natural entity when it was produced by the procreation of its parents. Must the child itself then not count as an artificial human product? However, it is clear from my point of view that this is a rather artificial and hair-splitting paradox, because the procreation of human beings is obviously different from other forms of human craftsmanship and industry, and pregnancy follows indeed its 'natural' program (in the sense of the old Greek verb *physein*, which meant "to grow"). Another argument concerning the artificiality of human existence is brought forward by authors who expect the further development of humanity as so-called *cyborgs*, i.e. mixed creatures consisting of human / biological and technical elements. Some of these authors argue that with simple instruments, like for example glasses, we have already transgressed the boundary between man (in the original sense) and *cyborg* (Warwick 2003: 131). However, this argument is equally not very convincing, because it misunderstands the natural predisposition of humans to use tools and technical implements and defines the *cyborg* in a very minimalistic sense. In this view, the Bronze Age Man would also count as a *cyborg*, when he uses knife and spear. The term *cyborg* would then have no different meaning than *man* or *civilized man* (ergo, it would be superfluous). On the other side, it is also not so easy to decide when, due to artificial prostheses, somebody loses his human identity and gains the character of a machine rather than of a human being. Given that there are at present already some very invasive interventions (like the implantation of artificial hip joints, cardiac pacemakers, or computerized and motorized lower legs), the common perception of human identity is obviously rather flexible and not so narrowly defined as one might expect.

The discussion about the moral estimation of the natural (in the sense of 'something which comes or stems from nature') is complicated by the fact that in many languages the unnatural is coupled with the immoral and the unethical. This statement is true for all kinds of human behaviour. For example, in English bad parents are called 'unnatural parents' although what it means to treat a child well or badly is rather a cultural and societal question and not a question of 'naturalness'.[1] Some moral positions also rely heavily on the notion of the natural, as for example the Catholic morals. *Persona Humana*, the declaration of the Vatican Sacred Congregation on *Certain Questions Concerning Sexual Ethics* from 29th Dec. 1975, states that:

> [...] Divine Revelation and, in its own proper order, philosophical wisdom, emphasize the authentic exigencies of human nature. They thereby necessarily manifest the existence of immutable laws inscribed in the constitutive elements of human nature and which are revealed to be identical in all beings endowed with reason. [...] In this regard the [Second Vatican] Council declares that the moral goodness of the acts proper to conjugal life, acts which are ordered according to true human dignity, "does not depend solely on sincere intentions or on an evaluation of motives. It must be determined by objective standards. These, based on the nature of the human person and his acts, preserve the full sense of mutual self-giving and human procreation in the context of true love." (Vatican Sacred Congregation 1975; emphasis added).

This results in a kind of moral naturalism, where human nature as such (as far as it is in the right order and disposition) is in accordance with religious doctrines (and therefore the divine will) as well as the laws of nature. When this relationship between religion and nature is correct, there must be objective (i.e., natural) standards regarding the right behaviour in the handling of the human body and the exercise of sexuality. For example, sexual activities which do not aim at human procreation can then be classified as *unnatural* or perverse. One of the conclusions the Vatican draws from this moral point of view is the condemnation of homosexual practices, which, according to a declaration of the Congregation for the Doctrine of the Faith from the year 2003, violate the 'natural moral law' (Frankfurter Allgemeine Zeitung 2003). Equally, according to the doctrines of the Catholic Church, the use of condoms for birth control and the application of *in-vitro-fertilisation* for reproduction are unnatural, and, because of the high moral value of human procreation and respectively of marital intercourse, also unethical. However, such a moral position is clearly weakened by the fact that there are at the same time other social arrangements (like for example celibacy) which are also very unnatural, but despite this are morally highly valued by the Catholic Church. From the mere biological point of view, the fixation on human procreation could be seen as a misinterpretation of human sexuality, because the existence of the two sexes does not necessarily imply that human sexuality must be exercised on every occasion with the aim of procreation. Even when one accepts the survival of the

[1] Webster's Dictionary of the English Language (1996: 1558) gives as synonyms for *unnatural* the words *inhuman, heartless*, and *brutal.*

species as a kind of 'objective' natural goal, this does not exclude sexual practices beyond heterosexual intercourse.

Nevertheless, psychiatry and medicine in the nineteenth century had very similar views on deviant sexual practices, as for example the *Psychopathia sexualis*, a well-known textbook of that time, from the German psychiatrist and forensic medical expert Richard von Krafft-Ebing shows:

> There seems to be a pathological phenomenon, where organ and function are in discrepancy, like in the case of the exclusive perverse male. The Membrum virile is foreseen for the Introductio in vaginam, this is demonstrated by its position and form, which is adjusted – in the erect state – to the vagina. When instead of the normal instinct another desire exists, which is not in accordance with the genital's anatomical design, then is there a discrepancy, which is not only abnormal, but pathological. (Krafft-Ebing 1924: 648, translation by the author)

According to this school of thought, homosexual intercourse is unnatural because of the existence of two human sexes and the form and position of the male and female private parts. When contrasted with both learned and popular contemporary opinion, the example also shows that many are nowadays much less willing to accept direct conclusions from the sphere of the *natural* to the sphere of ethical and cultural values (i.e., the so-called 'naturalistic fallacy').

We can conclude from these examples that human beings do a lot of things which are unnatural, but that this is independent from the question whether these things are also immoral or unethical.

At the same time, it cannot be excluded that there are things or human actions which are indeed not only unnatural, but also unethical. For example, it is unnatural to pollute the environment with fumes and chemicals (they do not 'come or stem from nature'), and it is at the same time immoral to destroy the natural resources of future generations. I would argue, that those things are not unethical *because* they are unnatural, but because they are dangerous and inconsistent with the health of the population. And the fact that they are dangerous has also something to do with the fact that they are unnatural (here rather understood like in the next paragraph of this article, in the sense of the 'normal'). As we have seen, the old opinion (especially in the context of the human body and sexuality) that the unnatural is always intertwined with the unethical is mostly wrong. However, we will see in the next paragraph why the deviation from the normal (for example, in the case of human enhancement) can have some ethical relevance.

3.4 The Natural As the Normal

The notion of the *natural* is often used as a cipher for the *usual*, for something which is already known and perceived as relatively harmless. Therefore, I will now refer to this second meaning of the natural described above. Given that there is a multitude of surprising, dangerous and incalculable phenomena in the natural world, it is far from clear why in many languages we use the adverb 'naturally' for the usual, harmless, foreseeable and even reasonable. Maybe this (linguistic) coalition between the natural and the reasonable has its roots in the ancient perception of nature as a well-ordered and good arrangement (like the old greek word for *cosmos* means not only *order*, but also *jewellery* or *adornment*). Additionally, enhancement is also difficult to classify as unnatural, because the modification of human qualities and changes of the individual appearance belongs to human nature. For example, animals do not manipulate their hair, skin, body shape, etc., but humans traditionally do.

Many people – and not only adherents of the ecological movement – expect that organic food is 'natural' and therefore also relatively harmless. After a long discussion between food industry and food activists about whether artificial food colours and other food additives influence the behaviour of children, in 2007 a study in the medical journal *The Lancet* demonstrated that artificial food colours in sweets and food do indeed trigger attention-deficit-hyperactivity-disorder in children, while natural colours do not (McCann et al. 2007). Somebody who wants to avoid all possible risks from nutrition and lifestyle can choose the strategy only to use organic food and has then the security that this food (which is very 'natural') has no unknown side effects. This is not because *all* organic material is risk-free (there are many highly toxic materials in nature), but rather because this traditional and organic food is known to have no unknown and dangerous side effects. In this sense, the natural can then also have an ethical meaning, i.e. that we have a lot of knowledge about it and evaluate it as widely or even completely risk-free. I found a nice example for this relationship myself when I was invited as a research fellow to the Welsh *Swansea University* where you can order *Ethical Apple-Juice* in the student restaurant. This may then be a major reason why people find it unethical not to be informed about transgenic ingredients in their food, in the strict sense not because it is unnatural, but because they fear unknown side-effects from such foodstuffs. By analogy, it may follow that the enhancement of human beings is not unethical *because* it is unnatural (in the sense of the unnormal), but because of other qualities (i.e. possible risks or harms) which are conjoined with the fact that the enhancement is an artificial intervention.

Therefore, human enhancement can very well be unethical (i.e., producing harm to oneself or other persons) when it is dangerous or has uncontrollable side effects, and perhaps this follows from the fact that a number of possible enhancement measurements are 'unusual' or demand extreme interventions into the human body and mind. For example, it can be very dangerous to add growth hormone to the human organism when there is no lack of such hormones in the body or it is dangerous to inject EPO because this leads indeed to an 'unnatural' viscosity of the human blood and sometimes to the blockage of blood vessels, i.e. thrombosis. In the case of the most frequent doping substances, anabolic steroids, a number of severe and sometimes deadly side-effects have been described: hypertrophy of the heart, stroke, myocardial infarction, gynaecomastia, impotence, damages of the liver and the kidneys, hepatic tumours, and aggressivity (Kistler 2006). Kohler et al. (2008: 179) mention in the context of the abuse of anabolic steroids the case of the famous sprinting star

Florence Griffith-Joyner, who had already suffered a stroke in 1996 before she died in 1998 at the age of 39 years. Comparable cases reported by Kohler et al. (2008: 179) are the death of the Eastern German heptathlete Birgit Dressler (multi-organ failure in 1987) and the Austrian bodybuilder Andreas Münzer (multi-organ failure in 1996).

These examples show the connection between enhancement interventions and the 'normal' or 'natural' status of the human body in the sense of the biostatistical theory of C. Boorse.[2] As a representative of 'naturalist' theoretical approaches, Boorse argued in a number of articles beginning in the 1970s, that according to the orthodox medical perspective, the *normal functioning* of the human body (on several levels, for example genome, cells, tissue, organs, and so on) is usually named as *health*. It is a rather complex task to describe this normal functioning for all relevant parts of the human body; however, Boorse's argument is that this is exactly what the science of medicine is doing, starting from the historical example of the description of the heart's function and the discovery of blood circulation until today's work of genetics and proteomics. Concerning enhancement interventions (i.e., medical interventions which improve the outcome of bodily or mental functions over and above the normal range), Boorse had the opinion that such interventions move the body out of its natural state (in the sense of 'normal functioning') to a state influenced by different values, thereby sometimes producing dilemmatic decisions:

> If the pursuit of positive health forces a choice between incompatible excellences, it requires an evaluative decision – by client, physician, or society – about what life goals are worthy of pursuit. What it would be for a person to become healthier is no longer fixed by the concept of health until someone's values are added on. This value-ladenness is the most striking difference between positive health and the traditional negative variety. Our conception of disease required no value judgment about what forms of human life are admirable or desirable. Diseases were interferences with an empirically discoverable species design. Thus what it is to eliminate disease is uniquely describable in advance of normative decisions. Health does not depend on values when it is freedom from disease, only how far one ought to pursue it. By contrast, ideals of positive health are not discoverable, but only advocable. Their advocacy raises familiar ethical dilemmas about the good life for man, yet no medical procedure can possibly resolve them. A major attraction of positive mental health is precisely the vain hope of giving personal or social values the objectivity of traditional medicine. (Boorse 1977: 571)

One can also say that the body's normal state enables the respective person a wide range of activities, while the overemphasis of some bodily or mental traits will

[2] Boorse (1977: 554f). "From our standpoint, then, health and disease belong to a family of typological and teleological notions which are usually associated with Aristotelian biology and viewed with suspicion. Often this suspicion is excessive. Informal thinking in the life sciences constantly uses typological and teleological ideas with profit, and much recent philosophical work has been done on concepts of function and goal-directedness in modern biology. This work suggests that aseptic substitutes can be found for ancient notions that continue to have a scientific use. […] Our version of the nature of the species will be a functional design empirically shown typical of it." For a more detailed analysis of Boorse's position, see also Lenk (2002), chapter III.

lead to a disequilibrium of the human body or mind. From an ethical point of view, the natural condition of the human body is not a value as such, but only a secondary value which is important for human health. It is not *as such* of ethical importance, how many blood cells somebody has in 1 ml of his blood, but only in relation to the consequences that has for him and other concerned persons. It is not *as such* problematic to confuse the natural balance which exists in the human body, but only when this has negative consequences for the health of the person concerned. And this is indeed what happens in the aftermath of some doping interventions.

3.5 The Natural As the Essential

We can understand the natural also in the third sense which was described at the start of this text, in the sense of the 'essencc' or 'human nature'. Enhancement can then be interpreted as something which would transcend human nature (as it is propagated by the so-called transhumanist movement). For example, Nick Bostrom, one of the transhumanist movement's representatives, demands 'morphological freedom' as a (post-) human right, the right to decide on and change freely one's own biological shape and identity:

> Transhumanists promote the view that human enhancement technologies should be made widely available, and that individuals should have broad discretion over which of these technologies to apply to themselves (morphological freedom)[.] (Bostrom 2005: 203)

One of the close companions of the World Transhumanist Association (nowadays: Humanity+), the artist Stelarc, has acted out this new freedom and engaged a surgeon who transplanted him a third ear onto his the forearm (BBC 2007; Danish Council of Ethics 2009). However, public reactions towards such an illustration of transhumanist thought were rather mixed (not only because the additional ear at this rather unusual – and equally unnatural – place has no physiological function; BBC 2007). Equally, it would be unnatural to implant somebody a third arm, because humans normally have only two arms. Or it would be against human nature to give somebody a blue skin colour, because the human skin is normally white or brown or olive-coloured. It is also unnatural in this sense to bereave a person of the ability to make autonomous decisions. However, I doubt that the moral problem to bereave somebody of this capacity results from the fact that it would be unnatural. Rather, it seems to be the case because we find it unethical to do so. We would also say that it is unethical to give randomly selected persons a blue skin colour without their consent, not because it is unnatural, but rather because they did not accept this change of their appearance and we don't have the right to interfere with their bodily integrity. In sum, enhancement is in this sense indeed 'unnatural' when we change essential human qualities. However, also in this case, the fact that something is unnatural in this sense of the word does not necessarily imply that it must also be unethical.

In the case of therapeutic interventions, it is commonly held that it is in the best interest of those who cannot decide for themselves (for example babies or toddlers) to receive such an intervention when it suspends the harm and suffering which is caused by an underlying disease. The existing harm for a person is then the justification for the therapeutic intervention, even when it is not possible to gain the informed consent of the person concerned. A purported non-therapeutic benefit which is not part of the normal, healthy human nature is much more problematic to justify an interference with the bodily or mental integrity of someone who cannot consent for him- or herself. This line of reasoning would, for example, exclude giving artificial qualities for specific sports competitions to non-competent minors. However, this is much more difficult in the case of adults who wish to transcend their own human nature, even when this would dramatically alter their own identity. At present, this does not seem to be a very common desire, but it cannot be excluded that, for example, some members of the Humanity$^+$ movement would wish such interventions for their own body and mind. Somewhat comparable to the discussion on euthanasia and assisted suicide, we could then have a public discussion about whether there exists a kind of human right for the transgression of natural human qualities (the above-mentioned 'morphological freedom'). However, the cases of euthanasia and assisted suicide also show that national and international courts can reject demands to be allowed to undergo procedures which harm the individual. But, also in this case, it seems not to be decisive that such interventions would transcend normal human qualities, but rather that society does not accept interventions which are too risky for the individual and where the targeted goal is too controversial.[3]

Some thoughts of the Swiss philosopher Anton Leist (2003) indicate that an enhancement by transgressing the limitations of the human biology in the sense of a *perfection* of the human body would be as such problematic, because in comparison to our tools and machines, it is not so clear to say, what makes the human body and mind more perfect.

> In the face of these contemplations, two questions arise. Firstly, does this mutual dependency not also mean that we are unable to perfect the body, simply because we can possess no yardstick of perfection independent of the body? For truly trenchant alterations this appears to be crucial. To the extent that we can perceive structural alterations, as have been illustrated in diverse books and films for example, they usually assume an animal-like specialisation: human beings with gills, with wings, with the teeth of a wolf, with the eyes of a fly, etc. In order to think in terms of other body forms, we would have to make fundamental changes to our human purposes of action and in life – something that seems impossible without possession of these new forms already, and jumping into some of them would be arbitrary. (Leist 2003: 205)

In other words, we could change human nature in a radical way but it would be fundamentally open whether we would also reach our intended goals with these changes. As the example with the third ear, Leist's examples rather make it seem plausible that the posthuman future would perhaps not be so futuristic as expected

[3] When a person demands a medical intervention which is refused by society and law, the problem is often discussed in terms of human dignity, not in relation to personal autonomy. For a more detailed discussion of this important topic, cf. Andorno (2009), Beyleveld and Brownsword (2002).

but rather oriented towards already known biological elements. This would indeed be unnatural and could also violate ‘human dignity’, because the concept of dignity was at least until now never expanded to such hybrid life forms. We could follow Kant’s arguments who claimed that all rational live forms also have dignity (derived from their ability to autonomous and rational decision-making), but there would be a definitive danger to produce a new class of human-like creatures which are not accepted as possessing equal rights, as argued by Jürgen Habermas (2001: 110ff.) in *The Future of Human Nature*.

3.6 Conclusions for the Moral Value of the Natural

One of the conclusions we can draw from the arguments presented here, is that it is particularly difficult to examine the moral value of the ‘natural’ in the case of the human body, mind, and behaviour. However, these are exactly the categories which are the most important for the ethical discussion on human enhancement. While it is normally not so difficult for us to distinguish between natural and artificial objects in the external world, we cannot easily place our own body and mind within these categories. The ‘nature’ of the human body and mind could be seen as a well-defined construct (like in the case of other biological species, where we normally do not have problems to explain what, for example, a ‘horse’ is); however, in the case of humans, there is a multitude of interpretations of human nature. As Ludwig Siep described, this is the logical consequence of the fact that we cannot perceive our own nature in a neutral way (from a normative perspective), but only accompanied by certain values and beliefs:

> Apparently human beings have regarded and treated their bodies in a great variety of different and opposing ways. The body has been considered the prison of the human soul or the presence (Dasein) of man’s personality (Hegel), the image of god or the instrument of the devil, the source of vexation or happiness. As ‘flesh’ it belongs to a world, to which the human being’s true spiritual essence does not belong and of which it longs to get rid of – death is the best thing that a philosopher or a true believer can strive for. But as the source of pleasure it can also be considered as man’s actual eternity (nunc stans), his only way of achieving immortality. (Siep 2003: 172)

Current debates about abortion, euthanasia, sex-changing surgery, self-mutilation and a possible freedom for self-enhancement equally show a multitude of sometimes opposing views on the human body. And even the value-neutral description of the human body (should this be the case) from the side of medicine or biology does not help us in solving ethical dilemmas concerning the biopolitics of body and mind. Ethical values do not flow out of the mere description of the human body and mind, but have to come out of the spheres of ethics and law. For example, the imperative to protect the integrity of the human body and mind does not follow from the nature of that body, but from ethical and legal principles (which are rather unnatural).

We should understand the natural in general as morally neutral. It does not follow from natural conditions *as such* that something would be either positive or problematic from an ethical point of view. However, this is not to say that the natural would be irrelevant for ethical considerations. For example, the natural environment is extremely important for the health and quality of life of people, and therefore also has an ethical value as an essential and enabling resource for this quality of life. Equally, every action which threatens or limits the access to such important resources then has to be justified very carefully. I would interpret the moral preference for the natural – which we often find in the public discussion – rather as a kind of precautionary rule in the handling of unknown techniques and innovations. Such an attitude is regularly described as anti-progressive and overanxious. But maybe this is a misinterpretation of a maxim which can be very reasonable in the handling of risks and unknown side-effects. When a person is generally satisfied with the status quo in one area of his or her life, for example nutrition, it is only reasonable to be careful with innovations which can also produce unintended side-effects, as is for example the case with transgenic plants.

A similar position can be developed for enhancement or doping in sports. Therapeutic interventions such as the implantation of artificial heart valves are extremely unnatural. However, due to their therapeutic value, this is hardly ever criticized. It is therefore not the unnaturalness as such which is criticized in the case of enhancement interventions, but rather the wrong aim which is targeted with such measurements (not a person's health, but the artificial improvement of performance). From my point of view, the decisive meaning of the 'natural' in the ethical evaluation of doping is in the sense of Boorse's biostatistical theory, i.e. that the natural condition of the body (in the sense of the 'normal') is the healthy condition and that the artificial deviance from the natural homeostasis regularly leads to unintended side-effects and harmful consequences for the athletes. For example, it is possible to give an extreme boost to the muscle growth for weightlifters, but this will change the original condition of the human body and lead to restrictions in other areas and activities. Similarly, the change of the body's physiological qualities, like in the use of EPO for doping activities, can lead to functional restrictions and serious complications as an unintended side-effect (Elliott 2008: 531). Therefore, in my interpretation, the 'natural' has no secret metaphysical meaning, but is rather a reasonable and pragmatic benchmark for the evaluation of the acceptability of changes in the human body.

A possible critique of my approach may be that this argumentation makes it very easy for those proponents of human enhancement arguing that we should make doping legal and safe first, and subsequently make it commonly available in sports competitions. I have outlined in another article why such a liberalisation of doping practices would be not very convincing from my point of view (Lenk 2007). For example, rules in sport competitions normally demand a basic equality of opportunity between the different participants. It would follow then that we would have to apply an official regulation in sport competitions on what kind of doping interventions are acceptable and to what degree. This would hardly change our present situation – with the big disadvantage, that we would accept the problematic consequences of doping practices for the athlete's health. I would conclude from the arguments above that the reference to nature is not pointless in our discussion on the acceptance of doping in sport, but that it is rather a secondary ethical argument, whereby the harm as a result of doping practices, the health and well-being of the athletes, and the fairness of sport competitions are the issues that really matter.

References

Andorno, R. 2009. Human dignity and human rights as a common ground for a global bioethics. *The Journal of Medicine and Philosophy* 34(3): 223–240.

Bayertz, K. 2003. Human nature: How normative might it be? *The Journal of Medicine and Philosophy* 28(2): 131–150.

BBC. 10.11.2007. Performer gets third ear for art. *BBC News online*. http://news.bbc.co.uk/go/pr/fr/-/2/hi/health/7039821.stm. Accessed 27 Mar 2012.

Beyleveld, D., and R. Brownsword. 2002. *Human dignity in bioethics and biolaw*. Oxford: Oxford University Press.

Boorse, C. 1977. Health as a theoretical concept. *Philosophy in Science* 44(4): 542–573.

Bostrom, N. 2005. In defense of posthuman dignity. *Bioethics* 19(3): 202–214.

Danish Council of Ethics. 2009. *Homo artefact. Art between body and machine*. http://etiskraad.dk/Temauniverser/Homo-Artefakt/Artikler.aspx. Accessed 20 Mar 2012.

Elliott, S. 2008. Erythropoiesis-stimulating agents and other methods to enhance oxygen transport. *British Journal of Pharmacology* 154: 529–541.

Frankfurter Allgemeine Zeitung. 21.7.2003. *Homosexuelle Beziehungen verstoßen gegen das Sittengesetz* [Homosexual relationships violate the moral law]. www.faz.net/artikel/C30190/vatikan-homosexuelle-beziehungen-verstossen-gegen-das-sittengesetz-30111212.html. Accessed 20 Mar 2012.

Habermas, J. 2001. *Die Zukunft der menschlichen Natur* [The future of human nature]. Frankfurt am Main: Suhrkamp Verlag.

Kistler, L. 2006. *Todesfälle bei Anabolikamissbrauch. Todesursache, Befunde und rechtsmedizinische Aspekte* [Fatalities under misuse of anabolic drugs. Cause of death, findings, and forensic aspects]. Dissertation, University of Munich.

Kohler, M., M. Thevis, W. Schänzer, and K. Püschel. 2008. Gesundheitsschäden und Todesfälle durch Doping [Damage to health and fatalities from doping]. *Rechtsmedizin* 18: 177–182.

Krafft-Ebing, R. 1924. *Psychopathia sexualis*. Stuttgart: Ferdinand Enke.

Leist, A. 2003. What makes bodies beautiful. *The Journal of Medicine and Philosophy* 28(2): 187–219.

Lenk, C 2002. *Therapie und Enhancement. Ziele und Grenzen der modernen Medizin* [Therapy and enhancement. Goals and limits of modern medicine]. Münster: Lit-Verlag.

Lenk, C. 2007. Is enhancement in sport really unfair? Arguments on the concept of competition and equality of opportunities. *Sport, Ethics and Philosophy* 1(2): 218–228.

McCann, D., A. Barrett, A. Cooper, et al. 2007. Food additives and hyperactive behaviour in 3-year-old and 8/9-year-old children in the community: A randomised, double-blinded, placebo-controlled trial. *The Lancet* 370(9598): 1560–1567.

Siep, L. 2003. Normative aspects of the human body. *The Journal of Medicine and Philosophy* 28(2): 171–185.

Vatican Sacred Congregation. 1975. Declaration on certain questions concerning sexual ethics. www.vatican.va/roman_curia/congregations/cfaith/documents/rc_con_cfaith_doc_19751229_persona-humana_en.html. Accessed 20 Mar 2012.
Warwick, K. 2003. Cyborg morals, cyborg values, cyborg ethics. *Ethics and Information Technology* 5: 131–137.
Webster, N. 1996. *Webster's encyclopedic unabridged dictionary of the English language*. New York: Gramercy Books.

Chapter 4
Dignified Doping: Truly Unthinkable? An Existentialist Critique of 'Talentocracy' in Sports

Pieter Bonte

4.1 What Doping *Is* – And What It Need Not Be

4.1.1 The Need for Rigorous Intrinsic Inquiry

In the spirit of this volume, this chapter seeks to cut to the chase of the doping debate. Having picked the thick cluster of doping-related problems apart, many scholars conclude that the problem posed by performance enhancing interventions runs deep, and that this depth is not adequately probed by addressing the many acute circumstantial problems they pose, such as health risks, rule-breaking, unequal access, direct and indirect coercion, abuse in perfectionist and hypercompetitive cultures, etc. (for two sustained analyses, see President's Council on Bioethics 2003: 101–158; Sandel 2007: 25–44) The World Anti-Doping Code recognizes that doping does not only pose such circumstantial issues which are ultimately remediable – at least theoretically so, even if they might prove to be intractable in practice. Doping is seen as wrong *in and of itself*. In this vein, the Code's Fundamental Rationale section stresses that "doping is fundamentally contrary to the spirit of sport" (World Anti-Doping Code 2009: 14) and the World Anti-Doping Agency's baseline reads not "play healthy" nor "play fair" but, rightly so: "play *true*". *No matter what*, WADA says, doping will always be wrong.

A strictly pragmatic, circumstantial rendering of the doping imbroglio carries the risk that nothing *specific* about doping will be called into question. As such, it leaves one at liberty to conclude that there may be nothing wrong with doping in itself, and that doping might therefore have to be freely allowed as soon its circumstances are tidied up. Such a 'thin' laissez-faire stance I hold to be frivolous and in disregard for a

P. Bonte (✉)
Faculty of Arts and Philosophy, Bioethics Institute Ghent (BIG),
Ghent University, Blandijnberg 2, Room 120, B-9000 Ghent, Belgium
e-mail: Pieter.Bonte@UGent.be

J. Tolleneer et al. (eds.), *Athletic Enhancement, Human Nature and Ethics*, International Library of Ethics, Law, and the New Medicine 52, DOI 10.1007/978-94-007-5101-9_4,

'thick', virtue-ethical spirit of sport. Alarmingly, this thin view is even to be found among WADA experts themselves. Harm Kuipers, for instance, who previously served as a member of WADA's doping commission, argues: "If a substance enhances performance and does not damage one's health, for me it can be used." (Kuipers in Starckx 2008: 118, my translation from Dutch) As I will argue here, such a 'health-only' view is thoroughly misguided and ethically reckless. To remedy such apparent oblivion for intrinsic issues, in the debate on doping we need to start talking more, and more clearly, about *doping*. In philosophical parlance: we need to address, in a direct and sustained manner, doping's *categorical* and *intrinsic* traits, by which I mean respectively the traits which are *common to all possible instances* of the category denoted by 'doping' and those that are *peculiar and distinguishing* to doping.

At the same time, however, I subscribe wholeheartedly to the important argument made by John Hoberman (2009) and Thomas Murray (2009a) that we should be very wary about dangerously naïve theorizing about performance enhancing interventions in the abstract. Such naiveté can indeed readily spring from at least these three distinct forms of 'theoretical blindness': (1) a too short-sighted engineering perspective ("the intervention works in the lab, so why not introduce it in society?"), (2) a too short-sighted libertarian rights perspective ("it's my body, so why should some regulator get to decide what should and what shouldn't be in it?"), (3) a too short-sighted virtue perspective ("The truly good and wise can find a way to use this intervention with dignity, so who are we to pass judgment and deny anyone a try?"). Trying to add to received wisdom, I will argue here that in dealing with doping we should *also* be wary of the common bias of *circumstantial blindness:* the inability to see beyond the predicament one – or one's society – is in at a specific time and place. To come to grips with doping we must neither be starry-eyed nor blinded by the floodlights of the stadium as it stands anno 2012. Both the conceptual and the contextual analyses are indispensable, and we should paddle back and forth between the two to realize a strong reflective equilibrium. Seeking to add to that collective discursive effort, I temporarily turn away from the 'broad', circumstantial side of doping and turn towards the 'deep', intrinsic end.

A tried and tested philosophical tool to sift out circumstantial traits and allow intrinsic and categorical traits to surface, is to think through what 'doping' would *persistently consist of* if it were to exist in a (number of) 'best of all possible worlds'. All that persists in such worlds indicates *intrinsic* traits, and all wrongs that there persist indicate intrinsic wrongs. For instance, Aldous Huxley's Brave New World (2006 [1946]) served to show how a pure 'happiness enhancing technology', the designer drug Soma, seems to be something intrinsically, deeply wrong even if it would be healthy, abundantly available and voluntarily engaged in. In fact, Soma provokes such profound indignation *precisely because* it demonstrates the issue of 'alienated happiness' in such an undiluted form. Robert Nozick (1974: 42–45) famously refined such arguments in his thought experiment of the 'Experience Machine', arguing that if a machine would be able to induce the perfectly realistic *experience of* the most stimulating and satisfying life imaginable, this should be rejected. As I will argue, a choice for 'mental state welfare' based on self-deception is profoundly undignified, and, curiously, certain pro-talent, anti-doping philosophies seem to advise using one's given human nature as such an experience machine: to generate the (illusionary) experience that one was purposefully created, which promises substantial existential relaxation.

Huxley and Nozick show that with all preconditions of autonomy, health and fairness met, things can still take a turn for the worse, if not for the worst: persons being perfectly happy and healthy, while at the same time intolerably alienated, inauthentic and dehumanized. In a similar exercise, throughout this chapter I will suspend my judgment on the countless circumstantial problems that surround doping-as-we-know-it-today and attempt a rigorous inquiry on doping's intrinsic, defining traits. The question is whether an athlete who dopes herself into a perfectly healthy and happy state of high performance thereby inadvertently degrades herself into a state of alienation, inauthenticity or dehumanization.

4.1.2 *What is Doping?*

To set apart doping's intrinsic traits, I propose the following as a working definition of doping. Doping can be conceived as:

> the use of an exogenous substance, device or method / that enables enhanced physical capacity / by altering a person's bodily make-up / so that a physical capacity level can be reached beyond (a) that which the non-disabled doper possessed before this use (bypassing or 'short-cutting' doping), (b) that which the non-disabled doper might come to possess if she were to apply maximized effort and enjoy optimal social and environmental circumstances (individual surpassing doping), and/or (c) that of the biological species *Homo sapiens* (species surpassing doping) (Fig. 4.1).

In crafting this definition I have taken care to filter out the following elements which are regularly presented as intrinsic traits of doping but which, as I would argue, are neither intrinsic traits nor necessary implications. Rather, they are *possible* (perhaps highly probable) contingencies connected to doping. A specific doping practice might entail several or even all of the following traits, but as I believe we

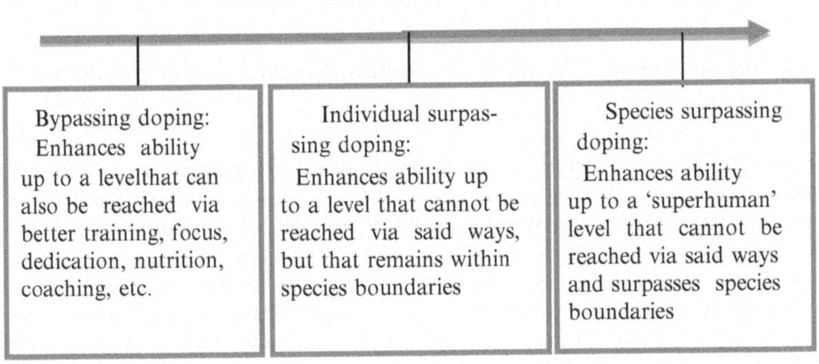

Fig. 4.1 Three degrees of doping

can construe *realistic*[1] alternative doping practices in which all these entailments would *not* be the case, the following traits are not categorically nor intrinsically tied into the concept of 'doping' itself:

1. Doping is unhealthy
2. Doping operates via *artificial* substances, devices or methods
3. Doping is unfair (as in being against the rules, as in there being insufficiently equalized access, or as in placing impermissible coercive pressures on those who do not wish to dope)
4. Doping involves an *intention* to enhance or a 'drive towards mastery'
5. Doping is a sufficient cause for enhanced performance, i.e. effectuates *by itself* or 'automatically' the enhancement of performance, thus bringing about a 'drift towards mechanism'
6. Doping diminishes the need for engaged and effortful involvement by the athlete

4.1.3 A Structured Search for Doping's Intrinsic Wrongs

The strongest contenders as (seemingly) intrinsic, categorical arguments against doping can be structured along the following three main lines, respectively based on the conviction that a proper athletic accomplishment should: (a) originate from proper origins, such as natural talent; (b) take place via proper, intentionally directed processes; and (c) result in proper, recognizably human outcomes (President's Council on Bioethics 2003; Sandel 2007; Murray 2009b; Loland and Hoppeler 2012). On these three fronts, doping threatens to turn that proper athletic accomplishment into something *debased*, *mechanistic* and *dehumanizing*, respectively (Fig. 4.2):

As the debate on doping is maturing and unreasoned intuitions are increasingly being taxed and asked to be replaced by sustained ethical argument, the ethical arguments against doping-itself revolve around concerns about the proper appreciation of one's natural endowment, about the proper cultivation of that endowment, and finally about the proper conservation of that endowment. Thomas Murray, chair of WADA's Ethical Issues Review Panel, has powerfully captured this talent-based Spirit of Sport which drives the global 'war on doping' and which often presents itself as *the* fundamental Spirit of Sport, allowing for no plurality on this fundamental level. (See Chap. 16 by Brownsword, this volume) According to Murray, Michael Sandel and the many who

[1] For a strict philosophical inquiry to be maximally revealing, thought experiments should be restricted by nothing more than logical possibility – a classic example of which, also relevant in the debates on human enhancement, would be Derek Parfit's *Reasons and Persons* (1984). If one demands realism, or even *actual occurrence* or *feasibility in a foreseeable near future*, one risks 'ontological parochialism', obscuring conceptual clarity. That said, the more realistic and contemporary we can construe a situation in which doping would not be intrinsically problematic, the more fruitful such findings will be for the eventual goal of the development of *practicable* policies, as it would allow us to identify, should they exist, (1) *actual* doping practices which are *now* receiving undue denunciation and persecution, and/or (2) *feasible* or *foreseeable* doping practices which are permissible and perhaps laudable to prepare and pursue.

Performance	Natural	Doped/Enhanced
Proper Origins	From 'given', natural origins Praised as 'gifts', providing a sense of objectively given place, purpose or predestination Proper, for sports should display 'god-given talents', or, secularly, 'innate nature-given talents'	From self-styled, artificial origins Denounced as 'debasement', eroding our sense of given place, purpose or predestination Improper, for it distorts the display of 'real', natural hierarchies
Proper Processes	Through one's inherent, endogenous bodily processes and one's active intentional effort Praised as authentic accomplishments	Through intrusive, exogenous means and by passively undergoing their influence Denounced as effortless & mechanistic
Proper Outcomes	'Vitruvian' Towards a perfected optimum within the normal, species-typical range Praised as perfected humanity	'Promethean' Towards a distorted excess over and beyond the normal, species-typical range Denounced as alienating dehumanization

Fig. 4.2 Three components of proper athletic accomplishments

concur, the Spirit of sport is about *the virtuous perfection of natural talent* (Murray 2009a; Sandel 2007 and Chap. 10 by McNamee, this volume). Sharing the commitment to virtuous sports, I now take to task these additional requirements of perfectionism and natural talent.

4.2 Proper Origins. May the Best, or May the Blessed Man Win

4.2.1 Talent As Robustness and Doped Performances As Flukes

Most doping practices we know today seem to enable only a *temporary boost* of performance levels, whereas the presence of natural talent indicates a more durable, longer lasting potential for high-level performance – a *predisposition* proper. For this reason, when comparing capacities to perform, talent-based performances can be believed to be categorically superior to doping-based ones. They provide proof for the existence of a deeply ingrained, robust capacity to perform, whereas doping-based performances only provide proof for a superficially induced, fleeting capacity to perform. Take away the dope, and the capacity level would drop significantly.

This may have limited validity if one restricts one's view to the effectiveness of contemporary doping technologies, but it would be an exaggerated simplification to say that, come what may, only natural talent can ever count as a truly reliable marker for deeply ingrained, robust potential. We must not let the image of today's pills, syringes and injection needles and the often fleeting effects they bring about obscure the fact that a plethora of future *and current* doping practices will not all follow the lines of this 'Popeye caricature': doping right before the performance is to be performed, bringing the body in a temporary high, resulting in a extraordinary strong performance, after which the enhancing effect fades away and it is revealed how the athlete without the spinach/doping is, 'in reality', a less able athlete who could never have performed her feat 'on her own'.

What this Popeye caricature misses, is that robust potential may also be obtained via certain doping practices, for instance the introduction of a doping agent that would secrete chemicals over a long space of time (contrast Popeye to Spiderman, Asterix to Obelix), or a permanent enhancing intervention such as today's Lasik eye surgery that golfers undergo to provide them with better than 20/20 vision. Theoretically we can conceive of a heart-widening procedure which would allow a cyclist to obtain the large heart (and great advantage for cyclists) which others may found implanted in them via genetic luck (Mehlman 2009). Ergo, deeply integrated doping practices such as the very real Lasik eye lasering or the still very theoretical heart-widening procedure seem immune to the 'fluke' critique of doping. What is more, even if – counterfactually – not a single type of doping would ever be able to provide a potential for physical performance as profound or durable as natural talent, this alone does not suffice to categorically depreciate the acquisition of abilities via doping – it would only mean that natural talent could be appreciated somewhat more in this regard as it would mark a somewhat more robust potential. In sum, it is an erroneous overgeneralization to hold that doping could only ever induce fluke performances and that natural talent is and always will be the best proxy for deeply ingrained and durable predispositions.

4.2.2 *The Talented As the Authentic and Dopers As Phonies*

Precisely by resolving the fluke objection, the deep integration of doping can raise a new set of objections on an altogether different and perhaps more fundamental plane. On this plane, succeeding all too well in endowing the athlete with a predisposition to perform at a level equivalent or even superior to natural talent becomes the problem. The deeper cause for concern is this: by implanting such novel (perhaps more enabling and satisfying) predispositions, one may betray the (perhaps more incapacitating and frustrating) predispositions that are properly one's own, that make up 'the essence' of who one is. The more permanently and profoundly one intervenes in one's own birth suit – and doping can do exactly that – the more one 'tries to be somebody else', the more one turns into a 'phony'. Doping, therefore, might deeply undermine personal authenticity.

In light of this deeper danger posed by human enhancement interventions in general, critics like Carl Elliot and Howard Baillie call for an ethic of authenticity, more precisely an ethic of *affirmative* authenticity (Elliot 2003; Baillie 2005). In such an ethic, self-exploration is conceived of as self-discovery: drawing out what is already inside of you, as opposed to drawing in alien things from the outside. This general authenticity argument can be invoked with extra vigour in the field of sport and doping. Articulating the more fundamental reasons why sport may be of great ethical value, several philosophers have characterized sports as a 'spiritual exercise' of self-discovery. While sport may often seem to be all about *Citius, Altius, Fortius* – that is: about transgressing given physical boundaries and striving towards "superhuman performance" (Savulescu et al. 2004) – this apparently *transgressive* practice can also be understood as an on-going *approximative* discovery of the eventual, ultimate boundaries of one's given potential – talent which nature had already implanted there but needed tending to ('virtuous perfecting') to be made manifest. The reward of an athletic accomplishment based on one's natural talent then lies not only in the pride one can take in the objective performance, but also in the 'revelatory experience' of finding out what you had in you all along. This perception of athletic exertion as a way of gaining valuable existential self-understanding seems to resonate in such widespread sporting slogans as "show what you are made of", "find out what you have in you", "stretch yourself to the limit", etc. Any sport hobbyist might thus engage in his private athletic soul-search, as for instance the environmentalist Bill McKibben describes the deeper meaning of his personal jogging routine in his anti-enhancement book *Enough* (2004: 1–66) or as Thomas Murray describes his cycling runs (Murray 2009b). Moreover, as Darian Meacham as well as Tara Magdalinski describe in their contributions to this volume, in the top performances of the global elite athletes this dynamic is played out on a species-level.[2] At the Olympics, those elite athletes reveal what 'mankind' has in itself – what mankind truly is. Between the gifted, perfected athletes and the less-gifted, less-perfected members of the audience, a deep form of bonding can take place provided that two premises (are believed to) hold: (1) the top athlete and the audience member share a species constitution, and (2) talents are distributed within the species in a random way, via 'the natural lottery'. As such, an audience member (as the Everyman) and the audience as a whole (as the People) can witness a top athlete accomplish superior feats and deeply identify with those accomplishments, thinking 'yes, look at what my/our mankind is capable of', and, 'yes, had fortune dealt its cards differently it might just as well have been me/anyone of us starring the show'.[3] For such revelations and identifications to take place, it is imperative that no-one is

---

[2] See the distinction made between individual surpassing doping and species surpassing doping.

[3] This dynamic of identification based on recognition + random elevation is, I believe, a major reason of the immense popularity of hero stories such as Spiderman, Popeye, Asterix & Obelix etc. The basic narrative is that of an Everyman, with which the reader of the story can readily identify, who has the luck of experiencing an extraordinary intervention (being bitten by a mutated spider, eating super-spinach, drinking a druid brew or being drenched in it as a child), after which he comes to obtain special, superior capacities and his life is elevated to that of a superhero. Should the same fortune ever befall the reader (and it is only random that it has not), she may start leading the life of

meddling with the communal constitution of human nature, nor with the dynamic of arbitrary allotment of talents via the natural lottery.[4]

Doping seems to be exactly such meddling, and this may provide good reason to categorically reject it. Indeed, from this perspective of self-discovery and group identification, doping presents itself as a diametrically opposed practice of self-*alteration* and group-*dissociation*. I will reserve a further discussion of group-dissociation for the section on proper outcomes and will now focus on the problem of individual self-alteration. Instead of showing what someone is made of, doping makes that someone anew. By redrawing one's given physical boundaries instead of approximating them, doping blurs precisely what the ethic of affirmative authenticity wanted to bring into sharp focus. Such arguments help to explain why doping can be considered as a form of cheating not in the superficial sense of breaking a conventional agreement that no one is to use stimulants just like no football player is allowed to carry the ball over the field in his hands, but cheating in a more profound sense as cheating oneself in becoming a fake, a phony, a fraud – denying and corrupting who one 'really is' or 'was cut out to be'.

However, the ethic of authenticity as presented so far seems too crude and uncritical of its own presuppositions. Identifying ourselves with our biological determinants and affirming those determinants as the constituents of our authentic selves yields fundamental problems too – problems pertaining to biological determinism, essentialism, the naturalistic fallacy and in extremis: social Darwinism. This is an instance in which the virtue ethical critique must cut *both* ways: true, technological alterations of one's biological constitution may well present grave challenges to the maintenance of one's authenticity, but an excessive conservation and normative affirmation of one's given biological constitution may do the same. In the timeless Nature-Nurture debate, there is an overwhelming contemporary consensus that no matter how much our biological underpinnings may determine our psychological and motivational identity structures, the *facts* of our biological existence have no say in determining the *values* of our *moral* existence. No matter how frustrating it may be in our search for a straightforward moral compass, there stands a wall of separation between 'is' and 'ought'. Our nature may tell us some things about what makes us *feel* good,[5] but is mute as a tomb about what might make us *be* good. In the abstract, one's biological nature and the natural processes by which one receives one's particular biological

a superhero, too. Similarly, should the natural lottery have given the audience member the genetic hand now dealt to the top athlete (and it is only random that things turned out otherwise), the now audience member who goes unnoticed would have been the glorified star in the arena. Oddly, we seem to adore 'doping sinners' in these cartoon stories.

[4] For an insightful critique of common mistakes in thinking through the 'natural lottery', applying among other things the non-identity problem to these issues, see Hurley (2002).

[5] As such it may have a *secondary* role in helping to determine what might improve human welfare (see for instance Bayertz 2003), but even then it can just as well be that all manner of 'unnatural' practices may make us feel even better.

endowment are a-moral and existentially pointless. Considering the specifics of the natural evolutionary forces that made us, they are quite pitiless too. Therefore, in posing the question whether we are in any way ethically obliged to honour our biological constitution, the conclusion seems rather inverse. We must "rebel against the tyranny of the selfish replicators [i.e. our genes]" (Dawkins 2006 [1989]: 201) From a understanding of the actual trappings of our evolutionarily cobbled together *Homo sapiens* nature, and from an affirmation of our *ethical* and *existential* authenticity, we do right to, in a sense, "establish an identity which lies at an ever-increasing distance from our organic nature[.]" (Tallis 2007) In sum, the notion that our natural constitution – our talents or our lack of talent alike – is something we should morally affirm and conserve on account of it being the authentic thing to do has at best a severely constrained validity. In fact, such affirmation and conservation may regularly be the morally and existentially *inauthentic* thing to do.

What does this imply with regards to doping? If one engages only occasionally in the aforementioned 'fluke doping', few authenticity concerns seem to rise. But if one engages in *habitual, chronic use* of such brief booster doping, or if one elects to undergo a form of *deeply integrated* doping, then such chronic or constant reconfigurations of one's embodiment and one's capacities may indeed bring about substantial reconfigurations of one's sense of self, too. In itself, however, such self-altering practices need not raise authenticity issues of 'no longer being oneself' or 'becoming a phony'. The recognition of this possibility does, however, make the maintenance of essentialist and teleological conceptions of the self difficult, and seem to require the adoption of some 'narrative' conception of personal identity (DeGrazia 2005; Levy 2011). In the abstract, such an assertive decision to take full responsibility over one's randomly obtained endowment – provided that that decision is earnestly recognized and undertaken *as such* – may even be virtuous acts of humanistic *Bildung* and existentialistic courage.

In practice, however, such decisions to substantially reconfigure and to intentionally re-create oneself in accordance with one's values and preferences, cannot be taken lightly (Frankfurt 1971; Schechtman 2004). Not only is there always the danger of dubious norms and motives informing the decision on how to alter oneself (the most outspoken concern of Carl Elliot); such self-alterations are likely to require intensive labours of deep reflection and deliberation – both within oneself as within one's community, as they have the potential to fracture the fragile fabrics of both one's own historical self-understanding and of one's myriad communitarian ties (the most outspoken concern of Harold Baillie). Even if deeply integrated or habitual doping may in general bring about less drastic 'upheavals of identity' than, say, sex reassignment surgery or the integration of means which directly alter one's emotional, perceptional or cognitive constitution, such doping practices still seem to form a particularly dramatic category of self-change activity. Undertaken in a wrong way, for wrong reasons (which may be very probable), deeply integrated doping would be deplorably undignified. Undertaken in a right way, for right reasons (which may be very unlikely), doping (of the deeply integrated or the fleeting variety) may be positively dignified – and a worthy, albeit risky aim for athletes of a humanistic or existentialistic bent.

4.2.3 *Natural Endowment As the Gift of Place and Purpose, and Doping As Its Loss*

Continuing the argumentative cascade, the very resolution of the authenticity concerns can give rise to a new set of objections on an even more fundamental plane than that of personal authenticity. That even deeper cause for concern is this: precisely by heeding the moral call of his personal authenticity only all too well, the 'self-made man' may come to neglect the respect he is due to not to his self, but to the forces that made him. Becoming oneself the decider on how one is going to be constituted, as doping arguably does, overrules the original plans as encapsulated in one's innate biological blueprint. This overruling can be seen as a moral transgression: at a certain level of depth – the specifics of which can be debated – one should curtail one's own sphere of legitimate agency and stay true to something *even more originary* than one's 'ego'. Perhaps this is where the intuition comes from that the 'gifted athlete', like the 'natural beauty', deserves special reverence: she got her special gift by nature, whereas the doper helped herself to a gift – as an usurper, as an ingrate, or as someone with an excessive *"drive towards mastery"* (Sandel 2007: 27).

In this regard Michael Sandel writes in the chapter "The Bionic Athlete" of his influential anti-enhancement essay *The Case Against Perfection*:

> The deeper danger is that [human enhancement technologies] represent a kind of hyperagency, a Promethean aspiration to remake nature, including human nature, to serve our purposes and satisfy our desires. [...] To acknowledge the giftedness of life is to recognize that our talents and powers are not wholly our own doing, nor even fully ours, despite the efforts we expend to develop and to exercise them. [...] It is, in part, a religious sensibility. But its resonance reaches beyond religion. (Sandel 2007: 26–27).

Ultimately, however, I fail to discern in Sandel's writing an actual *principled* argument to ground the claim that our (lack of) talents should be kept as-is on account of their giftedness. I do, however, find a number of *prudential* arguments on discomforts that are best avoided:

> [T]he real problem is the explosion, not the erosion, of responsibility. As humility gives way, responsibility expands to daunting proportions. We attribute less to chance and more to choice. [...] One of the blessings of seeing ourselves as creatures of nature, God, or fortune is that we are not wholly responsible for the way we are. (idem: 87)
>
> There is something appealing, even intoxicating, about a vision of human freedom unfettered by the given. [...] But that vision of freedom is flawed. It threatens [...] to leave us with nothing to affirm or behold outside our own will. (idem: 99–100, the book's closing paragraph)

I wholeheartedly agree with Sandel that the growing number of practical possibilities to make decisions about how we are constituted burdens us with new levels of responsibility, of indeed daunting proportions.[6] And I also agree that a vision of human freedom unconstrained by predetermined givens leaves us with

[6] On the contemporary problem of responsibility explosion in general, see Schwartz (2004).

little more to affirm or behold outside our own will. To be so 'foundationlessly free and ruthlessly responsible' can indeed be a despairingly unmoored predicament to find ourselves in. As Allen Buchanan and colleagues note: "If we can no longer convince ourselves that human nature provides significant constraints on the pursuit of individual or social good, we may feel cast adrift in a sea of possibilities." (Buchanan et al. 2000: 93) If we could legitimately escape such a predicament, that would indeed be a blessing.

Sandel clearly believes we can escape such a predicament: we should start "seeing ourselves as creatures of nature, God, or fortune" – that way "we are not wholly responsible for the way we are". Yet no argument is provided how we might *legitimately* see ourselves that way. Allusions are made, however, to the "blessing" (87) of the psychological comfort gained by believing oneself to be a creature of some originary creative force. In one convenient move, Sandel circumvents both those daunting degrees of responsibility as well as the aporic challenge of creating meaning for yourself from the vantage point of a godforsaken, nature-forsaken freedom.

With proper argument wanting and only a 'vision' of the lure of a life with lessened responsibility and a pre-existing meaning and purpose to our lives, I can only return to sender Sandel's argument: there is something appealing, even intoxicating, about a vision of *a given human nature* unfettered by *freedom and personal responsibility*. But that vision of *giftedness* is flawed, as it threatens to leave us with nothing to affirm or behold outside *pointless, pitiless nature*.

With Sandel I denounce the excessive "drive towards mastery" that seems to motivate some – perhaps many – advocates of human enhancement. But the ethics of virtue and authenticity equally compel me to denounce the excesses of a drive towards the comforts of an *amor fati* which seems to motivate some –perhaps many – to outsource responsibilities that are *de facto* theirs to entities such as a normatively charged conception of "nature, God, or fortune" – the existence of which remains to be argued for, whereas the arguments for their *mere desirability* are ready to come by. As such, it seems plausible to understand such lines of reasoning as a choice for (false) comfort and certainty over the dignity and virtue of daring to accept – perhaps dauntingly uncomfortable – facts of our existential predicament. In this sense I believe the widespread experience of "moral vertigo" (Sandel 2007: 9), outrage and indignation in the face of deep possibilities of self-creation may in part be the result of the fact that such "unwelcome liberations" (Bonte 2008) are increasingly making an existentialist self-understanding hard to deny or brush aside. I quote Jean-Paul Sartre's Existentialism is a Humanism *in extenso,* as it reads like a direct, point-per-point contestation of Sandel:

> We are left alone, without excuse. That is what I mean when I say that man is condemned to be free. [E]very man, without any support or help whatever, is condemned at every instant to invent man. [T]o begin with he is nothing. He will not be anything until later, and then he will be what he makes of himself. Thus, there is no human nature, because there is no God to have a conception of it. Man simply is. Not that he is simply what he conceives himself to be, but he is what he wills, and as he conceives himself after already existing [.] That is the first principle of existentialism. [T]he first effect of existentialism is that it puts every

> man in possession of himself as he is, and places the entire responsibility for his existence squarely upon his own shoulders. [M]an is in consequence forlorn, for he cannot find anything to depend upon either within or outside himself. [O]ne will never be able to explain one's action by reference to a given and specific human nature; in other words, there is no determinism – man is free, man is freedom. (Sartre 1989 [1946] – my italics)

Doping makes this existentialist self-conception of finding ourselves condemned to be 'foundationlessly free and ruthlessly responsible' strike right at the heart of sport, that cultural institution which is deeply predicated on both investing deep meaning in our nature-given embodiments as well as on fetishizing passionate and self-assured human action driven by prefixed rules to live by, crystal-clear goals to strive towards, and gratifyingly instant and absolute standards and judgments on behaviour that is right or wrong, excellent or failing. Sport thus fetishizes, albeit ironically in a wholly artificial environment constituted by man himself, the Nature-driven, Purpose-driven life. Precisely in this dreamland of teleology and given purpose, doping is providing proof of concept that a *telos* can be constructed at will. If there is a 'given fate' to be accepted, it is that man is 'the fateless animal'.

It is emotionally understandable that a person regularly seeks relief from such a disorienting self-understanding and rather leaves such thoughts unthought-of. A pervasive way in which this happens is by self-deceptive storytelling about how blessed it would be if we would be meaningfully created beings, with a self-evident natural mold to fall back on and be carried away by. Nobel-Prize winning poet Wislawa Szymborska masterfully strikes this chord in her evocative poem *In Praise of Self-Depreciation* (Krynski and Maguire 1981):

> The buzzard has nothing to fault himself with.
> Scruples are alien to the black panther.
> Piranhas do not doubt the rightness of their actions.
> The rattlesnake approves of himself without reservations.
>
> The self-critical jackal does not exist.
> The locust, alligator, trichina, horsefly
> Live as they live and are glad of it.
>
> The killer-whale's heart weighs one hundred kilos
> but in other respects is light.
>
> There is nothing more animal-like
> than a clear conscience
> on the third planet of the Sun.

Animals can indeed be said to be blessed with a "clear conscience" and they can seem to "live as they live and are glad of it". As mentioned above, sports (as well as religion) can perhaps be seen as an immensely popular 'therapeutic fabulation' to satisfy this deep yearning to recuperate a lost, self-evident life propelled by an animalistic sense of purpose. Doping, as a sobering testament to our 'natureless' existence, ruins the animalistic flight of fancy that is sports.

Ultimately, however, it is vital that these flights of fancy remain recognized as such: as therapeutic fabulation, as (rationally-)irrational 'coping practices' in function of our comfort and satisfaction. On an existentialist conception of human dignity, that dignity compels us not to get too carried away by athletic or religious make-belief about living a Nature-driven, Purpose-driven life. Such practices are to be understood as artful 'play' of the *Homo ludens* (Huizinga 2008 [1938]; Suits 1978).[7] Indeed, there is a sense in which doping could drive home the 'scandalous' thesis of both Huizinga and Suits that human culture, once it has overcome the preliminary business of overcoming obstacles to welfare and free self-determination, is eventually 'play' or 'game' – that is: "the voluntary attempt to overcome unnecessary obstacles" (Suits 1978: 41). Suits speculates that a life in which the need for instrumental labour is overcome – a life in which one finds oneself 'no longer of any use' – would commonly be considered superfluous and pointless, and people would resolve this frustration by resorting to self-deception, creating for themselves all manner of purposeful activity which they believe to be of vital necessity. The whole pseudo-instrumental enterprise would, however, be a tragedy, if not a farce, of self-deception. In this sense then, doping can appear as a vindication of the dignity of the *Homo ludens*. In direct opposition to both the animalistic and religious conceptions of sports (non-artistically understood), via doping one can affirm and embrace the existentialist self-understanding that eventually, human beings are bound to live a life beyond utility and need: playing 'useless' games of their own devise, in 'useless' bodies of their own devise.

It seems quite apt to consider human life, the human body included, as one's own work of art, as an artifice of one's own will. Human nature may be exactly this: to find, at the outset, "nothing to affirm or behold beside one's own will". Human dignity then lies in accepting and affirming this Ourobouros-like predicament we find ourselves in: being fundamentally circular 'strange loops' (Hofstadter 2007). Contra Sandel, such a vision of freedom is not at all flawed. But he may be right in finding it uncomfortable to the point of desperation.

4.2.4 Talent As a Signal of Fitness and Doping As Misleading Mimicry

Turning away from 'high minded' objections of (superficial) authenticity and (false) humility, a final contender for the categorical preference for talent over doping is of a primitive and lowly nature. Both sports and beauty pageants are regularly suspected of being perilously primitive, a relapse into more brutish ways of interacting with and appreciating one another. There may be something to that suspicion.

There is one fundamental dimension in which doping cannot provide what natural talent can. Doping lacks the same fundamental thing a silicone-filled bosom lacks: contrary to naturally given phenotypical traits, the deeply integrated athletic

---

[7] In my rendition of the *Homo ludens*, I add the dimension of the self-made body to the existentialist understanding presented by Suits, who had already radicalized the existentialist dimension of the *Homo ludens* when compared to Huizinga.

or aesthetic enhancements are no *signals of hereditary potential*. They can only mimic such a signal.[8] So long as the enhancement technologies do not induce *hereditary* enhancements (which is, however, not theoretically impossible), their effects are in this hereditary sense categorically more superficial and fleeting. Even if they would be so deeply integrated that they exert their enhancing influence flawlessly throughout the entire lifespan of the doper, perhaps even more reliably and robustly as natural talent ever would, their enhancing influence would never live on in their offspring.

To the extent that sport competitions still are in part atavistic relics of the ceremonial competition between animals to demonstrate who is the alpha male or female (a point strongly argued for by De Block and Dewitte 2009), doping would dramatically corrupt this primal 'Spirit of Sport': to demonstrate who is 'truly', genetically superior with respect to physical prowess. To the extent that we still are hardwired to seek out such reliable markers of hereditary physical superiority, doping, like silicone, frustrates. From the perspective of this primal evolutionary-psychological craving, which is itself highly dubious and dangerous as a normative stance ('fascistoid' even, according to bio-ethicist Torbjörn Tännsjö (2000)), they can be regarded as superficial, 'fake' and categorically inferior abominations that corrupt this primal point to sports. This may be a way to flesh out Eric Juengst's suggestion that deeper reflection on the widespread categorical objection to doping may reveal how many still turn to sport to "glorify a genetic prejudice that the world is working hard to evolve beyond in other spheres of human life." (Juengst 2009: 176–177)

In this sense, 'natural sport' and 'natural beauty pageants' may sometimes in part be *social Darwinist* institutions that take it one step further than classic *laissez-faire* social Darwinism. Here, an interventionist brand of social Darwinism may sometimes be operating, installing 'talentocratic' measures that actively keep the playing field tilted in favour of the in-bred privilege of the talented against those who would implant equal or superior abilities with which they might match or outperform the talented.

As they have often been in the past, in their response to human enhancement interventions both athletic competitions and beauty pageants may once again be at an increased risk of falling prey to *eugenic* modes of thinking. Curiously, the risk of eugenics looms at *both* ends of the spectrum in the doping debate. In pro-enhancement writing, one can regularly find the naturalistic fallacy being brazenly flaunted, for instance when 'transhumanists' believe themselves to be carrying out the imperatives of Evolution.[9] In doing so, they fallaciously upgrade the abstract evolutionary principle of 'survival of the fittest' into a moral and existential directive. Where some overbearing transhumanists feel they must obey the evolutionary directive that the 'the *de facto* strongest must prevail', some overbearing 'talentocrats' might feel that 'the *naturally* strongest must prevail' – that the point of sport is in part really to reveal, in a literal eu-genicist fashion, who has been endowed with the best genetic

[8] Indirectly and often via the process of sexual selection, the mere capacity to mimic the presence of a hereditary trait can sometimes suffice to become a fitness-increasing trait itself (De Block and Dewitte 2009).

[9] See the analysis of such lines of reasoning in Trijsje Franssen's (Chap. 6) contribution to this volume.

material. Too manifestly unnatural ways of becoming able-bodied are then to be actively made taboo and if possible, forbidden and persecuted. To the extent that either excessively atavistic evolutionary psychology or social Darwinism and perhaps literally gene-centered eugenicism would be found present in either a transhumanist or a talentocratic spirit of sport, this would need to be exposed in detail, rigorously critiqued and supplanted by a spirit of sport that escapes these excesses. In order to avoid such excesses and all the other issues hitherto raised, I think it best to forcefully discard the exclusivist requirement of natural talent.

4.3 Proper Processes. Just Do It, or: Let Nature Do It for You

If one accepts the conclusion that a proper athletic accomplishment does not necessarily have to originate in natural talent, the most fundamental problem seems resolved. Nevertheless, one can still object that the *process* of an athletic performance must be natural in some crucial way. When discussing proper origins, concerns revolved around doping threatening to uproot a person. The proper processes concerns are seemingly more mundane. No existential crises are feared of doping sinners becoming debased, hubristic and aporic 'hyper-agents' (Sandel 2007). On the contrary, the doping athlete is here experienced as a dud – a non-agent. The overt action is only *apparently* undertaken by the athlete. In reality, due to the aid, it actually becomes a non-act to a substantial degree. A doping athlete is then seen as one who fraudulently tries to pass as the agent accountable for her feats while in fact she relied on a 'hidden helper' which did a substantial amount of the work *for her.*

The issue seems to revolve around *accountability:* is a doped athlete really performing the feat herself, can she be held accountable for the outcome, does she deserve the pride and the praise, the position and the prize?

It is intuitively appealing to think that no, she doesn't deserve full credit for the performance, and if the context would be that an athlete A took recourse to such a 'hidden helper' while her competitors B to E all competed without it, 'on their own inner strength', then A cheated or is at least less deserving of her final position, even in the absence of an explicit ban: regardless of regulations and stipulations, it would be an offense against this intrinsic, essential characteristic of virtuous sport – that you yourself do the work.

We can even take it one step further. As the anti-doping author Bill McKibben notes while contemplating the ethical inner dynamics of his own recreational running experience, even in a *non-competitive* context would the insertion of such a foreign doping agent in your own body cause an unacceptable accountability problem. According to McKibben, this would upset the beneficial, natural way in which you can consider yourself the author of your own actions, and the interference of the performance enhancer with your natural capacities would send you reeling in "a spiraling self-doubt" (McKibben 2004: 55) of not being able to ascertain if it was *'really you' who is accountable* (to be distinguished from the dimension mentioned in the Proper Origins section, where a spiralling self-doubt lay in it becoming muddled *who 'the real you' is* – that is: of a person no longer knowing who she is anymore). For this reason, performance enhancing drugs may be found ethically impermissible in a categorical way, because as soon as you allow yourself to become 'mixed' like that, you will be at a loss to determine just what you have done 'on your own', and what was done by the dope.

Appealing as this alarmism may sound as a *prima facie* credible account of proper personal accountability, we must see how this view fares after taking into account the following three issues: (1) the possibility of *agency-enabling* doping, which lifts internal barriers so that one it put at liberty to engage in *increased* exertion; (2) the possibility of *baseline-lifting* doping, which increases the amount of feats one can perform without exertion only to put one at liberty to engage in *identical* levels of exertion when performing further feats at a higher level of complexity; and (3) the critical comparison of the possibility of passive reliance on doping to the pervasive reality of our passive reliance on (semi-)automatic assistive processes of our default biological constitution. We must thereby also ask the question why the relatively effortless natural grace of the naturally talented commonly inspires such awe whereas the realization of ease in performance via doping often inspires disgust and outrage.

4.3.1 Agency-Enabling Doping

If we move past armchair philosophy in these matters and look *empirically* at the diversity and detail of how the human body can be stimulated by all sorts of biotechnological interventions, we find that enhancement interventions needn't be *intrusive,* working you like a puppet. They may well be 'eruptive', lifting internal blockades on agency. In this sense, for instance anabolic steroids are not at all 'mechanizing' or 'passivity-inducing': they do not make muscle growth 'easy' or 'automatic' as a tenacious caricatural misconception has it. In fact, as Andrew Holowchak recounts in his contribution to this volume, it is the other way round: the actual influence of steroids on the body is that they make the body recover more quickly after very intensive exhaustion, so that you may exhaust the body anew at a quicker pace *provided* that you can muster the willpower to do so. Steroids do not do the work for you, on the contrary*:* they provide you with an opportunity to increase your active effort to exhaust yourself, which you still have to be motivated for. You still have to 'just do it' yourself. Ergo, powerlifting guru Louie Simmons is confident when he says: "the people who train with anabolic steroids train way harder than the people who don't." (Simmons in Bell 2008) As such, it seems to be an overstatement to think that all doping agents would 'do the work for you' and erode the degree of active engagement and willful effort beneath crucial thresholds. As the previous fluke and phoney critiques, this critique of passivity needs to be deflated from a categorical critique to a probabilistic, circumstantial one.

4.3.2 *Baseline-Lifting Doping*

What is more, even in the case of specific doping agents which do increase your athletic ability in some decisively 'mechanizing' or 'passive' way, such as blood doping or oxygen chambers seem to do, the most obvious effect, amply demonstrated in the history of sport when new facilitating support practices are being introduced (Van Hilvoorde et al. 2007) is that the athletes with enhanced baseline abilities will seek out new, more intensive and trying challenges. An all-encompassing conception of doping as an 'outsourcing of effort', setting us on a slippery slope towards a culture of slothful, passive technology consumers (McKibben 2004; Kass 2002) must be dismissed as hyperbolic.[10] As long as we are be able to retain our ability to relocate our locus of active engagement, the 'outsourcing of effort' – be it to some natural or some technological process, be it internal or external – is not intrinsically problematic: after we allowed fiber poles in pole jumping, we raised the bar and the spirit of sport lived on; and if for instance endurance enhancing doping such as blood doping, oxygen chambers and EPO-injections can be provided adequately safe and equitable, we could add an extra epic *col* to the Tour de France – or, alternatively, we could use such doping practices to *reign in* current excesses: to remake contemporary cycling into a somewhat more healthy and safe enterprise.[11]

4.3.3 *Passive Consumption of Natural Processes*

Again, it merits to redirect the critique directed at doping towards natural talent itself, and see how that fares. There seems to be little wrong with relying 'passively' on automatic or semi-automatic bodily processes, for that is the way the vast majority of our countless bodily processes operate, from blood flow to oxygen uptake over habituated skill and countless ways of subconscious mental pre-processing, by which we outsource enormous amount of effort so that our mind is freed up to be occupied with further, more complex activities. Using the inflated rhetoric of McKibben and Kass, we could say that if we are to some extent 'slothful, passive technology consumers', then we are to an incomparably greater extent 'slothful, passive *nature* consumers'.

Although the spirit of sport can at times be seen to revolve around notions of effort and exhaustive training, intriguingly, the athletes considered to be 'the greatest of them

[10] Although certainly, those susceptible to sloth (and this group may well be demographically very large) may indeed use such baseline-lifting doping as a way to lessen the need for their own active agency.

[11] Consider in this regard Denis Hauw's (Chap. 12) analysis in this volume of how contemporary elite cyclists often revert to dope simply in order to cope with the excruciating demands put on their natural constitutions by organizers and audiences.

all' are often those who to a certain but crucial extent exhibit *the exact opposite* of effort and exhaustive training: public exhilaration is heightened to a whole new level if an athlete can pass a heroic athletic test 'without even breaking a sweat'. The natural grace of a Usain Bolt or a Michael Jordan inspires the greatest awe:

> [S]triving is not the point of sports; excellence is. [...] No one believes that a mediocre basketball player who works and trains even harder than Michael Jordan deserves greater acclaim. (Sandel 2007: 28–29)

> [E]ven if we are prepared to admire people who have worked hard [...] I believe that we will have added admiration for a person who excels without having worked hard. If a middle-aged member of the audience who has never exercised unexpectedly walked down from the stadium and joined the Olympic 10,000 m race and, because of superior natural talent, defeated all the finalists, the success would be formidable. Our admiration for this person would be unlimited. So, basically, it is talent (which can be genetically explained), not achievement, that we admire above else. The point of the contest is to show who has the most superior talent. (Tännsjö 2000: 18)

But this confronts us with two apparent contradictions:

1. The WADC-spirit of sport is often portrayed to consist crucially in *effort and training*, but if someone with great natural talent can perform the athletic feat *(comparatively) effortlessly*: even better.
2. If a *gifted* athlete performs effortlessly, that is good or even wonderful, but if a *doped* athlete would perform effortlessly, that is bad or even horrendous.

These contradictions can be made sense of in the following two ways. To explain the first apparent contradiction, imagine a 200 m sprint where two runners, A and B, cross the finish line at the same time, but A has had to invest all his effort and staggers unto the ground panting, whereas B (B for Bolt Usain, say) runs unexhausted toward the cameras to mimick a lightning bolt striking at the speed of light. The fully exhausted A may be greatly appreciated for having made the excellent time, and for having demonstrated the impressive character traits of extraordinary determination and willpower. But with regard to athletic skill per se, athlete B can be admired more because he has clearly not exhausted all of his running capacity: compared to A, Bolt still has a reservoir of untapped potential, and thus demonstrates by his lack of fatigue and his lesser effort that he has a potential for running skill superior to A's. They may have crossed the line at the same time on this occasion, but should Bolt choose to also invest the extraordinary perseverance of A, his performance would outmatch A's.

The second contradiction, an apparent double standard, can be made intelligible by referring back to the social Darwinist and eugenicist understandings of natural sport. Today's class of the effortless naturally talented can be seen as the heirs of the radically different 'gentleman-amateur' spirit of sport that heavily influenced the Olympic Games and other sports milieus until mid-twentieth century, and was its official doctrine until the late 1980s (Guttman 2002). In that spirit of sport, sport was supposed to be a *leisurely* activity, certainly not something to spend all your waking hours on, let alone to engage in as a *profession* – that would be disgracefully obsessive and beside the point of sport as a wholesome display of readily available bodily capacity. Certainly in its nineteenth century beginnings, modern sport was meant to exhibit a spontaneous 'natural nobility'. The 'amateur gentlemen' athlete was to prove his noble pedigree via good sporting capacities, which provided proof 'of being of good stock' in an (often implicit, but occasionally explicit) eugenic sense. A sport performance, though very energetic and exhaustive, was nevertheless meant to exhibit, if it was to be truly outstanding, a measure of cool off-handedness – to come across as spontaneous demonstrations of stallions who have 'plenty more where that came from'. Today sport may have been turned into an industry filled with people *working* as athletes and spending their entire (young) life in specialized training facilities, and viewed from the old amateur spirit of sport, this would be seen as an unsavoury medico-scientific conditioning of people like rats in a cage (Young 1996, a critique I subscribe to myself), which only threatens to skew the view on who belongs to 'the *natural* aristocracy' (Jefferson 1988 [1813]: 387–391; Mehlman 2009). Interestingly, however, even in the contemporary high-intensity sport industry, sport fans commonly continue to reserve a modicum of extra appreciation to those athletes who still manage to exhibit that Old World regal sense of being 'nature's aristocracy' (for instance the flair of Michael "His Airness" Jordan) or being, in a more straightforward way, pure 'forces of nature' (for instance the frolicking Usain "Lighting" Bolt). This, then, is how the double standard about effortlessness can be made sense of: such 'off-handed' effortlessness of Jordan, Bolt and consorts is something completely different than the 'hands-on' effortlessness induced by an intrusive intervention in one's body with the explicit intention to maximize ones athletic performance. Not only could such endeavours of biotechnological self-change bespeak a fundamental character flaws of over-eagerness (a probabilistic critique I again subscribe to myself), moreover they are a corruption of sport as a transparent testing ground for talent-centered, genetic hierarchies (a categorical critique I would, on the contrary, staunchly oppose as indicative of a social Darwinist if not eugenicist interest).

In sum, as discussed here, these three concerns about proper processes fail to provide compelling reasons to demand that the processes of athletic accomplishment be *natural.*

Concerned about the neglect of the proper processes of human activity, Leon Kass writes: "there is a sense that the "naturalness" of means matters. It lies not in the fact that the assisting drugs and devices are artefacts, but in the danger of violating or deforming the deep structure of natural human activity" From this concern, he goes on to famously conclude that "the engaged and energetic being-at-work of what nature uniquely gave to us is what we need to treasure and defend." (Kass 2003) From my existentialist virtue-ethics perspective, I share much of this concern: we indeed need to treasure and defend our engaged and energetic being-at-work, and self-abdication into slothful passivity and fatalism must be strongly resisted. But ironically, it is for that exact same reason that I cannot accept the exclusivist clause that this must be done and can only be done with "what nature uniquely gave us". Not only does this neglects (or in the worst case: denigrates) those who are engaged and energetically-at-work *with what culture and technology gave them.* For instance, the double amputee and 'blade runner' Oscar Pistorius provides a powerful example of the dignified way in which a person can deeply submerge himself in a partly artificial bodily predicament, and our spirit of sport should honour such dignified interactions with deeply integrated artifice. (See also Chap. 13 by Magdalinski, this volume) Moreover, it neglects the fact that the virtue of temperance should also be applied to the extent in which we adapt a passive consumerist stance with regard to how (semi-)automatic natural processes do the work for us, and to the extent in which talentocrats would improperly claim intentional accountability and personal merit for beneficial natural endowments that are not their own doing, but rather their arbitrarily obtained good fortune. To conclude, these considerations seem to result in the following general virtue-ethical criterion about doping practices with regards to the issue of Proper Processes taken in isolation:

> Further possible issues suspended, insofar as a doping practice only *provides a constitution* from which all worthwhile aspects of engaged and energetic being-at-work are safeguarded, it is permissible. Insofar as a doping practice enlarges the extent in which one can be engaged and energetically-at-work, it is commendable and may be positively dignified.

4.4 Proper Outcomes. Sporting Towards a Blank Slate or To Showcase a Blueprint

If one accepts the conclusion that a proper athletic accomplishment does not necessarily have to draw on natural origins nor take place via natural processes, the two most formidable[12] problems have been resolved. One might, however, still object that the end states reached in a proper athletic accomplishment must always strive towards a (perfection of) the natural human figure and functions. that is: it must result in (the perfection of) an embodiment and performances that remain within the phenotypical range of *Homo sapiens*.[13]

Modern sport can be seen as a twin project: the perfecting of man, together with the *purification* of man. When Pierre de Coubertin rekindled the Olympian Flame in 1894 there was a lively sense that human biology still harboured vast amounts of untapped potential. Human living conditions were to a large extent unsanitary and unscientific, so there were countless medical, hygienical, nutritional, scouting and coaching advances lying ahead. By scouting for the most well-endowed individuals of the human race, improving their hygiene and nutrition, and enlisting them in a

---

[12] As I believe these concerns are relatively less profound and as space does not permit further extensive argument, I will deal with this third set of intrinsic objections, those on defiguring 'dehumanization', in a more brief manner, and several intriguing twists and turns will therefore be left unpursued here.

[13] See, for instance, Tara Magdalinski's (Chap. 13) contribution to this volume on the 'natural romanticism' that fueled the creation of modern sports as a wholesome activity. In addition, however, John Hoberman describes in his contribution the long history of the *Leistungsprinzip* or Performance Principle as a motor of modern sport. For more on the long history of human intervention in nature in the name of increased productivity and efficiency, see Claude Olivier Doron's introduction to Missa and Nouvel (2011).

meticulously planned and monitored scheme of training, the 'purified and perfected man' would be approximated ever more closely. From that historical vantage point, the Olympics could be given the motto *Citius, Altius, Fortius* (Faster, Higher, Stronger) from the best intentions: it was a clarion call for spreading the liberal dream of public health programs, the integration of physical exercise is a well-balanced art of living, open meritocracy and humanism predicated on the unity of mankind (Coubertin 1992 [1913]). *In tempore non suspecto*, it was not yet read as an implicit invitation to perform-at-all-cost and obsess over outward results, the naturalness and the virtue of the athlete be damned. The project of Coubertin and his successors was that displaying virtuoso expressions of the universal 'blueprint of humanity' seen as common heritage of mankind that is to be cherished, conserved, purified and perfected. The spirit of sport was one of 'Vitruvian perfectionism', after the image of Leonardo da Vinci's blueprint drawing of the perfectly proportioned, well-created Vitruvian Man. (See also Chaps. 7, 10, 13 and 14 by Meacham, Magdalinski, McNamee and Hoberman, this volume).

4.4.1 Reshaping the Human Figure As Straying from the Original Plan

The belief that humans are the more or less imperfect instantiations of an underlying pure template (a Platonic Form, a divinely ordained blueprint, an essentialist 'natural kind', etc.) has been decisively discredited by an overwhelming body of scientific evidence. Human biology is in no way a fixed natural kind containing a distinct essence of humanity. Nor can mankind be seen as some sort of pinnacle of the natural world – a highest or most perfect species. Such beliefs, and refined variations thereupon, nevertheless remain widely held and are occasionally explicitly articulated and argued for, most noticeably in theologies of sport (Weir 2011, see also John Hoberman's contribution to this volume). I cannot go into the relevant details of evolutionary science in the space of this chapter. Moreover, both Andreas De Block and Eric Juengst already present some basic facts to debunk such notions of species essentialism in their contributions to this volume. Suffice it to say that if the Vitruvian dimension of modern Olympism would imagine itself to be in pursuit of a matter of (naturalistically demonstrable) fact, it is deeply mistaken.

4.4.2 Reshaping the Human Figure As Repugnant Deformation

Alternatively, one could argue for the categorical denunciation of doping on the grounds of a deeply felt repugnance. Such profound sentiments of repugnance in the face of too aberrantly 'unnatural' entities and practices have been intensively studied in (evolutionary) psychology. Doping and the deviant embodiments it may bring about can indeed ignite deeply ingrained aversions. For instance, the practices of ingesting, injecting or implanting some enhancing substance may spark aversion based on the avoidance of 'contamination'-like interactions, even if, in reality, the substance would be perfectly healthy or even health-improving (as, for instance, the widespread 'enhancement' practice of vaccination does). Also, introduction of enhancing substances drawn from other animals may spark aversions based on the avoidance of species intermingling, as it might also frustrate the deeply ingrained (but all too 'rough and ready') cognitive mechanisms to classify entities into distinct and essentialistically understood 'natural kinds'. To conclude, doping that effectively 'deforms' the natural human form is also likely to jar the manifold ways in which the human mind is preconfigured to react to the particularities of characteristically human features – such as characteristically human body shapes, facial features, movement patterns, odours, etc. The deviant embodiments doping can bring about might run counter to these deeply entrenched perceptive and evaluative systems, creating possible intolerable levels of confusion, frustration and aversion, which may in turn solidify into a stance of 'allergic' repugnance, intolerance and, in the most extreme cases, outright hatred and aggression against those who dope and turn themselves into 'entartete' deviants of the normal, natural human mold.

Whether they are shaped by these quirks of evolutionary psychology or not, and whether they cannot be helped or instead can be mitigated or fully overcome, in principle everyone is entitled to have the tastes and distastes one has – no questions asked.[14] As such, a value community that finds doping categorically repugnant could be allowed to organize sports activities in which doping athletes will not be allowed because that community finds them repugnant – similar to how a swing dance troupe may enforce an internal ban on cha-cha should they for instance find cha-cha intolerably campy and a ruination of the brittle and unique internal spirit of swing, which they perceive as the only way to dance like one should.

This, however, works both ways. Certain (minoritarian) value communities may find very attractive, on account of their own aesthetic idiosyncrasies, certain 'unnatural' bodily figures and functionalities (such as for instance enormous amounts of bulging muscle mass and an 'inhuman' weightlifting capacity, hyper-flexible limbs and an 'inhuman' snake-like flexibility, surgically split tongues, abnormally large and spherical breasts, scarred, pierced or tattooed skin, etc.) or figures and functionalities that are not 'naturally', normally found attractive (baldness, flabbiness, a combination of male and female sexual traits, etc.). Even if some such practices may be met with widespread or near-universal repugnance, in a liberal, human rights abiding society we must be watchful that such majoritarian repugnance does not turn into an officially enforced 'oppressive taste'. (See Chap. 16 by Brownsword, this volume) Moreover, there may be ethical reasons to encourage the aesthetic

[14] For instance, in his contribution to this volume Andreas De Block (Chap. 8) examines the possibility to ground the categorical objection to doping on such widespread, evolutionary rooted distastes. The anti-doping project would then no longer have to argue for the increasingly contested position that doping is categorically *unethical*, but instead could take the more robust line of argument that doping is categorically *un-aesthetical*.

appreciation of the two types of deviance mentioned, as they may lessen the pressure put on people to conform to (possibly suspect and oppressive) norms of 'naturalness' and 'normality'. For this reason, some pro-enhancement movements align themselves with the disability rights movements, with the liberation from uniform beauty ideals, etc. as the way forward for accepting deep societal diversity (see for instance Chap. 13 by Tara Magdalinski, this volume, Carrico 2009)

4.4.3 Reshaping the Human Figure As (Mutual) Alienation

The celebration of the peaceful cohabitation of mutually deviant lifestyles and embodiments can, however, easily turn into an overly naive neglect of the enormous practical difficulties that are raised by constant and pervasive societal friction between differently natured communities. Consider, for instance, the difficulties to manage the peaceful cohabitation of differently *cultured* communities in multicultural political constellations. Moreover, such celebration might tip over into a somewhat nonsensical appreciation for diversity and deviance for their own sake, in which the maintenance and construction of communitarian ties threatens to be neglected. As Darian Meacham argues in his contribution to this volume, perhaps not a biological but a more symbolical, *phenomenological* sense of species recognition and belonging seems to be a fundamental human need. This, however, can be historically negotiated and constructed. Indeed, in line with Dale Carrico's thoughtful though informal writing in this regard (2009), I equally wish to underscore, contra libertarian and technophile transhumanism, the primordial importance of civic and communitarian negotiation in accommodating disturbingly deviant, 'queer' (embodied) lifestyles. To take a sports-specific example, there is a legitimate demand that the activity of differently embodied athletes can be compared in a meaningful and practicable way. For instance, it is legitimate for organizers of running meets (1) to investigate whether Oscar Pistorius, when he is taking great strides on prosthetic blades, is engaged in an activity that can still be categorized under the negotiated rubric of 'running' – and not, for instance, under a substantially different rubric of 'blading'. (2) Should substantial differences be found, a further negotiation should be engaged in to decide whether the kind and degree of the difference can be adequately assessed and compared. (3) If, after a final study, corrective circumstances can be added so as to redress irrelevant inequalities between the differently embodied athletes, it may still prove possible to conserve both the "sweet tension of uncertainty of outcome" (Fraleigh 1984) as well as the competition as a test of relevant capacities, most importantly as a test of character and will power, even if athletes are substantially 'other' or 'alien' towards one another. As the specifics of such methods of inclusive organization entail largely 'logistical' issues instead of ethical ones, I will not pursue them further here. However, for such humane, inclusive negotiations to take place, it seems imperative that the demands for dominance of innate natural talent and for *biological* instead of phenomenological species similarity be relaxed, and as I would argue together with Max Mehlman ground-breaking writing (2009) on these issues, actively discarded.

4.5 Conclusion. The Unbearable Lightness of Being a Self-Made Man, Out There Playing Games

On the one hand, doping – constructing a substantially artificialized embodiment and drawing on that embodiment to realize athletic performances – has been analysed to be neither debasing, mechanistic or dehumanizing. On the other hand, the demands for the necessary presence of natural talent, for its dominant presence, and for taking protectionist measures to ensure that natural talent remains a necessary and dominant requirement have been analysed as 'talentocratic' demands that can become riddled with dubious comfort-seeking, overly superficial conceptions of authenticity and agency, oppressive prejudice against deviants and in the worst case scenario: share with certain transhumanisms the flirtation with social Darwinism and literal eu-genics: holding in higher esteem those who happen to have some genetic endowment deemed better.

This is not to say that we should turn a blind eye to talent. Our talents, together with the cravings and needs that emanate from the constraints of human biology, can legitimately play a dominant role in 'fleshing out' the specifics of what is and what is not wholesome to pursue, as they are make up a fair part of the tangle one finds oneself thrown in (Bonte 2011). They are largely what make us get up in the morning and what keep us (pre)occupied throughout the day. But existentially, their role is secondary (Bayertz 2003): having a talent, or being natured in a certain way, does not make the cultivation of that talent or the conservation of that nature meaningful. One's talents, one's nature, and all the comfort and satisfaction one may draw from it, may still be both ethically dubious and existentially absurd. We deny the depth and difficulty of human existence if we con ourselves into thinking that there is a 'given' human mold we all share, together with a mold with specific traits, talents and afflictions 'given' to each individual, that is meaningful in and of itself. We are not 'meant' to have the improvised, ramshackle nature given to us by the 'Blind Watchmaker' of natural selection (Dawkins 2006; Buchanan 2011), nor are we 'meant' to continue the business of evolution, as some transhumanists would have it. Rather, we are free and forlorn – that is the (perhaps poisoned) 'gift' of life as a human, which spoils and makes impossible the self-evident, passive life of an animal, no matter our nostalgia for the lost comforts of that Arcadia. This existentialist self-understanding does not deliver some final blow to religion. Indeed, it can itself be a religious sensibility, such as in the hallmark 1486 text of renaissance humanism, Pico della Mirandola's *Oratio on the Dignity of Man*, where the allegorical gods inform Man of his plight:

> Adam, we give you no fixed place to live, no form that is peculiar to you, nor any function that is yours alone. According to your desires and judgment, you will have and possess whatever place to live, whatever form, and whatever functions you yourself choose. All

> other things have a limited and fixed nature prescribed and bounded by our laws. You, with no limit or no bound, may choose for yourself the limits and bounds of your nature. (Mirandola 1999 [1486])

It can also, of course, be a secular sensibility, borne from a sobering understanding of the randomness of biological evolution and an uncompromising adherence to the force of the naturalistic fallacy.

I believe we cannot, without self-deception, *choose against self-creation*. Standing on the tipping point of the tense cognitive dissonance between *finding ourselves* foundationlessly free and ruthlessly responsible on the one hand, and *wanting ourselves* to be self-evidently driven by a nature that was meaningfully given on the other hand, we cannot abdicate from choosing some resolution of the dissonance. But we can make a fundamental choice on how we will rise to that existential occasion. We can either create a counterfactual belief that we are *not* the creator and carrier of our own life project, or we can (with courageous humility even) accept and affirm our self-creating predicament. Though apparently dubious, counterfactual beliefs in nature's normativity can be a highly wholesome coping practice – as invigorating mythmaking to provide repose from existential gravitas. I have argued that this can be an enjoyable opiate integrated in a virtuous art of living, when dosed in good measure. But as a constant belief, held in earnest and motivationally rooted in a "fear of freedom" (Fromm 1960 [1942]), it becomes a cause of great ethical concern.

I cite in full agreement that "[a]n untroubled soul in a troubling world is a shrunken human being" (Kass 2003) and that we must not "remake nature, including human nature, to serve our purposes and satisfy our desires" (Sandel 2007: 26–27). Yet my argument leads me to conclude that Kass' and Sandel's proposals may ironically root in the very indignities they rail against. Building on humanist-existentialist strands of thought, the violation of human dignity occurs not by transgressing supposedly ethically charged natural constraints on our existence. In a sense, to the contrary. A person violates her dignity when she falls beneath a treshold of maintaining moral character and realizing virtues, such as truthfulness and moral courage. Now, when a person denies our foundationlessly free and ruthlessly responsible predicament, and grounds her denial not in rational rebuttal but in a belief for which no argument has been given and which is conspicuously convenient (*in casu*, that one has been given a meaningful nature in which one can find true purpose and authentic fulfillment), then indignity seems nigh.

I have sought to identify and critique a cluster of 'talentocratic' assumptions on which much intrinsic, categorical anti-doping thinking seems to be premised, and have pitted it against a positive existentialist account of human dignity. Pending further scrutiny, I would wager that it is better – not in terms of contentment and happiness but in terms of dignity and the love of truth – to first gauge our god- and nature-forsaken condition to the fullest, and then from that vacuous and circular predicament, testify of our will to live in truth, beauty and goodness by improvising something out of ourselves – body and all – in a 'virtuous exploration of bodily virtuosity'.

Perhaps, beneath the governing taboos and odium in public life today, at intersections of virtue-ethical undercurrents with clandestine doping underworlds, such virtuous explorations have already been undertaken in more or less erratic, confused, conflicted or understated ways. Even if contemporary doping practices would all prove to be deeply corrupted by the thicket of *other* problems non-intrinsic to doping itself (and should space permit I would argue that such corruptions do indeed run deep and wide given the perfectionist, puritan, individualist, domineering and exploitative values running through contemporary capitalist culture, sports culture surely not excluded), that would still not allow us to categorically conclude against doping. Instead, it would call on us to combat those corruptions not from 'anti-doping' grounds, but from an understanding of all that needs to be put in place before doping can be done in dignity.

References

Baillie, H. 2005. Aristotle and genetic engineering: The uncertainty of excellence. In *Is human nature obsolete? Genetics, bioengineering, and the future of the human condition*, ed. H. Baillie and T. Casey, 209–232. Cambridge, MA: MIT Press.

Bayertz, K. 2003. Human nature: How normative might it be? *The Journal of Medicine and Philosophy* 28(2): 131–150.

Bell, C. 2008. *Bigger, stronger, faster. Is it still cheating if everyone is doing it?* (Documentary). New York/Austin: Magnolia Home Entertainment.

Bonte, P. 2008. De verbeterkunde: een ongewenste bevrijding? *Ethiek en Maatschappij* 11(2): 18–35.

Bonte, P. 2011. Why should I be natural? A fivefold challenge to the supposed duty to 'be natural' as grounds for outlawing human enhancement. In *Technologies on the stand: Legal and ethical questions in neuroscience and robotics*, ed. B. van den Bergh and L. Klaming, 215–248. Nijmegen: Wolf Legal Publishers.

Buchanan, A. 2011. *Beyond humanity. The ethics of biomedical enhancement*. Oxford: Oxford University Press.

Buchanan, A., D. Brock, N. Daniels, and D. Wikler. 2000. *From chance to choice: Genetics and justice*. New York: Cambridge University Press.

Carrico, D. 2009. A condensed critique of transhumanism. Using technology to deepen democracy, using democracy to ensure technology benefits us all. *Amor Mundi*. http://amormundi.blogspot.com/2009/01/condensed-critique-of-transhumanism.html. Accessed 27 Mar 2012.

Coubertin, P. 1992 [1913]. In *Essais de psychologie sportive*, ed. Rioux Jean-Pierre. Paris: Editions Jérôme Million.

Dawkins, R. 2006a. *The blind watchmaker: Why the evidence of evolution reveals a universe without design*. New York: Norton.

Dawkins, R. 2006b. *The selfish gene*, 30 Anniversaryth ed. Oxford: Oxford University Press.

De Block, A., and S. Dewitte. 2009. Darwinism and the cultural evolution of sports. *Perspectives in Biology and Medicine* 52(1): 1–16.

DeGrazia, D. 2005. *Human identity and bioethics*. New York: Cambridge University Press.

Della Mirandola, P. 1999 [1486]. Oratio on the dignity of man. In Reading About the World, Volume One, ed. Paul Brians, Mary Gallwey, Douglas Hughes, Azfar Hussain, Richard Law, Michael Myers Michael Neville, Roger Schlesinger, Alice Spitzer, and Susan Swan. Fort Worth: Harcourt Brace Custom Publishing.

Elliot, C. 2003. *Better than well: American medicine meets the American dream*. New York: W. W. Norton and Company.

Fraleigh, W. 1984. *Right actions in sport: Ethics for contestants*. Champaign: Human Kinetics Publishers.

Frankfurt, H. 1971. Freedom of the will and the concept of a person. *Journal of Philosophy* 68(1): 5–20.

Fromm, E. 1960 [1942]. *The fear of freedom*. London: Routledge.
Guttman, A. 2002. *The Olympics, a history of the modern games*, 2nd ed. Champaign: University of Illinois Press.
Hoberman, J. 2009. Putting doping into context: Historical and cultural perspectives. In *Performance-enhancing technologies in sports. Ethical, conceptual, and scientific issues*, ed. Thomas H. Murray, Karen J. Maschke, and Angela A. Wasunna, 3–27. Baltimore: Johns Hopkins University Press.
Hofstadter, D. 2007. *I am a strange loop*. New York: Basic Books.
Huizinga, J. 2008 [1938]. *Homo Ludens. Proeve Eener Bepaling Van Het Spel-Element Der Cultuur*. Amsterdam: Atheneum.
Hurley, S. 2002. Luck, responsibility and the 'natural lottery'. *Journal of Political Philosophy* 10(1): 79–94.
Huxley, A. 2006 [1946]. *Brave new world*. New York: Harper Perennial Modern Classics.
Jefferson, T. 1988 [1813]. *The Adams-Jefferson letters: The complete correspondence between Thomas Jefferson and Abigail and John Adams*, ed. Lester Cappon. Chapel Hill/London: University of North Carolina Press.
Juengst, E. 2009. Annotating the moral map of enhancement: Gene doping, the limits of medicine and the spirit of sport. In *Performance-enhancing technologies in sports. Ethical, conceptual, and scientific issues*, ed. Thomas H. Murray, Karen J. Maschke, and Angela A. Wasunna. Baltimore: Johns Hopkins University Press.
Kass, L. 2002. *Life, liberty, and defence of dignity: The challenge for bioethics*. San Francisco: Encounter Books.
Kass, L. 2003. Ageless bodies, happy souls. *The New Atlantis* 1: 9–28. http://www.thenewatlantis.com/publications/ageless-bodies-happy-souls. Accessed 29 Mar 2012.
Krynski, M., and R. Maguire. 1981. *Sounds, feelings, thoughts: Seventy poems by Wislawa Szymborska*. Princeton: Princeton University Press.
Levy, N. 2011. Enhancing authenticity. *Journal of Applied Philosophy* 28(3): 308–318.
Loland, S., and H. Hoppeler. 2012. Justifying anti-doping: The fair opportunity principle and the biology of performance enhancement. *European Journal of Sport Science* 12: 347–353.
McKibben, B. 2004. *Enough. Staying human in an engineered age*. New York: Henry Holt and Company.
Mehlman, M. 2009. *The price of perfection. Individualism and society in the era of biomedical enhancement*. Baltimore: Johns Hopkins University Press.
Missa, J.-N., and P. Nouvel. 2011. *Philosophie du Dopage*. Paris: Presse Universitaire de la France.
Murray, T. 2009a. Ethics and endurance-enhancing technologies in sport. In *Performance-enhancing technologies in sports. Ethical, conceptual, and scientific issues*, ed. Thomas H. Murray, Karen J. Maschke, and Angela A. Wasunna, 141–159. Baltimore: Johns Hopkins University Press.
Murray, T. 2009b. In search of an ethics for sport: Genetic hierarchies, handicappers general, and embodied excellence. In *Performance-enhancing technologies in sports. Ethical, conceptual, and scientific issues*, ed. Thomas H. Murray, Karen J. Maschke, and Angela A. Wasunna, 225–238. Baltimore: Johns Hopkins University Press.
Nozick, R. 1974. *Anarchy, state, and utopia*. Oxford/Cambridge: Blackwell.
Parfit, D. 1984. *Reasons and persons*. Oxford: Oxford University Press.
President's Council on Bioethics. 2003. *Beyond therapy. Biotechnology and the pursuit of happiness*. Washington: Dana Press.
Sandel, M. 2007. *The case against perfection. Ethics in the age of genetic engineering*. Cambridge/London: The Belknap Press of Harvard University Press.
Sartre, J-P. 1989 [1946]. Existentialism is a humanism. Translated by Philip Mairet. Reproduced from *Existentialism from Dostoyevsky to Sartre*, ed. Walter Kaufman. Meridian Publishing Company. http://www.marxists.org/reference/archive/sartre/works/exist/sartre.htm. Accessed 29 Mar 2012.
Savulescu, J., B. Foddy, and M. Clayton. 2004. Why we should allow performance enhancing drugs in sport. *British Journal Of Sports Magazine* 38: 666–670.
Schechtman, M. 2004. Self-expression and self-control. *Ratio* 7: 409–427.

Schwartz, B. 2004. *The paradox of choice: Why more is less*. New York: HarperCollins Publishers.
Starckx, S. 01/07/2008. Interview ex-schaatskampioen en sportwetenschapper Harm Kuipers. 'Doping is voor een groot stuk placebo.' *EOS Magazine*: 116–119.
Suits, B. 1978. *The grasshopper: Games, life and utopia*. Edinburgh: Scottish Academic Press.
Tallis, R. 2007. Enhancing humanity. *Philosophy Now Magazine*. http://www.philosophynow.org/issues/61/Enhancing_Humanity. Accessed 29 Mar 2012.
Tännsjö, T. 2000. Is it fascistoid to admire sports heroes? In *Values in sport: Elitism, nationalism, gender equality, and the scientific manufacturing of winners*, ed. Tännsjö Torbjörn and Claudio Tamburinni. London: Kluwer Academic Publishers.
Van Hilvoorde, I., R. Vos, and G. de Wert. 2007. Flopping, klapping and gene doping: Dichotomies between 'Natural' and 'Artificial' in elite sport. *Social Studies of Science* 37(2): 173–200.
Weir, J.S. 2011. Theology of sport – historical overview. *Verité Sport*. http://www.veritesport.org/downloads/Theology_of_Sport_An_historical_review.pdf. Accessed 27 Mar 2012.
World AntiDoping Code. 2009. *World Anti-doping Agency*. http://www.wada-ama.org/World-Anti-Doping-Program/Sports-and-Anti-Doping-Organizations/The-Code/. Accessed 27 Mar 2012.
Young, D. 1996. *The modern Olympics: A struggle for revival*. Baltimore: Johns Hopkins University Press.

Part II
Transgressing the Limits of Human Nature

Chapter 5
Subhuman, Superhuman, and Inhuman: Human Nature and the Enhanced Athlete

Eric T. Juengst

5.1 Introduction

At the core of some critiques of performance enhancement in sports lies a negative account of human nature. It is not negative because it is a pejorative account; in fact, its proponents are committed to protecting and preserving the human nature that this account describes above all other goods. But it is an account that defines human nature negatively, in terms of what it is not, rather than through a positive vision of what it might be. It boils down to the recognition that human nature is not, by definition, subhuman, superhuman, or inhuman. The subhuman is less than human, like a beast; the superhuman is more, like a god; and the inhuman something simply different, like a machine. From this contrast-definition flows a simple argument: if we agree that preserving the human nature of athletes is morally important, we should strive to protect them against interventions that would cross any of those boundaries. To the extent, then, that performance enhancing interventions in athletes do risk such transgressions, they are ethically suspect. This logic will indict transgressive enhancement interventions even if other ethical concerns about the interventions' physical safety, the fairness of their distribution, and the social consequences of their acceptance can be addressed. As a result, the critics can conclude, the only critical challenge in condemning any particular form of performance enhancement is to identify the ways in which it 'dehumanizes' athletes against one of these boundaries: if that can be accomplished, the day is won.

This chapter reflects on this argument by examining the ways in which it is fleshed out and deployed in the contemporary literature on performance enhancement in sports. I argue that, while seductive, the content-less vision of human nature employed in these critiques leaves the critics in a dangerous trap: in their efforts to

E.T. Juengst (✉)
UNC Center for Bioethics, University of North Carolina, 333 MacNider Hall,
Campus Box 7240, Chapel Hill, 333 S. Columbia Road, 27599-7240 Chapel Hill, NC, USA
e-mail: ejuengst@med.unc.edu

J. Tolleneer et al. (eds.), *Athletic Enhancement, Human Nature and Ethics*, International Library of Ethics, Law, and the New Medicine 52, DOI 10.1007/978-94-007-5101-9_5,

be fashionably agnostic about the content of human nature they are forced back into a hierarchical genetic essentialism for sports much like those that have so repeatedly proven oppressive in other spheres of human activity.

5.2 The Appeal of Philosophical Boundary Work

Appeals to human nature have a strong and ancient history in moral philosophy, and in contemporary practical ethics that tradition remains vibrant. In order to interpret our relationships with one another and the world around us, it makes good sense to begin with an understanding of what we cherish about being human. Doing so can, in theory, help identify features of the human experience we should always strive to preserve, and set aspirational goals for our actions with respect to the rest. In the practical contexts of applied and professional ethics, however, developing or adopting a universal, normative account of human nature as a prerequisite to decision-making is never an option, since such an account is perennially 'in process.' Indeed, one part of human nature seems to be the perpetual contemplation and re-interpretation of our self-understanding as our experience of ourselves, each other, and our world evolves. The best that practical ethicists and policy-makers can hope for is to snatch from the on-going project of philosophical anthropology insights and ideas about particular aspects of human nature – like the one in the preceding sentence – that can be helpful in their tasks at hand. In doing so, of course, they import into their policy-making deep, competing, and even incommensurable philosophical assumptions, on which no public or philosophical consensus may exist. As a consequence, the arguments of those who appeal to human nature for guidance in matters like performance enhancement are easy to demolish, on both logical and political grounds (see also Daniels 2009; Buchanan 2009; Harris 2011; Lustig 2009; McConnell 2010).

Against this background, appealing to the limits of human nature seems like an attractive alternative strategy. Even if we argue over the ideal way to be human, we may be able to find common ground in identifying ways to fail to be human at all. In that project, moreover, the three traditional distinctions between the human and the subhuman, the superhuman, and the inhuman seem like obvious places to start, as the boundaries that frame the contested ground of our positive accounts of human nature. By policing those boundaries, critics can remain agnostic in their philosophical anthropology, and still draw on a commitment to protecting human nature in practical decisions about, for example, which performance enhancing biomedical interventions to permit in sports, and which to prohibit on moral grounds. But it is worth paying close attention to the language that critics use in accusing particular interventions of distorting athletes' human nature across each of these boundaries. All three modes of 'dehumanization' are embedded in rich histories of moral and political discourse, which each offer important cautionary lessons for this line of argument.

5.3 Stooping to the Subhuman

In sports ethics outside of the (increasingly overwhelming) discussion of performance enhancement, claims that sports is degrading to the humanity of athletes are usually aimed at athletic practices accused of brutality. The archetypal analogy in these critiques is the Roman gladiatorial arena, as when sports ethicists dispute the proposition that sports physicians should attend ring-side at boxing matches by arguing that "the prospect of suturing lion wounds in the coliseum was not why Galen moved to Rome in the second century, and patching modern gladiators is no more a part of the physician's mission today." (Bernstein et al. 2000: 55) In addition to its violence, the gladiatorial arena still resonates as a paradigm of sports that even sports physicians should eschew, because of its literal '*brut*ality': the dehumanizing consequences of forcing humans to compete for their lives with each other and other predators, as if they were animals in the wild. By the same token, any practice that treats athletes in the ways we allow ourselves to treat animals – 'breeding' them, buying and selling them, coercing their participation, subjecting them to harsh and demoralizing training, involuntarily 'putting them out to pasture' when their performance declines – is susceptible to this critique. We should not treat athletes as if they were less human than the rest of us.

Of course this analogical form of the argument only works when there is no question that the athletes involved in these practices are human beings like the rest of us. What is interesting about the performance enhancement context, however, is that it is this crucial assumption that the critics seem to be doubting. For example, in its discussion of athletic performance enhancement, the U.S. President's Council on Bioethics wrote:

> The runner with genetically enhanced muscles is still, of course, a human being who runs. But the doer of the deed is arguably less obviously himself and less obviously human than his unaltered counterpart. He may be faster, but he may also be on the way to becoming 'more cheetah' than man. (President's Council 2003: 144)

Here, the concern is not that the use of performance enhancements will lead others to treat the runner *as if* he were a cheetah. Rather the President's Council seems worried that in fact an enhanced runner may become cheetah-like, perhaps after the fashion of the human/cheetah hybrid featured on the cover of Andy Miah's 2004 book, *Genetically Modified Athletes* (Miah 2004). Making reference to the geneticist and evolutionary biologist J.B.S.Haldane (1892–1964), John Hoberman ends his book on the 'dehumanization of sport' with a similar concern:

> Haldane also dreamed of gene-grafting techniques that would permit the crossing of men and beasts, of legless astronauts and other specially adapted creatures – a vision perfectly suited to the development of athletes who would be monsters as well. The supreme biological question confronting mankind today is whether Haldane's vision of the pursuit of organismic efficiency will prevail over the human image that appeared in the Old Testament thousands of years ago. (Hoberman 1992: 290)

This rhetoric suggests that these critics are involved in a different game than those who decry the brutalizing effects of sports. At stake is a deeper concern to protect what are perceived to be important ontological categories. Drawing from theological natural law traditions, Aristotelean essentialism, and nineteenth century Romantic sensibilities, this concern gives high normative weight to the biological kinds produced by the 'wisdom of evolution,' and their relative ranking in a hierarchical 'great chain of being'. Under this ontological caste system, protecting the 'integrity' of natural kinds becomes imperative, at two levels. First, like importing new species into an established ecosystem, 'splicing life' in the creation of transgenic organisms, interspecies tissue chimeras, or hybridized embryos, it is always dangerous enough on this view to at least trigger the need for the "precautionary principle." (Robert and Baylis 2003) Second, blending these categories risks the creation of 'abominations': the "monsters" that Hoberman cites who threaten the established order of our world by undermining the boundaries it is built upon (Stout 1988). Where it is our own place in that order – the integrity of the human species – that is put at risk by these boundary crossings, the intervention becomes morally suspect as 'dehumanizing'. The fear is that by polluting the constellation of traits that humans have inherited from our ancestors – our evolutionarily (or divinely) given 'human nature' – with nonhuman attributes we will inevitably degrade the elements of human identity we find morally important, whatever those might be.

Of course, it is not just the blurring of human/nonhuman lines that people may find abominable: conflating the natural kinds of humans – as men and women, for example, or children and adults – evokes the same repugnance in sports ethics. For example, sports ethicists report that:

> While anecdotal, the members of our undergraduate Sport Ethics class were generally revulsed at the sight of the Chinese women swimmers. These swimmers displayed a body type that did not fit into the socially constructed category of female…. They looked like men, but were labeled 'women.' Interestingly, the class was not similarly disgusted with Ben Johnson…. His sin was against a lesser god: he merely cheated. The problem with the Chinese swimmers was that they were gender freaks first, and cheaters a distant second. (Burke and Roberts 1997: 103)

Unfortunately, as Davis and Delano point out, even female athletes who look like women can evoke similar reactions when they simply attempt to compete with men, creating barriers that keep mixed gender sports rarer than our bodies necessarily dictate (Davis and Delano 1992). And of course, in the past, competitions between athletes from different racial groups – or even different social classes – have raised similar concerns about disturbing the natural order and its degrading effects on the most privileged players (Hoberman 1992).

These allusions to the history of sexism, racism and social class bias in sports are important to note here, because they suggest that there are dangers to beware in using the 'lower boundary' of human nature in order to criticize performance enhancement. The notion that nature is normatively ordered into a hierarchy of natural kinds, each with their inviolable essence and fixed moral status, is a distinctly premodern view, increasingly rejected over the last two centuries by the natural sciences as a matter of fact and by the political sciences as a guide to human rights, and by theology as an adequate vision of Creation. In most other spheres of human activity, the use of this hierarchical worldview to declare some people less fully human than others in order to rationalize unjust social practices has led to its wide-spread rejection. Against that historical backdrop, resurrecting this approach for the practice of sports would be an uphill challenge that most sports ethicists would rather avoid, particularly if there are other arguments to employ.

5.4 Aspiring to the Superhuman

The second boundary on human nature that critics of performance enhancement police is the 'upper boundary' between human beings and their superiors. Here, the critics signal traditional concerns about *hubris*: concerns that enhancement practices flow from and reflect a dangerous aspiration to transcend natural human limits that, in seeking the superhuman, transgresses the supernatural. Miracles are the province of gods and magic is the province of demons: attempts to perform either by ordinary mortals (saints and witches, by definition, excluded) reflect the moral vice of *hubris*, and usually backfire in punishing ways. This, of course, is not just a premodern set of ideas, but positively ancient: it resonates with the lessons of Greek and Norse mythology and Old Testament Judiasm. Interestingly, it remains very alive in the rhetoric of performance enhancement, where allusions to Prometheus, the Garden of Eden, and Ragnorak (the apocalyptic war of the Norse giants against the gods), are stock metaphors (Zylinska 2010).

Here the concern is not so much based on violating or debasing human nature, but on ignoring our limits and attempting to transcend our nature. The distinction is often grounded in (or refuted by) views of the proper relation of humans and the divine: to use biomedical means to improve ourselves is to 'play God' in a dangerously over-reaching way, risking *hubris* in ways that humbly accepting the limits of natural achievement does not. As Michael Sandel says:

> I do not think the main problem with enhancement and genetic engineering is that they undermine effort and erode human agency. The deeper danger is that they represent a kind of hyperagency -- a Promethean aspiration to remake nature, including human nature, to serve our purposes and satisfy our desires. [...] And what the drive to mastery misses and may even destroy is an appreciation of the gifted character of human powers and achievements. (Sandel 2004: 54)

The best exposition of the roots and contours of this interpretation comes from Erik Parens. In his trenchant comparison of how both critics and proponents of enhancement technologies appeal to the moral ideal of authenticity (Parens 2005). Parens contrasts two 'ethical frameworks' in the debate, 'creativity' and 'gratitude' frameworks, which support different conceptions of what it means to be true to oneself or to human nature. Under the 'creativity framework', human authenticity resides in our freedom to remake the world in service to our own ends. By contrast, Sandel's concerns exemplify the 'gratitude framework': the view that "we human beings are not the creators of life; we are creatures, whose job is to remember that life is a gift. It is our responsibility to express gratitude for the mysterious whole, which we have not made." (Parens 2005: 37) In sports ethics, this is often rendered as the 'admiration' and 'celebration' of the talents of 'gifted' athletes, and a criticism of the "ethic of willfulness", "mastery", or "striving" (to use Sandel's terms) that would motivate a turn to performance enhancing technologies.

On the other hand, Parens' main point about the two 'ethical frameworks' in the enhancement debate is that they inevitably live in tension with each other, as complementary expressions of the ideal of authenticity. Performance enhancement in sports is a perfect example of this tension, because so much of the spirit of sport is captured by the motto of the Olympic movement: "faster, higher, stronger." It is not a coincidence that "Olympian" names both elite athletes and the Greek gods: to a large extent, as many have noted, sport does seem to be about transcending the 'records' of human experience, about self-mastery, and about creative striving. Without a way to accommodate this Promethean dimension of sports, the 'gratitude framework' is not going to suffice to ground a successful critique of athletic performance enhancement.

It is interesting to note how often the charge that performance enhancement is hubristic is invoked with apocalyptic overtones, as if our Promethean efforts at performance enhancement in sports had cosmological significance – at least for the universe of sports. For example, in criticizing the hubris they see motivating performance enhancement, the members of the U.S. President's Council on Bioethics write that:

> The desire to have a perfect body [...] is tantamount to a desire to transcend our embodiment altogether, to become as gods, to become something more-than-human. [S]uch longings risk becoming a full-scale revolt against our humanity altogether. Fueled in addition by a thirst not merely to excel but to defeat and surpass our rivals, the desire for superhuman powers easily becomes boundless. (President's Council 2003: 150)

Even more explicitly, one of the European editors of a 2005 volume on gene doping open his introduction with this declaration:

> Present genetic technology poses a challenge to sport so serious that it is hard to overestimate. What is at stake is the very ethos of sport, nothing less than an epochal confrontation between a model of human identity as spelled out in the Book of Genesis and a science-based libertarian model. According to the former, sports is a means by which we explore human nature, admire it at its peak, and gain self-understanding. It is not up to us to play God, or, to put it in a more mundane and secular way, to meddle with our evolutionarily determined human nature. (Tännsjö 2005: 2)

This will be worth returning to when we consider the axiology of sports below, because the editors may be right: performance enhancing technologies just may provoke the Ragnorak that spells the beginning of the End for sport as we have known it.

5.5 Engineering the Inhuman

The third boundary on human nature that critics of performance enhancement invoke, draws a different kind of distinction from the first two. The distinction still concerns important transgressions against a normative order, as Habermas signals in the beginning of this passage:

> Bodies stuffed with prostheses to boost performance [...] dissolve boundaries and break connections that in our everyday actions have up to now seemed to be of almost transcendental necessity. There is fusion of the organically grown with the technologically made, on the one hand, and separation of the productivity of the human mind from lived subjectivity, on the other hand. (Habermas 2003: 41)

Unlike the first two boundaries, however, the boundary here is not between grades of agency (subhuman or superhuman), but, as Habermas suggests in the last clause above, between agency and artifice. If athletic performances have artificial, rather than natural causes, the authenticity of the athlete's accomplishments are called into question and with it their agency as human beings. Since athletes' causal agency is critical to their claims to the credit for their competitive accomplishments, any dilution of that agency by artifice is worrisome. As Hoberman puts it:

> The scientization of the athlete, either through drugs or other techniques, also involves a crisis of identity. What sort of human being is competing? To what extent can the emotional experience of competition be truly shared by an athlete who has transformed himself with drugs? On this level, Ben Johnson revealed himself to be a modern reincarnation of Victor Frankenstein's artificial man. (Hoberman 1992: 21)

Allusions to this boundary are much more common in the literature on performance enhancement than either of the other two: in fact, the typical way to express concern about a particular intervention is to label it in a way that highlights this distinction. Thus, we are warned against "synthetic steroids," (Hoberman 1992) "designer doping" (Joyner 2004) "artificially-induced hypoxic conditions," (World Anti-Doping Agency 2006) and even "artificial genes" (Unal and Unal 2004). This suggests that this boundary has particular salience for sports, compared with worries about athlete's station in the Great Chain of Being, but those who employ these labels seldom explain precisely why they matter.

The distinction between the natural and the artificial is common in other bioethical debates as well, over issues ranging from the right to a 'natural death' to the moral merits of 'artificial reproductive technologies.' Interestingly, however, it is usually invoked when the process in question – death, conception, respiration, hydration, aging – are basic biological processes which we share with other living organisms. The distinction's point is to draw moral attention to biomedical attempts to 'artificially' replicate, control, or improve upon these processes, by emphasizing that these 'natural' processes are not human inventions in the first place. As a call to arms in sports, unfortunately, this contrast falls flat in the face of the fact that sport itself is a highly constructed human practice governed by invented rules and dependent on multiple technologies of human manufacture. This leads some to dismiss this interpretation quickly as incoherent (Mehlman 2003: 95–101) perhaps too quickly, since in fact its salience to sports can help illuminate a deeper set of issues, as we shall see below.

Some of those concerned to flesh out the moral meaning behind this distinction argue that criticizing performance enhancement in terms of its artificiality is to express the concern that the achievements it would make possible would be not be achievements authored by the athlete, and for which he or she could legitimately take credit. Under this view, enhanced victories would be fake, hollow, or inauthentic because they literally would not be the athlete's victories (President's Council 2003: 124–134). Even the World Anti-Doping Association's Ethics Committee takes this approach, when it argues that, for example:

> Our analysis of artificially induced hypoxic conditions to modify performance alerted us to an important distinction: between technologies that operate on the athlete and in relation to which the athlete is a merely passive recipient, versus technologies with which the athlete actively engages and interacts as part of the process of training and competing in order to enhance performance. From the individual athlete's point of view, my responsibility for my performance is diminished by technologies that operate upon me, independent of any effort on my part. [...] The human athlete utilizes, masters and controls the technology, not the other way round. (World Anti-Doping Agency 2006)

The problem with this interpretation, of course, is the difficulty in disentangling the many conscious decisions made by athletes that are accepted as legitimate, even praiseworthy contributors to their responsibility for their accomplishments in the normal course of sports from a decision to use a performance enhancing technology. If accepting a particular diet can be a legitimate feature of an athlete's 'victory narrative,' why not the decision to accept a performance enhancing intervention? A doped victory may not be morally praiseworthy, but it would still be authored by the athlete. Cole-Turner compares athletes with writers using computer technology:

> The fact that I write at a computer makes writing easier by eliminating retyping and other frustrations, but writing itself is an intense struggle and it will remain so under any technological condition. The technological advance does not eliminate the struggle so much as relocate it; indeed, it makes it possible to eliminate secondary aggravations and focus attention on the core struggle at hand, namely, expressing new insights in just the right words. Even if technology increased our cognitive ability itself, [...] so that we could calculate or write or think more clearly, these activities would still be a struggle in the face of even greater intellectual challenges to which we human beings inevitably set ourselves. (Cole-Turner 1998: 156)

Claudio Tamburrini goes even further to suggest that performance enhancement interventions may even allow us to focus more explicitly on the features of athletic performance that do deserve our praise and admiration:

> We will no longer admire sport stars for their outstanding physical traits (they will be genetically designed). Rather, we will continue to admire them for all the sacrifices endured to actualize their genetic predisposition. In the same way that today's natural talents cannot do without hard training, the genetically transformed athlete will have to devote herself to her discipline in a goal-directed and professional manner. [...] Our admiration for sport heroes will, to a much higher degree than today, concentrate on their dedication and efforts, rather than on a fortuitous physiological predisposition. I cannot see anything unsound in such a development. (Tamburrini 2002: 266)

Of course, unless the naturally talented are somehow prohibited from using performance enhancements (returning us to today's situation), the availability of such interventions might never have this levelling effect. But even if it could, others can see, *contra* Tamburrini, something unsound in such a development. What makes a biomedically engineered victory hollow, they argue, is not that it renders individual athletes inauthentically human, but rather that makes them inauthentically athletic, by missing the point of sports as a social practice (Loland 2002).

It is through this account that the salience of appeals to the boundary between human agency and artifice for sports ethics begins to become clear. On this view, even if the explicit rules of a sport are changed to allow new enhancements, their use would erode the athletic value of the achievements they make possible, by undermining the features of athletic agency that make the achievements valuable in the first place. This line of thinking is becoming increasingly influential in sports ethics, and seems to offer a stable moral platform from which to evaluate the performance enhancement issue and the fundamental sports policy choice it raises. It means that either athletic games must be redesigned to find new ways to celebrate agency that are not affected by available enhancements, or they must prohibit the use of the enhancing shortcuts. However, knowing which way to go means having at least the sketch of a positive theory of the forms of human agency that sports as a social practice celebrates. Fortunately, the WADA Ethics Committee itself offers such a sketch, which is instructive to examine.

5.6 The Meaning of Athletic Agency

If in fact the problem with performance enhancement is that it misses the point of sports, what intrinsic values seem to be at stake? In a recent statement on hypoxic training chambers, WADA, echoing Sigmund Loland, Tom Murray, and others, says that the spirit of sport lies in "the celebration of natural talents and their virtuous perfection," which they explain in the following way:

> In one important sense, sport is a celebration of human variation. Not all of us have the physiology or anatomy to be a great swimmer, hurdler, skier or discus thrower. Yet, biology is not all that matters: the most naturally gifted athletes must work to perfect their talents. […] The spirit of sport, as we understand it, celebrates natural talents and their virtuous perfection. We say 'virtuous' in this context because virtues are qualities of character admirable in themselves, the qualities that outstanding athletes develop and embody in their quest for excellent performance. Some means we respect and want athletes to employ exemplify aspects of character that we admire in people more generally, such as fortitude, dedication, self-discipline, the willingness to suffer in the service of a worthy cause, courage, and strategic wisdom. […] So, for any particular means for enhancing performance […] the crucial test will be whether it supports or detracts from sport as the expression of natural talents and their virtuous perfection. (World Anti-Doping Association 2006)

This sketch has several important features that bear scrutiny. First and foremost, sport celebrates human variation: that is, the differences between individual athletes. But not all human differences are of interest in sports. Sport is concerned with celebrating differences in natural talents and in their competitive development by athletes. Moreover, not all natural human talents are relevant in sport. Sport, we learn, is concerned with celebrating the variation in human natural talents that lead athletes to their goal of "excellent performance."

What constitutes excellent athletic performance? WADA's opening question offers three scales for judgment: a performance's moral virtues ("honorableness"), aesthetic qualities ("beauty") and its place in a hierarchy of human accomplishments ("admiration"). Honorableness, as WADA suggests, is a necessary condition for performance excellence, and to a lesser extent beauty may be as well. But neither seem sufficient conditions, alone or together. As other philosophers of sport have pointed out, an athletic performance can be beautiful and well-played, but if it does not win, either by beating the athlete's own record or the competition, it is not as excellent as an equally beautiful, well-played event that does produce a champion. Performance excellence in sport, in other words, seems to be inevitably and intrinsically about the comparative ranking of performances. Thus, Sigmund Loland concludes:

> In spite of great diversity in sport-specific goals, then, it is possible to formulate a general goal that characterizes sports competitions as such: the goal of sports competitions is to measure, compare and rank two or more competitors according to athletic performance. This goal seems to be common to all sports however diverse their ethos. It defines sport's characteristic social structure and I shall therefore call it the structural goal of sport competition. (Loland 2002: 10)

The key role of hierarchical ranking in sport is often ignored in the official rhetoric of sports organizations, but as sports ethicists like Coakley acknowledge, fixation with hierarchical ranking –with competition, contest, score-keeping, record-breaking, championship, victory and defeat – is pervasive in the everyday practice of sport (Coakley 1998). In ordinary discourse, sports remains true to its historical roots in the war games of ancient and medieval 'arenas', 'tournaments' and '(battle)fields'. As Tännsjö has provocatively pointed out, the admiration of champions can easily become contempt for the losers: to 'beat', to 'own', to 'dominate' the competition is to metaphorically subjugate them, just as conquering champions have always enslaved those they overcome (Tännsjö 2000). While the hierarchies established by sport are usually more benign than the language of sports (and Tännsjö, for that matter) suggests, the aptness of our bellicose metaphors shows a similar human urge at work: the urge to establish interpersonal and intergroup hierarchies and locate ourselves within them.

Moreover, as we have seen, the value of comparative ranking also insinuates itself into – and helps explain the appeal of – other attempts to use the boundaries of human nature to criticize performance enhancement. Thus, it is a concern for complicity with suspect social tendencies to use hierarchical human ranking in stigmatizing and oppressive ways that, in limit cases, causes physicians to eschew particularly brutal forms of sport (like boxing, or feeding Christians to lions) and to worry about the social consequences of performance enhancement. Similarly, the value of hierarchical taxonomies for sport echoes through the worries that performance enhancement might 'degrade' or 'dehumanize' athletes by infecting their identities with the attributes of lower classes of entities, like animals or machines, or because they violate 'our place in nature' in the other direction, by willfully encroaching on the gods' domain.

Returning to WADA's description of the spirit of sport, then, this analysis suggests an important clarification of their claim that sports "celebrates human diversity." Perhaps one could say that the many different forms of sport accommodate human diversity in an appreciative and equitable way. But within any one sport, the "celebration of human variation" is always comparative and hierarchical. The spirit of sport, then, is the celebration of the differences in the human talents and virtues that allow athletes to accomplish achievements that place them higher in a hierarchy of human excellence than their competition.

Again, the point of analyzing WADA's description of the spirit of sport was to determine if that spirit is betrayed by performance enhancement so badly that any achievements it might facilitate would be hollow victories, artificial in the sense of inauthentic, and therefore 'inhuman' in a way that is morally problematic. So far, nothing about WADA's account seems to be betrayed by the prospect of safe and openly available performance enhancements. As the apologists for performance enhancements insist, biomedical enhancements would not diminish the need for effort, courage, or self-discipline in sports. Across a universally enhanced field of competitors, these virtues would still be required to differentiate the champions (Miah 2004; Tamburrini 2002). In fact, enhancing interventions could help produce even greater champions to celebrate, by elevating the physical and mental talents that are the starting points for their accomplishments.

But there is still another feature of WADA's description that has not been examined: the fact that in its invocation of the spirit of sport, WADA's ethics committee restricts the kinds of talents sports celebrates to "*natural* talents." I think that this is a new use of the word 'natural' yet again, and this is where the real problem lies for gene doping.

The term is being used here to qualify the kinds of talents which athletes attempt to virtuously perfect and which authentically excellent (honorable, beautiful and admirable) athletic performances celebrate. We have already put aside the possibilities that this use of 'natural' might be intended to exclude subhuman or superhuman talents, or talents aided by technological artifice *per se*, for the same reasons that brought us to this point in the analysis. Only one other plausible interpretation of 'natural' in this phrase presents itself: 'natural talents' are the talents you were born with: i.e., inherited characteristics over which athletes have no control and which are ultimately traceable to particular combinations of ancestors and their genes. As Sigmund Loland concludes:

> Now I can say more precisely what I mean by talent in sport. It is an individual's genetic predisposition to develop phenotypes of relevance to performance in the sport in question. The distribution of talent in the natural lottery is a random process. (Loland 2002: 69)

On WADA's interpretation, in other words, the spirit of sport is in part a celebration of differential genetic endowments, distributed through the 'natural lottery' of geneology. Sport creates a system of values, virtues and practices designed to hierarchically grade people in terms of their (virtuously perfected) inherited traits and glorify the best specimens as champions. This is a turning point for our analysis: biomedical interventions at the genetic level would miss the point of the sports if this view is correct, because gene doping would undermine the ability of sports to distinguish those who passively inherited their talents from their progenitors from those who actively acquired them from their physicians. As Sandel says:

> Which aspect of the athletic ideal – effort or gift – would be more deeply offended? Some might say effort: the problem with drugs is that they provide a shortcut, a way to win without striving. But striving is not the point of sports; excellence is. And excellence consists at least partly in the display of natural talents and gifts that are no doing of the athlete who possesses them. [...] No one believes that a mediocre basketball player who works and trains even harder than Michael Jordan deserves greater acclaim. The real problem with genetically altered athletes is that they corrupt athletic competition as a human activity that honors the cultivation and display of natural talents. (Sandel 2004: 55)

At last our tour of human nature's boundaries has brought us to an interpretation of the problem of performance enhancement that its critics might use to theoretically ground its prohibition and motivate the search for detection techniques. Performance enhancement is wrong for athletes to pursue and sports medicine to provide, we can say, because it compromises the ability of athletes to segregate themselves into genetically defined classes and the ability of sports to elevate genetically advantaged athletes from their disadvantaged competitors. This will be true even if performance enhancement is proven safe and effective, and even if it could be provided equitably to every competitor in non-coercive ways.

5.7 Conclusion

After all our conceptual orienteering, why is reaching this destination so disturbing? Perhaps it is because of the uncomfortable light that the problem of performance enhancement throws back on the spirit of sport itself and the broader questions it raises for sports ethics. The very prospect of performance enhancement forces sports to notice that the 'natural lottery' of genetics is getting less 'random' all the time as our biomedical knowledge grows, and as that happens the moral justification for celebrating human genetic hierarchies as facts of nature begins to falter. It is one thing to acknowledge, as perhaps unfortunate but not unfair, that human talents vary across individuals in ways that allow some to do some things that others cannot. To admire those inborn differences as beautiful and to reward those who seek to perpetuate and increase them seems arbitrary, but no less pernicious than other idiosyncrasies of social taste, like celebrity worship. But to glorify those genetic disparities to the extent of prohibiting their abatement when biomedicine provides the ability to do so, exposes a fundamental tension with the spirit of sport itself: while the variable genetic distribution of human talent may be a natural fact, the creation of social hierarchies out of that distribution is a distinctly human project, and never a random one. For sports ethicists, this raises a challenging question for further discussion: as a human project, what differentiates the hierarchical ranking of genetic talents by sports institutions from the reciprocal devaluing of genetic defects by insurance risk underwriters? In both cases, the internal logic of the practice is clear, but their use of inherited genetic identity as a social ranking criterion raises large questions about the fairness of the enterprise as a whole.

This may be the reason for the apocalyptic tone with which sports authorities frame the issue of performance enhancement. After all, in most discussions of the ethics of enhancement outside the sports context, a key concern is to avoid the creation of a biological caste system and the errors of biological essentialism and risks discrimination that would entail (Mehlman 2003). In a world which is struggling to realize a commitment to treating people as moral equals despite their biological differences, a social practice that creates and glorifies hierarchies of genetic endowment seems anachronistic and slightly ominous. As Sandel acknowledges about his defense of the role of sport's commitment to genetic stratification:

> This is an uncomfortable fact for democratic societies. We want to believe that success, in sports and in life, is something we earn, not something we inherit. Natural gifts, and the admiration they inspire, embarrass the meritocratic faith; they cast doubt on the conviction that praise and reward flow from effort alone. (Sandel 2004: 55)

To discover that the principal problem with performance enhancement's threat to the human nature of athletes is that it threatens sport's commitment to the creation and promotion of human genetic stratification should give us pause. On one hand, of the many ways humans use inherited traits to create interpersonal hierarchies, athletic competition is amongst the most benign. When it is 'just a game,' comparative interpersonal ranking in terms of genetic identity in sports is a welcome substitute for blood feuds, racism, and genocide. On the other hand, what the performance enhancement debate suggests is that sports may have more in common with these extreme human vices than we would like to think. When sports becomes a matter of national pride and a source of economic opportunity, athletic losers risk more than simply admiration and social status: like insurance applicants with genetic susceptibilities, less naturally talented athletes risk access to important social benefits and potential life plans. In this regard, the apocalyptic challenge that performance enhancement poses to sports reduces to a simple question that echoes Cole-Turner's concern to "relocate the struggle" of human achievement on what really matters: are there ways to enjoy, appreciate and even show off our bodies and abilities without requiring someone else to lose social standing on genetic grounds?[1]

[1] This chapter expands on themes, arguments and examples to be found in Juengst (2009).

References

Bernstein, J., C. Perlis, and A. Bartolozzi. 2000. Ethics in sports medicine. *Clinical Orthopaedics and Related Research* 378: 50–60.

Buchanan, A. 2009. Human nature and enhancement. *Bioethics* 23: 141–150.

Burke, Michael, and Terrance Roberts. 1997. Drugs in sport: An issue of morality or sentimentality? *Journal of the Philosophy of Sport* 24: 99–113.

Coakley, J. 1998. *Sport in society: Issues and controversies*, 6th ed. Boston: McGraw-Hill.

Cole-Turner, R. 1998. Do means matter? In *Enhancing human traits: Ethical and social implications*, ed. Parens Erik, 151–161. Washington, DC: Georgetown University Press.

Daniels, N. 2009. Can anyone really be talking about ethically modifying human nature? In *Human enhancement*, ed. Nick Bostrom and Julian Savulescu, 25–43. Oxford: Oxford University Press.

Davis, L., and L. Delano. 1992. Fixing the boundaries of physical gender: Side effects of anti-drug campaigns in athletics. *Sociology of Sport Journal* 9(1): 1–19.

Habermas, J. 2003. *The future of human nature*. Cambridge: Polity Press.

Harris, J. 2011. Taking the "human" out of human rights. *Cambridge Quarterly of Healthcare Ethics* 20: 9–20.

Hoberman, J. 1992. *Mortal engines: The science of performance and the dehumanization of sport*. Caldwell: The Blackburn Press.

Joyner, M. 2004. Designer doping. *Exercise and Sport Science Reviews* 32(3): 81–82.

Juengst, E. 2009. Annotating the moral map of enhancement: Gene doping, the limits of medicine and the spirit of sport. In *Ethics, genetics and the future of sport: Implications of genetic modification and genetic selection*, ed. Thomas Murray, 175–204. Washington, DC: Georgetown University Press.

Loland, S. 2002. *Fair play in sport: A moral norm system*. New York: Routledge.

Lustig, A. 2009. Are enhancement technologies "unnatural"? Musings on recent Christian conversations. *American Journal of Medical Genetics. Part C, Seminars in Medical Genetics* 151C: 81–88.

McConnell, T. 2010. Genetic enhancement, human nature, and rights. *The Journal of Medicine and Philosophy* 35: 415–428.

Mehlman, M. 2003. *Genetic enhancement and the future of society*. Bloomington: Indiana University Press.

Miah, A. 2004. *Genetically modified athletes: Biomedical ethics, gene doping and sport*. New York: Routledge.

Parens, E. 2005. Authenticity and ambivalence: Toward understanding the enhancement debate. *The Hastings Center Report* 35(3): 34–41.

President's Council on Bioethics. 2003. *Beyond therapy: Biotechnology and the pursuit of happiness*. http://bioethics.georgetown.edu/pcbe/reports/beyondtherapy. Accessed 27 Mar 2012.

Robert, J., and F. Baylis. 2003. Crossing species boundaries. *The American Journal of Bioethics* 3(3): 1–14.

Sandel, M. 2004. The case against perfection: What's wrong with designer children, bionic athletes and genetic engineering. *The Atlantic Monthly* 293: 51–62.

Stout, J. 1988. *Ethics after Babel: The languages of morals and their discontents*. Princeton: Princeton University Press.

Tamburrini, C. 2002. After doping, what? The morality of the genetic engineering of athletes. *Sport Technology: History, Philosophy and Policy* 21: 243–268.

Tännsjö, T. 2000. Is our admiration for sports heroes fascistoid? In *Values in sport: Elitism, nationalism gender equality and the scientific manufacture of winners*, ed. Tännsjö Torbjörn and Tamburrini Claudio, 24–38. New York: Routledge.

Tännsjö, T. 2005. Genetic engineering and elitism in sport. In *Genetic technology and sport: Ethical questions*, ed. Tamburrini Claudio and Tännsjö Torbjörn, 57–70. New York: Routledge.

Unal, M., and D. Unal. 2004. Gene doping in sports. *Sports Medicine* 34: 357–362.

World Anti-Doping Agency. 2006. *WADA note on artificially induced hypoxic conditions*. http://www.mcst.go.kr/servlets/eduport/front/upload/UplDownloadFile?pFileName=Note%20on%20Hypoxia%20May%202006.pdf&pRealName=F362307.pdf&pPath=0404150000. Accessed 27 Mar 2012.

Zylinska, J. 2010. Playing god, playing Adam: The politics and ethics of enhancement. *Bioethical Inquiry* 7: 149–161.

Chapter 6
Prometheus on Dope: A Natural Aim for Improvement or a Hubristic Drive to Mastery?

Trijsje Franssen

6.1 Introduction

In the current debate on performance enhancement in sport, the athlete is often pictured in completely antithetical ways. Sometimes, he represents a creative "artist of self-design" (Young 2006: 306), whereas in other situations he embodies a "gullible and overcommitted" being who lacks "any kind of moral framework" (McNamee 2007: 190). The present chapter will closely examine arguments in favour of and against enhancement in sport in order to investigate these contradictory images of the athlete.[1] The examination of these images suggests that the arguments depend in large part on two main, contradicting, value-laden concepts of human nature. I will argue that these concepts represent a strong tension between two demands present in the way sport is practised and perceived today: on the one hand the athlete is expected to break world records, cross human boundaries, and achieve something that no one has ever achieved before, while on the other hand this same achievement has to remain within 'natural' limits – and thus no doping is allowed. This situation necessitates that the athlete deals with those two contradictory demands, as coaches, other contestants, public, media and official committees insist on both at the same time.

[1] I will not confine myself solely to texts which focus mainly on sport, as in general enhancement arguments the image of the athlete is also used because of its exemplary nature, and it is often in these examples in particular that the underlying demands and assumptions shine through.

T. Franssen (✉)
Department of Sociology and Philosophy, University of Exeter,
Saint German's Road, Byrne House, Room Number FF10, Exeter, UK
e-mail: tmf204@exeter.ac.uk.

J. Tolleneer et al. (eds.), *Athletic Enhancement, Human Nature and Ethics*, International Library of Ethics, Law, and the New Medicine 52, DOI 10.1007/978-94-007-5101-9_6,

Many of the texts I discuss in this chapter refer to the mythological figure of Prometheus. This myth will play a central role in my critique, as I think it offers a very useful tool to probe into the different images of the athlete. Thus, the myth can provide insight into the disparity within the rhetoric surrounding sport and even its practice. This chapter will first briefly narrate the Prometheus myth, after which I will present the arguments of three advocates and three opponents of (performance) enhancement. Subsequently I will analyse each of these, with the aim of providing insight into the debate and the practical issues of performance enhancement in sport.

6.2 The Myth

There are many versions of the myth of Prometheus, of which one of the best known is *Prometheus Bound* by Aeschylus (fifth/sixth century BC). After the older generation of deities had been conquered in the Clash of the Titans, Zeus threw all of them in the lowest depths of the underworld. Only one of them, Prometheus, was saved because he had helped Zeus to win the war by means of his cunning and advice. Soon, however, Zeus turned out to be a merciless tyrant. When he conceived a plan to wipe out the whole human race, Prometheus took pity on them. He stole fire for them from heaven, gave them wisdom and taught them all kinds of techniques and arts – from mathematics to reading and the art of prophecy. However, when Zeus found out, Prometheus was severely punished. Zeus chained the immortal god to a rock, where a vulture would eat his liver. Every night it would regenerate so that his torture could be repeated the next day, until many centuries later Hercules should free him.

Hesiod's earlier version of the myth[2] tells how, apart from punishing Prometheus, Zeus also penalised mankind. He sent them Pandora, the first woman, who carried the famous box that spread out evil, misery and disaster over the human race. Later, in Plato's *Protagoras* (fourth century BC), Prometheus is described as taking part in the creation of mankind. He and his not-so-clever brother Epimetheus were put in charge by the gods to assign to all the mortal races their defining attributes. Epimetheus, however, used up all the qualities on the animals, so that no powers were left when he came to man: "the human race was naked, unshod, unbedded, and unarmed" (Plato 1997: 757). To compensate for his brother's fault, Prometheus decided to give them fire, knowledge, and other civilizing arts and techniques which, again, he stole from the gods. Plato's story does not mention Pandora or the rock, but merely that later Prometheus was charged with theft. First, however, humans "were being destroyed by wild beasts because they were weaker in every way" (ibid.). Only after Zeus sent them the art of politics, justice and shame, were they able to establish order, found cities, and bond in friendship.

[2] In his *Theogony* and *Works and Days* (eighth century BC).

6.3 The Spirit of Sport: A Natural Aim for Improvement

What exactly makes the Prometheus myth relevant to the debate on enhancement in sport will hopefully become clear in the following sections. I will start by presenting the arguments of three advocates of (performance) enhancement who picture the athlete as a courageous, perfection-seeking and boundary-transcending being.

6.3.1 Gregory Stock

In his book *Redesigning Humans* (2003: 1) the biophysicist Gregory Stock, one of the major figures in the pro-enhancement camp, argues we are "on the cusp of profound biological change". Given emerging technologies such as cloning and human genetic modification it is only a matter of time before "these developments will write a new page in the history of life, allowing us to seize control of our evolutionary future" (ibid: 2). The trend has been set and further developments are inevitable. He suggests that some of us *are* already enhancing themselves by means of, for instance, performance enhancing drugs in sport: "Professional sports offers a preview of the spread of enhancement technology into other arenas. Sports may carry stronger incentives to cheat, and thus push athletes toward greater health risks, but the nonsporting world is not so different" (ibid: 8). In other words, according to Stock doping shows that in general, if people believe enhancement is to their advantage, as soon as they think the risks are worth taking they will be willing to use these new technologies.

Quoting James Watson, co-discoverer of the structure of DNA, Stock asks us: "if we could make better humans … why shouldn't we?" (ibid: 12) It is true that there will be concern, mistakes and even abuse, but that could be said about any radical new development or technology. Of course the application of the technologies should be done with care, he argues, but no serious scientist will start messing around with human genes until such interventions can be safely carried out. So "why all the fuss, then?" (ibid: 12) To try to stop the developments is unrealistic and simply impossible, Stock claims. Rather, we should think about how to minimise the risks and maximise the benefits.

> Humanity is moving out of its childhood and into a gawky, stumbling adolescence in which it must learn not only to acknowledge its immense new powers, but to figure out how to use them wisely. The choices we face are daunting, but putting our heads in the sand is not the solution (ibid: 17).

Instead, we should be brave and face the unknown dangers that inevitably accompany all radical new developments:

> Some imagine we will see the perils, come to our senses, and turn away from such possibilities. But when we imagine Prometheus stealing fire from the gods, we are not incredulous or shocked by his act. It is too characteristically human. To forego the powerful [enhancement] technologies […] would be as out of character for humanity as it would be to use them without concern for the dangers they pose (ibid: 2).

6.3.2 *Simon Young*

The transhumanist Simon Young does not even pose the question whether doping or enhancement should be used in sport. On the contrary, in *Designer Evolution* (2006) he simply states that we should, for instance, hold up bodybuilders' practices as examples, and shape our life according to their principles:

> What exactly is wrong with regarding the body as creative material ready to be carefully enhanced – as bodybuilders and aesthetic surgeons do already! [...] To see oneself and one's species as artworks in the making is not to devalue life but to venerate it. [...] There can be nothing bad about the desire to be the best one can be. So let us reclaim our respect and admiration for designers, and choose to be artists of self-design (Young 2006: 306).

According to Young there is nothing wrong with such creative enhancement or self-design, as humans are, by nature, all "imbued with the innate Will to Evolve – an instinctive drive of a conscious entity to expand its abilities in pursuit of ever-increasing survivability and well-being" (ibid: 19). Young's transhumanist ethic urges us to *"foster"* this Will, so that "by acting in harmony with the essential nature of the evolutionary process – complexification – we may discover a new sense of purpose [...] and come to feel ourselves *at home in the world* once more" (ibid: 19).

The Will to Evolve is symbolised by Prometheus, who represents "the innate human drive to increase knowledge and abilities, even at the expense of present pains" (ibid: 39) – the drive to progress, improve, *enhance*, that is. Young does recognise that a future of self-enhancement is not without risks. However, to reject the 'Prometheus Drive' will lead to stagnation or decline, for "that which ceases to grow, begins to decay" (ibidem). It would mean remaining forever in the power of our limitations and keeping on suffering from disease and death. Therefore, Young pleads:

> Let us be the New Prometheans. Let us unite in our commitment to boldly go where none have gone before in search of the knowledge by which to transcend the limitations of the human condition. Let us cast aside cowardice and seize the torch of Prometheus with both hands (ibid: 40).[3]

6.3.3 *Julian Savulescu, Bennett Foddy and Megan Clayton*

In their article 'Why We Should Allow Performance Enhancing Drugs in Sport' (2004), Savulescu, Foddy and Clayton present – as the title obviously suggests – a number of

[3] Note that 'to boldly go where none have gone before' is a reference to Star Trek, which Young calls a 'technomythology', characterised – just like, in my view, transhumanism itself – by an essentially humanistic philosophy that believes in 'the ongoing progress of the species through reason, science, and technology' (ibid: 39).

arguments in favour of athletic enhancement.[4] The paper opens with a paragraph which claims that "the athletic ideal of modern athletes" remains what it has always been since the ancient world: "superhuman performance, at any cost" (Savulescu et al. 2004: 666). Therefore, as performance enhancing drugs are becoming more effective each day, the chances to be tested are relatively small with the benefits of success incredibly high, many athletes are very likely to cheat. However, they argue, cheating exists because of the rules we define. A drug is illegal because the World Anti-Doping Agency claims it violates the 'spirit of sport'. "If we made drugs legal and freely available, there would be no cheating" (ibid: 666). Savulescu et al. have a different view of sport, a much less 'naturalistic' one, less a 'test of biological potential':

> Humans are not horses or dogs. We make choices and exercise our own judgment. We can display courage, determination, and wisdom […] Human sport is different from animal sport because it is creative. Far from being against the spirit of sport, biological manipulation embodies the human spirit – the capacity to improve ourselves on the basis of reason and judgment. When we exercise our reason, we do what only humans do (ibid: 666–7).

If athletic enhancement were to be allowed, winning would be less the result of the 'genetic lottery' and – apart from training – more determined by human creativity, determination, reason and judgement. One could claim performance enhancing drugs are unfair, but, they argue, so is nature: she simply deals some a better genome than others. The solution to sport's genetic discrimination would be to allow everybody to use performance enhancing drugs.[5] "Far from being unfair, allowing performance enhancement promotes equality" (ibid: 668). Similarly, economic inequalities could be removed by providing grants to poorer athletes. Therefore, Savulescu et al. conclude that the only serious limit to performance enhancement is safety; rules should be drafted to ensure that the drugs don't threaten the athlete's health and fitness. The welfare and safety of the athlete should matter to us, not whether performance enhancing drugs are natural or artificial. "Performance enhancement is not against spirit of sport; it is the spirit of sport. To choose to be better is to be human" (ibid: 670).

6.4 Performance Enhancement: A Hubristic Drive to Mastery

Of course, the opponents of enhancement have a rather different opinion when it comes to these issues. In this section I will present the thoughts of three critics of (performance) enhancement, who present the athlete as either epitomising the talents of nature, or, if enhanced, a hubristic and limits-ignoring being.

---

[4] This paper does not refer to the myth of Prometheus. However, its focus on doping in particular and the connection it makes between human nature and the role of the athlete make it relevant nevertheless.

[5] It is precisely such a view of sport that the cultural historian John Hoberman has in mind when he speaks of the 'progressive alternative' to the 'conservative ideal' of sport. He describes the former as a 'romantic Prometheanism that dreams of human transformations and demands unrestricted access to the drugs that might catalyze such metamorphoses' (1992: 28).

6.4.1 Michael Sandel

In *The Case against Perfection* (2007), the political philosopher Michael Sandel tries to articulate the unease which enhancement and genetic engineering present us with, and find answers to what is morally troubling about such practices and why many feel that they are morally wrong. In an attempt to answer these questions, he investigates what is admirable in athletic achievement. According to him, 'the point of sports' is *excellence*, which does not merely consist in effort and striving but "consists at least partly in the display of natural talents and gifts that are no doing of the athlete who possesses them" (Sandel 2007: 28). The problem with performance enhancing drugs thus is not so much that they undermine the value of the effort by means of which a goal is reached, but rather the way in which they depreciate the so-called 'gifted' character of our natural powers. Genetically altered athletes would "corrupt athletic competition as a human activity that honors the cultivation and display of natural talents. From this standpoint, enhancement can be seen as [...] a kind of high-tech striving" (ibid: 29). Games in which enhancement would be allowed would, Sandel claims, therefore be changed into something other than sport – the game would rather become a 'spectacle'.

> Spectacles, by isolating and exaggerating through artifice an attention-grabbing feature of the sport, depreciate the natural talents and gifts that the greatest players display. In a game that allowed basketball players to use a trampoline, Michael Jordan's athleticism would no longer loom as large (ibid: 37).[6]

In general, Sandel argues that the main problem with enhancement and genetic engineering is "that they represent a kind of hyperagency – a Promethean aspiration to remake nature, including human nature, to serve our purposes and satisfy our desires" (ibid: 26–7). Such a 'drive to mastery' will destroy the appreciation of our natural talents and gifts, of what Sandel calls the 'giftedness of life'. He suggests that it is this reverence for giftedness which keeps the Promethean tendency in check, regardless of whether we think our endowments proceed from God, nature, or fortune. "Appreciating the gifted quality of life constrains the Promethean project and conduces to certain humility. It is in part a religious sensibility. But its resonance reaches beyond religion" (ibid: 27). Complete mastery, however, would demolish this valuable humility, as it wouldn't recognise "that not everything in the world is open to any use we may desire or devise" (ibid: 27). Furthermore, he claims that it would make us responsible for everything we are: parents for their children – if they could control their genetic makeup, for instance – and athletes for their 'talents'. According to Sandel, such responsibility would undermine our solidarity with those who are less fortunate: if people's genetic endowments would become

[6] Sandel recognises that the line between natural and artifice, mere improvement and enhancement can be very unclear. In order to determine whether an innovation improves or corrupts a particular sport, one will have to look at 'the nature of the sport', and try to judge whether it manages to select the supreme players by emphasising their talents, or whether it rather overshadows them.

achievements or choices rather than gifts or fortune, the 'disadvantaged' at the bottom of society "would be viewed not as disadvantaged, and thus worthy of a measure of compensation, but as simply unfit, and thus worthy of eugenic repair" (ibid: 92). They would be considered to be lacking through a fault of their own – they (or their parents) could have chosen to be different. Sandel admits that the 'Promethean self-image' might be attractive somehow. "But that promise of mastery is flawed. It threatens to banish our appreciation of life as a gift, and to leave us with nothing to affirm or behold outside our own will" (ibid: 100).

6.4.2 *Michael McNamee*

Similar reservations appear in the article 'Whose Prometheus? Transhumanism, Biotechnology and the Moral Topography of Sports Medicine' (2007) by philosopher of applied ethics and contributor to this volume, Michael McNamee (Chap. 10). Concentrating on its application in sport in particular, he challenges "the hubris of modern biomedical science", and its "vertical ambition in transforming our very nature as humans" (McNamee 2007: 182–3). McNamee sees the traditional goals of medicine, such as healing and relieving suffering, being extended "not merely to intervening in, but also to re-envisioning the body" (ibid: 184). For example, hypoxic chambers, anabolic steroids or EPO – all regularly used in doping – "are often designed with medical therapies in mind. It is their use in the elite sports population that is problematic, not the nature of the products themselves" (ibid: 186). Broadening his criticism to the application of biotechnology to human enhancement, McNamee presents some arguments against transhumanism.[7] He dismisses it as 'undesirable utopianism' because it would lead to social inequality; further, it rests upon a very specific conception of the good; and finally (quoting Jürgen Habermas), it would deprive humans of their 'naturalness', and change their 'normative self-understanding'. "At TH's (sic) heart, it seems to me, is a view of technology at the mercy of scientists generally […] which is simply a case of Prometheanism" (ibid: 188). However, McNamee thinks that the charge of Prometheanism is often made but not properly explained, and thus he returns to two versions of the original myth of Prometheus: Hesiod's and Aeschylus's. On Hesiod's account, he says, Prometheus is a rather deficient character. He is "a cheat and a thief" (ibid: 189), whereas in Aeschylus' interpretation Prometheus is the courageous benefactor of mankind. After presenting a short overview, he cites D. J. Conacher, writing that "the Hesiodic Prometheus […] is presented as the indirect cause of all man's woes; the Aeschylean Prometheus, on the other hand, […] is presented as the saviour of mankind" (ibid: 190).

[7] Transhumanism is a philosophy that advocates the use of technology for unlimited human enhancement.

Although McNamee doesn't make any specific comparison between the different interpretations of the myth and transhumanism (or contemporary biotechnology), in the end he does imply that it is *Hesiod's* hubristic version which represents the transhumanist project – thereby warning against what would happen if unlimited enhancement were to be given free rein. He concludes by saying that in order to be able to evaluate the enhancement agenda in the context of sport, moral questions need to be raised to tease out "what I take to be the 'traditional' (natural?) work of medicine in the relief of suffering and the more recent (and artificial) goal of performance enhancement or the augmentation of natural abilities as opposed to the traditional therapeutic role of medicine" (ibid: 191). In other words, questions such as: Where to draw the line between therapy and enhancement? How much control over our bodies should be allowed? What limits should be observed? "We are mortal beings. Our vulnerability to disease and death, far from something we can overcome or eliminate, represents natural limits both for morality and medicine generally and sports medicine in particular" (ibid: 183).[8] However, McNamee says, especially elite sportsmen and women, their coaches and sport institutions

> all have an interest in surpassing limits. This denial of the necessity of limits in nature by some, the desire […] to control these human-limiting factors by the unfettered use of biotechnology is something that should concern us all in sports. I submit that philosophers of both sport and medicine begin to press such questions home in the public spaces of the media as well as the gymnasium and the university so that sports do not become the vanguard of Hesiod's Promethean project (ibid: 191).

6.4.3 Leon Kass

In an article entitled 'Ageless Bodies, Happy Souls' (2003) physician and medical ethicist Leon Kass investigates what he thinks is the deepest source of our concerns, fears and anxieties regarding biotechnology: the prospect that biotechnical powers will be used "to pursue 'perfection'" (Kass 2003: 10). One of the main worries is that this pursuit of perfection – of 'ageless bodies' and 'happy souls' – threatens human life and our very nature as such, and will lead to *dehumanisation*.

Kass tries to identify what exactly makes human nature worth defending. He argues that several objections that are regularly raised against enhancement – such as health and safety issues in the case of doping, or the unfair advantage of the enhanced athlete over others – still don't explain why a safe, fair and individual improvement by means of, for instance, genetic engineering remains disquieting.

---

[8] This passage actually comes from the beginning of the article in which he announces that he will conclude with this claim. Since it nicely sums up the last paragraphs of the article, I decided to quote the passage at this point.

Further, by criticising amongst other things Sandel's plea for giftedness, Kass claims that a mere appeal to human nature, the 'natural', the given etc., doesn't answer the question either. Such appeals can be a guide for what to do with regard to biotechnology "only if there is something inherently good or dignified about, say, natural procreation, human finitude, the human life cycle" (ibid: 20), etc. We have from the start been creatures that 'naturally' look for ways to improve or perfect their lives, so what could be wrong with enhancement, with perfection? It can't be mere 'unnaturalness', he claims.

Yet to achieve excellence in sport by means of athletic enhancement, is judged as cheating.

> We believe […] that people should work hard for their achievements […]. Even if one prefers the grace of the natural athlete, whose performance deceptively appears to be effortless, we admire those who overcome obstacles and struggle to try to achieve the excellence of the former (ibid: 21).

Such effort is a display of character, discipline and courage – traits which would disappear if the athlete's achievements would be all attained by means of drugs. Therefore, Kass argues, there must still be something wrong with the achievement of excellence by these 'artificial' means.

Indeed, "there is a sense that the 'naturalness' of means matters. It lies not in the fact that the assisting drugs and devices are artefacts, but in the danger of violating or deforming the deep structure of natural human activity" (ibid: 22). The inherent worth of the natural, Kass argues, lies in the fact that our activities aimed at self-improvement involve effort, choice, work and meaning. Biomedical interventions however, act directly on the human body, leaving the human himself no role in the process. Thus, the 'acts' of self-improvement and their effects will lose all their meaning, as there will be no intelligible connection anymore between means and ends. "Most of the best of [life's activities] […] are self-fulfilling beyond the need for praise and blame or any other external reward. In these activities, there is at best no goal beyond the activity itself. It is the deep structure of unimpeded, for-itself, human being-at-work-in-the-world, that we are eager to preserve against dilution and distortion" (ibid: 24). Although it sounds rather attractive to have an ageless body, in the end senescence and even finitude are indispensable for many of the most important human goods such as involvement, meaning, beauty and virtue.[9] And with respect to the 'happy soul' Kass claims that however awful events, memories and feelings can be, at the same time they make up our identity and teach us how to respond in the future. "An untroubled soul in a troubling world is a shrunken human being" (ibid: 27). Moreover, happiness and unhappiness are two sides of the same coin: one must desire and thus lack in order to be able to be fulfilled.

[9] Since Kass presents his arguments surrounding this issue in another essay ('*L'Chaim* And Its Limits: Why Not Immortality?' (2001) and since they don't add very much to the arguments relevant for this paper I chose not to explain this claim on finitude any further.

> A flourishing human life [...] is a life of aspiration, made possible by and born of experienced lack [...]. Not the agelessness of the body, nor the contentment of the soul, nor even the list of external achievement and accomplishments of life, but the engaged and energetic being-at-work of what nature uniquely gave to us is what we need to treasure and defend (ibid: 27–8).

6.5 Analysis

Interestingly, the arguments within all of the texts discussed above seem to rely on particular assumptions or claims about human nature, although some are a lot more explicit than others. Moreover, the way the athlete is pictured is inherently connected with these assumptions about mankind, whether the author argues in favour of or against (performance) enhancement and whether sport is the main subject of the text or not. Further, these assumptions – far from being neutral – often include a normative judgement as well.[10]

6.5.1 *The Advocates*

Stock makes a (admittedly rather inexplicit) connection between sport and human nature. To use performance enhancing drugs is a way to cheat, and of course it is dangerous, he argues. However, he also states: "[a]s difficult as the problem of doping is, it will soon worsen, because such drugs will become safer, more effective, and harder to detect" (Stock 2003: 8). Whereas at first sight it might seem that Stock is arguing against doping and enhancement, in fact he is not. He claims that these issues in sport offer a 'preview' of what will happen in society in general, for "the nonsporting world is not so different" (ibid: 8). He explains how, if the risks would diminish, many people – analogously to athletes – would be willing to use drugs, genetic enhancement or other ways of manipulation to improve their health, retard ageing, etc. It will be dangerous at first, there will be mistakes, and those who try out the newest technologies will probably be called imprudent, careless or crazy. "But as much could be said about any potent new development. [...] Our history is not a tale of self-restraint. [...] We may like to believe that primitive peoples lived in balance with nature, but when they entered new lands, they reshaped them in profound, often destructive ways" (ibid: 10). In other words, new developments always bring problems, misunderstandings, dangers and often even destruction – that is the way humanity explores, acquires its knowledge, and develops its techniques. "As always, we will have to earn our knowledge by using the technology

---

[10] Some of the arguments I make here are inspired by Michael Hauskeller's paper 'Prometheus Unbound' (2009), which argues that transhumanists refer to a value-laden concept of human nature when claiming what we should do with respect to enhancement. Hauskeller also pays attention to their references to Prometheus when making his arguments.

and learning from the problems that arise" (ibid: 11). According to Stock it is *human* to continue with the development of a new technology, even when we realise that there are risks involved. Therefore, we will not turn away once we realise what risks enhancement may have. As he states at the start of his book: stealing fire from the gods "is too characteristically human".

This reference to Prometheus – the courageous Titan who challenged the Olympians, faced danger and crossed boundaries, who brought civilisation, technological development and progress – is not merely a way of encouraging us not to be afraid. Stock is comparing the heroic Prometheus to humanity in general, implying that it is simply our *nature* to confront the new and the dangerous, to be rebellious, to seek to cross our natural limitations – even if this means taking part in our own creation. Stock also seems to suggest that to turn away from the – possibly dangerous – challenges and dismiss the new technologies would not merely be unrealistic – it would mean to *lack something fundamental*. Perhaps one would even *not really be human*, for as he put it, it would be "*out of character* for humanity" (emphasis added). A human being acting in character will dare to take the challenge of enhancement – it is only natural.

Underneath Stock's concept of the human lies a normative judgement. He claims a "profound biological change" is inevitable because of our nature. We've always tried to improve ourselves despite the dangers and destruction, so why stop now? "If we could make better humans … why shouldn't we?" (ibid: 12). In that sense, both the Promethean metaphor and the example of the athlete, as presented by Stock, indirectly lead to a moral imperative. Humanity has always explored, taken risks, developed techniques and learned from its mistakes – just as the heroic Titan and the enhancing athlete. This is what we have always done, what we still do, and thus, Stock reasons, what we *should* do. This argument seems vulnerable to the criticism any is/ought kind of reasoning receives – an issue I will address later.

Stock recognises that there are problems, concerns, and dangers surrounding enhancement. However, at the same time he uses some quite disapproving language when it comes to the non-enhanced human being. Apart from virtually calling it unhuman not to try and transcend our present limits by means of enhancement, Stock compares our current, non-enhanced state to 'childhood': "Humanity is moving out of its childhood and into a gawky, stumbling adolescence in which it must learn not only to acknowledge its immense new powers, but to figure out how to use them wisely" (ibid: 17). In other words: not to act according to the Promethean tendency would mean to never 'grow up' as a species. Obviously, what we should do according to Stock is follow our self-improving nature.

Young provides another example of a pro-enhancement debater who connects sport, human nature and Prometheus. Just as in Stock's text the example of the athlete is used to support the argument in favour of enhancement. We already enhance our bodies by means of clothes, makeup, etc., so why shouldn't we use other forms of enhancement? The bodybuilder just uses a different means to fulfil "the desire to be the *best* one can be" (Young 2006: 306). This, as Young explained earlier, is our nature, our inborn "Will to Evolve". Thus, his argument in favour of enhancement again relies on a – rather explicit – claim(s) about human nature. The bodybuilder simply represents a creative way to satisfy this natural desire: he is being propelled by the Prometheus Drive which makes him yearn for improvement, and search for ways to transcend his limitations. For Young, both Prometheus and the athlete exemplify the innate human drive, which provides the basis for a moral imperative: we *are* creative, daring, knowledge-seeking and limits-transcending beings, therefore we *ought to* enhance ourselves. If we don't, we refrain from acting "in harmony with *the essential nature* of the evolutionary process" (ibid: 19). Moreover, he suggests that there are awful consequences awaiting us if we do not cherish the Prometheus Drive, for "without the instinct to progress, humankind is doomed to remain forever at the mercy of disease, decay, and the limitations of the human body and mind" (ibid: 39). However, what seems to be most important of all is that according to him it is *ethically right* to act 'in harmony' with nature – "human beings were not meant to live in the moment like the lilies of the field – our purpose is to be all that we can be" (ibidem). One of Young's major arguments is – just as Stock's, though much more explicit – thus one with an is/ought-structure: this is what we are, this is our nature, and therefore the way it should be.

Savulescu et al.'s arguments in favour of performance enhancing drugs also rely on a very explicit view of human nature. "Humans are not horses or dogs" (Savulescu et al. 2004: 666) – they are *different* creatures, and therefore shouldn't be judged on the basis of their biological potential only. Unlike animals, humanity is endowed with creativity, reason, and judgement. These characteristics, they claim, enable us to choose our own way to reach our objectives, and to improve ourselves *whichever way we want* – even by means of doping or enhancement. It is human to long for amelioration.

Savulescu et al. are arguing against "the old naturalistic Athenian vision of sport" (ibidem), which merely tries to find those who nature selected as the chosen ones. However, interestingly, one could argue that, in a way, they too defend a naturalistic view, as – even though they reject the current discrimination against the 'genetically unfit' – their view of sport depends on their idea of human nature. In the end, athletic enhancement is not problematic for Savulescu et al., because creativity and the aspiration for improvement are *natural*. The fact that creative improvement defines human nature as such makes them conclude that performance enhancement doesn't violate 'the spirit of sport' – on the contrary. It is our natural creativity and urge for improvement that make performance enhancing drugs acceptable – as long as those drugs do not threaten the athlete's welfare, that is.

Thus, Savulescu et al. make fundamental claims about human nature and use these claims to support arguments in favour of athletic enhancement. Their concept of mankind, in other words, is not a mere description but includes a normative judgement: the human is characterised by creativity and the aspiration to be better, and that is *good*. They seem to argue that *since* it is our nature to create, reason, choose, and improve, we *should*. When they claim that "[p]erformance enhancement […] is the spirit of sport. To choose to be better is to be human" (ibid: 670), the only thing that is missing between the two sentences is the word 'because'. Apparently sport should be modelled upon these fundamental characteristics of our nature.

Interestingly, the pro-enhancement texts discussed here all present, in their own way, an is/ought-argument along lines that make it resemble a naturalistic fallacy: it *is* naturally human to take risks, transcend our boundaries, improve and create, and therefore we *ought* to do this (and thus enhance). The validity of this type of reasoning is dubious, of course: why should the mere fact that something is, or is natural, lead to the conclusion that therefore this is the way it should be? Moreover, to ground a claim on something, apparently value-laden, like 'human nature' in the first place, does not correspond with the emphasis most – if not all – of the proponents of enhancement put on objective (rational, and/or scientific) knowledge and truth as both the foundation of their reasoning and its purpose. Further, the naturalistic fallacy is precisely something many advocates accuse the opposing 'bioconservatives' (Francis Fukuyama, for instance) of committing, which, as we will see, is not unfounded.

6.5.2 The Opponents

In arguments of the opponents of (performance) enhancement, assumptions or claims about human nature play an important role as well, and again the manner in which the athlete is pictured is inherently connected to these assumptions. In his quest to verbalise what is morally wrong with enhancement Sandel focuses on what he sees as the point of sport – *excellence*. This excellence subsists in the exhibition and cultivation of "natural talents and gifts" (Sandel 2007: 28): the capacities in which the athlete has had no hand. Using performance enhancing drugs in sport is troublesome as it disturbs the appreciation of the 'giftedness' of the athlete's natural powers. Sandel is thus implicitly stating that the naturalcy of these talents lies in the fact that they are 'given – whether this is done by fortune, nature, or God.[11] This reveals an assumption about the nature of the human in general (without thereby defining human nature in its entirety): in his natural state, the human – not only the athlete – is characterised by the talents and attributes that he happens to possess. These talents and gifts are not his doing, but emanate rather from an origin that exceeds his control. In this way, the athlete implicitly epitomises what is considered to be essential about human nature. As in the preceding arguments of the proponents – though from a completely different angle – the athlete thus becomes a paradigmatic example of the human while being used to support the author's stance with respect to human enhancement.

Interestingly, Sandel also uses the image of Prometheus. Here Prometheus represents – instead of a courageous hero – the arrogant 'drive to mastery', that is: the hubristic tendency to want to control everything and anything (formerly) determined

[11] Note that 'natural' in Sandel's 'natural gift' refers to a broad, common sense concept of whatever is considered to be innate, inborn and outside our control, whereas nature is here *one* of the sources which – dependent on what people believe – may provide these natural gifts (which thus may also proceed from fate, chance, God, fortune, etc.).

by nature, God, or fate. For Sandel, the Titan God seems to epitomise a fundamental *threat* to what is essential about being human – the gifted character of our nature, our appreciation of this giftedness, and the humility it induces – as the Promethean tendency is the "aspiration to remake nature, including human nature, to serve our purposes and satisfy our desires" (ibid: 26–7). Moreover, Prometheus' role as creator is also emphasised: he becomes not only the presumptuous archetype of man's controlling urge, but also the symbol of the self-made man – *literally* self-made: moulding and transforming himself, crossing his limits while striving for mastery and perfection. Although Sandel doesn't claim that everything which is natural is good, to him the gifted character of the natural *is*. Therefore, again human nature – though less directly – is connected with a set of values that leads to a moral imperative. First, life as a gift has a deep moral value in itself; second, our appreciation of that gift and our humility towards the gifted are also valuable as such; third, the extreme mastery and self-moulding to which enhancement would lead would have bad consequences such as the 'explosion' of responsibility and the decrease in solidarity, while nothing would remain beyond our desires and design. Therefore, the Promethean ambition to master and even remake (human) nature is wrong, and thus Sandel argues that we *should not* engage in human enhancement.

In McNamee's paper there are not merely assumptions about human nature, but explicit statements. "We are mortal beings" (McNamee 2007: 183), he claims, and characterised by limits. By wanting to overcome these limitations , modern biomedical science – including sport medicine – hubristically ignores our vulnerability, mortality, naturalness, and the necessity of our limits. Such biotechnology is Promethean in that through this controlling and overcoming urge it is cheating nature and stealing its heavenly character, which will only have disastrous consequences. For McNamee, as for Sandel, Prometheus thus embodies the origin of all human misery – *Hesiod's* Prometheus, that is, as McNamee explicitly wishes that sport will not embrace the morals of this author's Titan'. Although McNamee is relatively inexplicit on the simile between the myth and biotechnology and sport medicine, it is clear that the Promethean enterprise threatens human nature as such.

McNamee focuses on the way the natural human is being threatened in (elite) sport, which is due to the deficit of morality from which the susceptible and overly devoted athletes suffer. Because of these athletes' urge to optimise their (physical) abilities and the need to enhance their performances every time, the natural seems to have lost all its value. This is why performance enhancing drugs are problematic for McNamee: they 'deny' the necessity of our natural limits, which is why it is vital that this denying and controlling drive be addressed in the world of sport. Therefore, McNamee (citing Charles Taylor) pleads for the establishment of a 'moral topography', that is: the development of an ethical map which draws the boundaries between the natural and the artificial within medicine, and between therapy and enhancement, so that the difficult questions regarding new technologies within sport may be addressed from an informed point of view.

In other words, according to McNamee sport medicine and sport as such seem to be exemplary spaces where these ethical boundaries seem to have lost their importance, or even disappeared. Where these boundaries need to be drawn exactly so that the artificial may be distinguished from the natural isn't something McNamee means to decide here. However, what is certain is that in this text the athlete's character as the boundary-crossing being again provides *the* way to emphasise what is natural about the human: precisely these boundaries, the (necessity of his) limits, his vulnerability and mortality, and his imperfection. Again, these characteristics are not merely natural, they are *valuable*, and so a strong moral imperative is present: this natural, limited humanness should be maintained and so no (performance) enhancement should be allowed – provided it is clear what exactly belongs to these categories. Once more, an argument that seems vulnerable to the naturalistic fallacy objection.

Kass also uses the image of the athlete to emphasise what is natural about the human, and what makes it valuable. Although we may favour the sportsmen and women with the tremendous innate talents, we nevertheless have great respect for the athletes who invest all their powers, effort, and energy in trying to compete with those who are better equipped from birth. According to Kass, humanity's 'naturalness' is inherently worthy precisely because human activities involve such effort, choice, and work. Differing than Sandel, he focuses on what lies *within* rather than what lies outside the athlete's – and thus mankind's – might. It is exactly the activity itself which makes the goal valuable, meaningful, and rewarding. This is why he argues that (performance) enhancement would disturb "the deep structure of natural human activity" (Kass 2003: 22): it is solely focused on the ends and goals, and thereby it takes away the fulfilment, satisfaction and meaning which result from the activity itself.

Obviously, Kass' example of the non-enhanced athlete serves to articulate his idea of (the value of) human nature: the athlete's effort, work, and struggle to achieve improvement of performance symbolise the character of natural human activity in general. Further, as in the preceding texts, his concept of the human includes a normative judgement: we need to conserve our naturalness. Kass argues that our desiring, engaged activities and the fact that we are lacking, mortal, and imperfect beings, make up our identity and character, and create the possibility of happiness. Therefore, we need to 'treasure' and 'defend' our naturalness – even though it also makes us struggle, limited and troubled – since in the end we need it in order to (be able to) desire, be fulfilled and lead a flourishing life. If we would manage to attain 'ageless bodies' and 'happy souls' this naturalness is what we would lose. It would dehumanise us, he says, and thus biotechnological enhancement should not be allowed.

Kass also refers several times to Prometheus, though not in the text at hand but in his book *Life, Liberty and the Defence of Dignity* (2002). As in the case of the other two critics, the Titan epitomises a hubristic disposition to rational mastery by means of technology, which could destroy what makes us human. Kass argues that if we would become able to even master the human mind, with all its desires, pains and pleasures, the result would be outright dehumanisation, reminiscent of the people in Huxley's *Brave New World*:

> Homogenisation, mediocrity, pacification, drug-induced contentment, debasement of taste, souls without loves and longings – these are the inevitable results of making the essence of human nature the last project for technical mastery. In his moment of triumph, Promethean man will become also a contented cow (Kass 2002: 48).

Again, Prometheus thus represents the destructive threat to human nature as such.

Just as those in favour – although from an opposite point of view of course – McNamee and Kass also make claims along naturalistic and is/ought lines: human nature, as it is, is inherently good, and therefore it ought to be preserved. Sandel's argument is slightly different as it doesn't value human nature because of its naturalness as such, but because of its giftedness. Interestingly, both sides of the debate thus argue according to a similar – although logically invalid – structure, while even using the same figure to depict their argument.

6.5.3 Ambiguity: The Myth and the Athlete

Although at first sight it might have seemed to be a mere illustration, the myth of Prometheus has turned out to be a tool for proponents as well as critics of enhancement to emphasise the importance and moral value of human nature. Moreover, as I will explain below, contrasting the different applications of the myth highlights an analogous contrast between the various ways in which the athlete and the point of sport are depicted.

On the one hand, Prometheus represents the courageous and boundary-crossing paradigmatic human who embodies mankind's natural creativity, while on the other hand he is the hubristic and controlling barbarian, who demolishes the human being's natural, gifted character. It is precisely this ambivalence which explains why *this* myth is used so often, as there are many more myths and stories dealing with topical subjects such as hubris (Daedalus and Icarus, for instance) or the temptation of knowledge (Faust). Not only do the stories in the original versions of this myth differ radically from chronicle to chronicle but – and especially – the *evaluation* of Prometheus' actions also varies. Hesiod depicts Prometheus as a trickster who brings almost nothing but misery to himself and humanity, while Plato portrays him as the heroic, courageous creator of mankind, whose punishment remains unmentioned.[12] As there is no one canonical text, the myth itself is extremely malleable, and the plot may be easily liberated from its original context. This enables each author to mould it into various directions and let it carry the meaning he prefers – dependent (amongst other things) on his account of the human. It is exactly this ambiguous nature that explains why the myth lends itself to such different applications in the accounts analysed – why Prometheus can symbolise humanity's ill-fated urge to master nature, as well as the promise of humanity's evolutionary and technological progress.

To ask which of the two main interpretations of the myth of Prometheus is correct would be unhelpful: there are different versions of the original myth, but they all present a story in which the main character is a figure with good and bad

[12] As said, all Plato writes is that he was accused of thievery.

characteristics, his deeds have good and bad consequences and the story carries both a positive and negative moral. In short: both interpretations make sense in context, because of the myth's ambivalence.

The characteristics of the athlete are very similar to those of Prometheus: he can be both a hero and a cheater, is courageously taking risks, facing dangers, and looking for perfection while at the same time he is pushing and sometimes hubristically crossing human limits, and transforming his body. Therefore, he can be depicted in (roughly) two completely different ways – he is just as ambiguous as Prometheus. For the advocates, the athlete embodies the essentially human and creative urge for improvement and perfection, while for the critics he represents either the display of natural talents or, if enhanced, a figure that crosses the limits of humanity and thereby destroys his nature and its giftedness. Which of these images is chosen is, just as in the case of the myth, inherently connected to what is understood to be human nature. As said, these are value-laden concepts of human nature, providing the basis for what ought to happen, be done, or exist (without, however, thereby holding all that is natural to be good). As David Hume already knew, deducing an 'ought' from an 'is' seems logically impossible, and so the logical validity of these arguments is questionable. This chapter has demonstrated that this type of (fallacious) reasoning is found at *both* sides of the debate – despite the fact that many pro-enhancement authors accuse the opponents of naturalistic fallacies. My objective here has been to reveal and contrast these two morally and naturalistically informed images of the athlete, and that this ambiguity of the athlete leads to a corresponding ambiguity when it comes to the way the spirit of sport is portrayed nowadays. I have not engaged in the normative discussion on which of the two visions of athleticism we should advocate: whether the point of sport is excellence through creative transcendence of boundaries, or through (the appreciation of) natural, gifted talents and activities within the limits of humanness. I have clarified how both interpretations of the athlete's role and the correlating spirit of sport can make sense up to a certain point, and that a deep tension underlies the way sport is practised and conceived today.

6.5.4 Contradictory Demands

The analogy with the ambivalence of the myth highlights a strained contradiction within the world of sport, which requires closer analysis. On the one hand sport is about crossing boundaries. An athlete who participates in the Olympics doesn't merely want to win the game or the gold medal – he probably also wants to run faster than anyone has ever run before, jump higher than anyone ever jumped. In many sports the goal is not merely to win against the competitors of this particular race or swimming course – it is about winning against *all others* who have ever competed in the same kind of game. It is about achieving the best or most outstanding performance ever made in the field, and breaking records. In other words, the objective is to achieve something that no other human has ever managed to accomplish.

It is not merely about trying to cross one's own limits or the limits of those around you; it is about transcending the boundaries of what is considered to be humanly possible. The important thing is that this isn't simply an individual aim – it is also something requested by the outside world. Whether it is mostly the pressure from coaches, the competitive nature of the sport itself or the combative behaviour of adversaries; whether it is the anticipation from family, the role of the media or the public's high expectations – the demand is insistent.

On the other hand, the superb achievements are expected to remain 'natural' – world records are news worthy, but so are doping scandals. Not only the World Anti-Doping Agency but public, media, and at least *part of* family, friends, and athletes themselves are against doping and enhancement. Apart from the fact that it is seen as cheating, unfair, and unhealthy, this disapproving attitude with respect to performance enhancing drugs is also a product of the fact that the self-enhancing athletes engages in, as the German physician Hans Seel already suggested in the 1930s, "deviations from a natural way of life" (quoted in Hoberman 1992: 132).

The result is a rather antithetical combination of two demands: the athlete is expected to transcend his human boundaries, while at the same time this has to be done by means of his human capacities, remaining within natural human limits. Moreover, these antagonistic expectations do not come from two different and opposing parties, but often from the same source: the world of sport in its entirety, from the athletes themselves to the media's representation of their actions. For the athlete, this presents a rather uncomfortable – if not a *lose-lose* – situation.

6.6 Conclusion

The myth of Prometheus has proved to be a useful tool to structure the bigger debate on (performance) enhancement and provide more insight within the practice of sport. First, the analogy between the ambivalent myth and the way the athlete and sport are conceptualised highlights the two main contradicting views which are present in the debate. Most of all, it emphasises their moral value and their connection with human nature. Second, the analogy also uncovers the deep contradiction between the two demands upon the athlete which exists in the practice of sport.

Instead of being 'just' a philosophical analysis, these two points are important for several reasons. Revealing the underlying (moral) concepts of human nature contributes to a better understanding of the priorities of each of the positions. It provides more insight into the inconsistencies within the rhetoric surrounding (doping in) sport, which could help in finding a resolution – not merely with respect to theoretical issues, but also to pressing normative issues within ethics, policy and legislation. Furthermore, the analysis makes us aware of the fact that the athlete finds himself continuously having to deal with these contradictory demands as he makes practical decisions – what means to use and what means to reject when trying to improve his performance. One can imagine this to be a rather difficult situation.

Being aware of this fact we should reconsider the way we perceive and approach the athlete. Moreover, we should acknowledge the fact that – even though it might be tiny – we all carry part of his responsibility, as together we make up his public.

References

Hauskeller, M. 2009. Prometheus unbound. *Ethical Perspectives* 16(1): 3–20.

Hoberman, J.M. 1992. *Mortal engines. The science of performance and the dehumanization of sport*. Caldwall: The Blackburn Press.

Kass, L. 2002. *Life, liberty and the defence of dignity*. San Francisco: Encounter Books.

Kass, L. 2003. Ageless bodies, happy souls: Biotechnology and the pursuit of perfection. *The New Atlantis* 1: 9–28.

McNamee, M. 2007. Whose Prometheus? Transhumanism, biotechnology and the moral topography of sports medicine. *Sport, Ethics and Philosophy* 1(2): 181–194.

Plato. 1997. Protagoras. In *Complete works*. Cambridge: Hackett Publishing Co.

Sandel, M. 2007. *The case against perfection*. Cambridge, MA/London: Belknap Press of Harvard University Press.

Savulescu, J., B. Foddy, and M. Clayton. 2004. Why we should allow performance enhancing drugs in sport. *British Journal of Sports Magazine* 38: 666–670.

Stock, G. 2003. *Redesigning humans*. London: Profile Books.

Young, S. 2006. *Designer evolution – A transhumanist manifesto*. New York: Prometheus Books.

Chapter 7
Outliers, Freaks, and Cheats: Constituting Normality in the Age of Enhancement

Darian Meacham

7.1 Introduction

In this chapter, I propose that attitudes in sports toward doping, technological enhancement of the body, and also what I call the 'natural unnatural' are influenced by an experience of the normal that is built up historically and can take on a normative force or significance. The unease surrounding enhancement in sport is not simply a concern about cheating, fair play, or the health and safety of athletes. Nor can it be attributed solely to a concern for the purity of sport. Rather, it must be understood as part of a much broader concern about the violation of the normal.

This concern is linked to, though not necessarily exclusive to, a concern for the Enlightenment categories of 'man' and 'humanity', categories that enhancement calls into question. Organised sports, and the Olympic movement in particular, serve as a kind of testing ground for the physical limits of man: a sphere where limits can be pushed in a regulated or rule bound manner. Olympic sports are in this sense a testing ground not only for the *nomos* of sport, but also for the *nomos* of man; *nomos* in a double sense, meaning an order or law based on convention, and a boundary or frontier.

The experience of normality can be explicated phenomenologically. Following Edmund Husserl, I contend that it is an essential element of perceptual experience *per se*, i.e. of our experience of the world. However, *what* is experienced as normal is a matter of historical contingency. In this context, I point to a distinction between the 'normally abnormal' and the 'abnormally abnormal' that is relevant to the sports enhancement debate in its relation to the categories of man and humanity. These are categories that Husserl himself seemed to take for granted, while furnishing us with the tools to interrogate them. The phenomenology of the normal scope of species

---

D. Meacham (✉)
Department of Arts, Faculty of Arts, Creative Industries and Education,
University of the West of England, St. Matthias Campus,
Room Number 1CK20, Bristol BS16 ZJP, UK
e-mail: darianmeacham@gmail.com

J. Tolleneer et al. (eds.), *Athletic Enhancement, Human Nature and Ethics*, International Library of Ethics, Law, and the New Medicine 52, DOI 10.1007/978-94-007-5101-9_7,

functioning and the categories that it reveals tells us something about the boundaries of the human or of man that sport both pushes and now seeks to defend.

Ultimately, I argue that enhancement as a violation of the normal threatens to render the notion of man or mankind obsolete, by splitting the identity of the species concept human and the ethical concept man. This would leave us (whomever that may end up being) in a fragmented situation where various sub-species of the human, or various descendents from the common ancestor, human or man, would co-habit the same space. The challenge would be whether these differently natured groups could share a world in a phenomenological sense: communicate and engage in common projects. In other words, humanity would have to be reconceived as a communicative project in a manner that would potentially incorporate more than one species or sub-species.[1]

I conclude by noting that developments in human enhancement technology will soon create a situation where international sporting institutions and the Olympic movement in particular will need to make decisions concerning the very ethos of international and Olympic sport. Modern sport has two choices. Hold dear to its Enlightenment ethos and become a celebration of the shared heritage of the (human) body and its traditional form, turning the Olympics into something resembling a UNESCO heritage site. Or it can become a sphere where the boundaries of the human are continually pushed, and perhaps where humanity finally exceeds or breaks free from both the Enlightenment idea of man—the Kantian 'empirical-transcendental doublet'—and its limitation to the taxonomic category of the human. This latter possibility is a transhumanist vision of the future of sport, wherein the project of humanity is pursued beyond the horizons of mankind or the human.

7.2 'Citius, Altius, Fortius', But Not That Fast!

When we watch elite sports, we expect to see extraordinary physical feats performed by extraordinary people. There is little doubt that even those world class athletes whose names never make the headlines are outliers of what can be called normal species functioning. The runner who finishes last in the quarter finals of the 100 m at the world championships, or the nameless *domestique*, who pulls his team leader up the *Alpe d'Heuz* in the Tour de France, only to fall back at the last moment exhausted and physically spent, regularly performs physical feats that the normal human being could only dream of. Yet, these extraordinary outliers are not our heroes. Their efforts and abilities go largely unremarked upon.

It is the true outliers that most captivate us. Those freaks who stand at the very pinnacle of our species' physical capacity. But is there a point at which a line is crossed, when the extra-ordinary is too much? When is an athletic performance so far

[1] A phenomenological species is a group of beings that share a world or could potentially do so. It remains to be explained then what a phenomenological sub-species would entail.

beyond the pale of the normal that it raises suspicions? Is there a point at which, we, the viewing public or the institutions that govern sporting competition say no! or enough! to the Olympic motto '*Citius, Altius, Fortius*' – faster, higher, stronger?

Yes. At a certain rather indeterminate point an athlete who functioned too far beyond the scope of what was perceived to be normal would begin to arouse a certain sense of uneasiness. I assert that most people experience a certain form of attachment to the normal, or more precisely, to what is perceived as normal. And this attachment exerts a kind of normative force that is also at the root of the uneasiness generated by the abnormal, in this case the enhanced as a type of abnormal.

In investigating our attachment to the normal and our uneasiness with its violation, there are two types of concrete cases that I am interested in exploring. The first concerns the prohibition of enhancement as a form of doping and uneasiness on the part of both sport's governing institutions and the sporting public with the use of technology to radically enhance sporting performances. The terms enhancement and doping are often used synonymously, but this is somewhat misleading. WADA defines doping as evidence of the use or attempted use of a prohibited substance in an athlete's body. Thus, various forms of enhancement, if prohibited, are doping, conceptually not very helpful. A more general understanding of doping as the addition of an impurity may be of more conceptual use here. To understand enhancement as doping is to see it as the addition of an impurity of some sort to what was previously taken as pure—the natural or traditional human body—for the sake of improving an individual's capacities and not preventing or treating a disease or condition. As I explain in note two below, I prefer a slightly finer grained understanding of enhancement. Nonetheless, the relevant point here is that equating enhancement with doping is to add a negative valuation having to do with an idea of the body as having a pure or natural state that should be positively valued and that the enhancement is a deviation from. Although the question of whether or not he is an enhanced athlete in the relevant sense is still open,[2] I nonetheless think that the South African 'runner', Oscar Pistorius, provides a good illustration of this type of case.[3]

The second type has to do with what I think we can call the natural unnatural: the perception of a person or a performance being unnatural or abnormal despite there being no artificial or technological intervention. This describes the media furore and many negative reactions around the performance of South African runner Castor

---

[2] The strong definition of enhancement refers to interventions that move the enhanced individual or population beyond the normal scope of species functioning, or at least to its outer limits in relation to one or more particular capacities. An enhancement within the normal scope of species functioning would be considered as weak. I readily concede that the Pistorius case is certainly a borderline or problematic one insofar as my strong definition of enhancement is concerned.

[3] Scare quotes here to alert the reader to a possible set of concerns that arise in this case and point to its difficulty: what is running? Does running require legs, and if so how much of a leg? What constitutes a leg? According to one study, Pistorius's prosthetics generate an energy savings of 25% over 'normal' (read: natural) limbs (Camporesi 2008). But the 'natural' limbs of most elite athletes have been thoroughly worked-over by technology, to the extent that the bodies of elite athletes must also be considered technological objects.

Semenya.[4] Despite quite radical differences in the two cases, one involving technological enhancement (carbon fibre prosthetic legs) and the other not, I think that they are both founded on the same type of experience: an experience or perception of normality in relation to what can be called a scope of typical species functioning. Enhancement in sport questions the experienced concept of normality, and perhaps also the boundaries of the species (understood phenomenologically), eliciting a deep uneasiness going beyond concern with cheating or breaking the rules.

7.3 Doping and the Nomos of Sport

I want to make a distinction between two different objections to technological enhancement in sport, understood as doping or not. The objection that enhancement violates the rules of sport, and is hence doping, should be separated from the objection that technological enhancement of the body violates the purity or naturalness of athletic accomplishment and is therefore morally objectionable in a more universal sense, and doping in the sense of adding an impurity, not merely using a banned substance. This second objection holds the concern for the ethos or spirit of sport in relation to the larger sphere of human activity.

The language of purity and naturalness is a manifestation of a concern for normality. Sport is one of the most appropriate domains for the manifestation of this concern as it is traditionally one of the most public spheres where the boundaries of normality are regularly pushed. So, sport emerges as a testing ground for the wider anthropological as well as ethical and political questions about the boundaries of the normal.

Cheating is a violation of the agreed rules of a sport and is objectionable from within the clearly defined world of that sport. The identity of any sport or game, its existence and delineation as a sphere separated from the rest of the world by way of its *nomos*, relies on this. A game of football or a chess match can only exist by virtue of the participants acknowledging the rules. But cheating is not necessarily a violation of the *nomos* of a sport or game. Many football players try to illegally bat the ball with a forearm or a hand. This may be bad sportsmanship, but it does not threaten the integrity of the game. What is important is that the violation, whether caught or not, is recognized as such. A player who denies that hitting the ball with her hand is a rule violation has called into question the *nomos* and with it the very integrity and identity of the game.

---

[4] Pistorius and Semenya are both South African athletes whose eligibility for competition has been brought into question: Pistorius because he has had both legs amputated below the knees and runs with the aid prosthetic lower legs, Semenya because her status as a women was questioned and she was asked to undergo 'gender testing' (a misnomer) following her victory in the 800 m at the 2009 IAAF World Championships. Semanya was ruled eligible to compete as a woman in July 2010. Pistorius was ruled ineligible for competition by the IAAF in January 2008, the decision was then overturned in May 2008.

If enhancement is banned, then it is cheating and ethically objectionable from within the rules of the game. In this sense, asking if enhancement should be allowed is something like asking if the offside rule in football should be changed. In order to avoid the redundancy involved in implying the legalistic definition of doping, I'll use the term enhancement from this point on, unless referring specifically to enhancement as rule breaking. However, doping in the sense of the addition of an impurity remains relevant to the discussion of enhancement as potential violation of normality.

Question about the rules of a sport may of course be affected by decisions outside of it strictly speaking; for example, worries about the long-term health of the players or, more relevant to our concerns here, what we want sport to be, i.e. how it should function in relation to the wider scope of activity it is situated within and related to. Doping as rule breaking is also different from other violations as the act of doping itself mostly takes place outside of the strict boundaries of the game or competition. However, this does not impact the general point about the distinction between the *nomos* of a sport and the role that sports play in regulating or modulating the *nomos* of the wider sphere of species activity that it is in, what in phenomenological parlance would be called the lifeworld.[5]

7.4 The Aesthetics of Sport

From a spectator's perspective, there also is an aesthetic dimension to violations of the *nomos* of a sport. But the aesthetic dimension of doping as rule violation illustrates the interrelatedness between concerns for the violations of the rules and concerns for the role of sport in relation to the wider sphere of the lifeworld.

I take it as a given that watching continual violations of a sport's rules sullies the aesthetic experience, likewise with learning that a performance was doped. The latter is more complicated because it is most often only after the fact that we learn of an athlete doping. Nonetheless, the spectator's aesthetic experience can be retroactively corrupted and made ugly in a similar fashion to learning that a painting is a counterfeit.[6]

The historical context or tradition within which artistic appreciation takes place can be shown to be somewhat analogous to the recognition and celebration of a shared natural and cultural heritage of the human body, which is one of the ways in which sport could be understood in the age of enhancement.

---

[5] Very generally speaking, the lifeworld is the world as experienced in an 'everyday' manner, that is, in relation to the interests, values, and practices of experiencing subjects. In this sense, it is the world as we encounter it intuitively in experience. For an elaboration, see, for example, Husserl's *Crisis of the European Sciences and Transcendental Phenomenology* (Husserl 1970).

[6] I am grateful to Andreas de Block for this comparison between doping and counterfeiting, see Chap. 8, this volume.

The famous case of Hans van Meegeren's Vermeer forgeries provides an example to illustrate the point about the aesthetic value of art, and by analogy a sporting performance. The French philosopher Maurice Merleau-Ponty argued that upon the discovery that the paintings were forgeries, their aesthetic and artistic value was diminished, even if prior to the discovery one held the works to be of great value. This is not simply art-world hypocrisy. Rather, our perception of a painting is through and through historical. We perceive a painting as responding to a certain question that is/was posed by the artistic tradition as a whole at the time of the painting. What the question is about i.e. is it about vision, or about painting itself, is part of the question itself. Every great painter in some way takes up the question of what the question of painting is about. The history of art is best understood as a question that is constantly being taken up and reformulated in each great work. A great painting responds to the question that is the tradition by reforming it from within. From this perspective, a van Meegeren does not hold the same value as a Vermeer. Even if most art experts at the time could not tell the difference, it does not come out of the same interrogative context as a Vermeer. Authorship in this sense matters a great deal. As a response to the question that is the tradition of painting, the van Meegerens have no value; they are entirely derivative. Upon the discovery of their true provenance they are reduced to a mere novelty sideshow in the history of art.

This is also the case when it comes to the experience of watching sport. Our appreciation of a sporting performance is not purely visual; or rather vision includes more than what is simply before our eyes in the normal sense. Vision and aesthetic appreciation of sport are historically contextualized. A great performance is only great in relation to the history of similar performances. The aesthetic experience of watching the world record being broken in the 100 m or watching a great climber pull away from the peloton on the Mont Ventoux acquires its greatness—and the frissons it provokes—in a historical context.[7] In the same manner that the history of painting is a question about painting posed to painters, the history of sport is a question or a challenge that is posed to athletes or to the body *per se*.

A distinction between the two is also needed here. In order to understand and appreciate the difference in aesthetic value between a Vermeer and a van Meegeren, one needs to know something about the history of Western art. To some extent this also holds true with sport. However, more generally speaking, sport is different, precisely because the tradition of the body is universal: it is the tradition of mankind, to use the term in the Enlightenment sense. The tradition of the human body as the taxonomical classification of Homo sapiens is taken to be synonymous with man or mankind, this identification itself being the product of the Enlightenment.

A great athletic performance is apparent to all because we all have a pre-reflective sense of what the normal is. The distinction between the normal and the extraordinary is apparent *in perception*, we see it immediately. The reason for this is that we all share in the tradition of the body. The pre-reflective understanding of the normal

[7] A similar point is made in Wasserman (2008).

does not rely on any understanding of biological classification or Enlightenment universalist philosophy. Human, taken in its biological taxonomic sense, is not a terribly helpful term here, which is why I think a phenomenological species concept is more useful in understanding the experience of the normal. However by identifying the normal with concepts of the human, human nature, or mankind we are linking it to an Enlightenment project of humanity understood as synonymous with the human or mankind. It is precisely this project that enhancement threatens. So while the experience of the normal and even the species, taken in the phenomenological sense, may be a pre-reflective part of our originary experience of the lifeworld, the concern or unease that is generated and institutionalised in response to certain violations of this norm (e.g. calls to ban enhancement) has a specific historical origin.

The analogy with art also helps explain why watching or remembering a great past sporting performance still creates a frisson of excitement and aesthetic pleasure even if the performance itself has been far surpassed by more recent ones. A performance is great in its historical context, as a response to the question posed to the (human) body at that time. Like the great work of art, it holds its value both because of and in its historical context.

Learning that an athlete's performance was technically enhanced in a manner that violates the *nomos* of the sport (nearly all elite athletic performance is today in some way technically enhanced in the weak sense) has a similar impact to learning that a painting emerged out of a totally different context than originally thought. Appreciation, be it for a work of art or an athletic performance, is contextualized within a history and within boundaries established by and within that tradition. If sport is a taking up of the tradition of the body understood as a question, then our aesthetic appreciation of that taking up, i.e. of the performance, is in many ways dependent on knowing how the question is taken up and the context in which it is taken up. This context is provided by the particular rules or *nomos* of the sport. In this way our perceptual appreciation for a performance is nullified upon discovery that the performance was enhanced in a manner that violates the *nomos* of the sport. As with forgery, our appreciation of sport is retroactively disappointed by being misled concerning the context of the performance. Here is where the question of aesthetic appreciation brings together a concern for the particular rules of a sport and a concern for the relation of sport to the rest of the lifeworld.

7.5 Enhancement and the Nomos of Humanity

Despite this interrelation, I think that it is helpful to distinguish the concern for the ethos of sport generally speaking (its role in the lifeworld) from a concern for the particular rules of a sport, (which may or may not prohibit enhancement). The former is often articulated in terms of purity or naturalness. I think that the concern for the ethos, purity, or naturalness of a sport has to do with an intertwining or overlap between the *nomos* of sport strictly speaking and what much more vaguely can be called the *nomos* of man or the *nomos* of humanity. In this sense, sport, like the arts, takes up the question of humanity in the broader sense. This question is itself historical and cultural. I do not wish to argue that posing of the question of 'what is man?' is part of an essential human nature, despite the experience of normality being an essential aspect of the lifeworld as experienced.

The question 'what is man?' is perhaps poorly posed and certainly sexist in its phrasing, but I think it important that it be distinguished from the questions of 'what is humanity?' and also 'what is the human?'. It is important that we are attentive to the breakup of this triplet, which has so seriously marked European ethical reflection since the eighteenth century. Therefore, 'what is man?' seems the clearest, articulation of the problem. One way to articulate the distinction between man, humanity, and human, is to say that man or mankind is the ethical correlate of the species concept human. I also think that the concept of man or mankind is only meaningful when taken in a universal sense: it must not only encompass all humans, but also points to an ideally unproblematic distinction between all beings that share our nature and all other beings. By this definition, the human sciences are not the study of the human but of man and mankind.

On this basis, humanity can then be understood in two senses. In its weak sense, it should be understood as the project of the self-realization of mankind, a growing awareness of our shared nature. In a second, stronger sense, humanity is perhaps best and most simply understood as what mankind should aspire to be. The term humanity entails both a normative and a (weak) teleological vision that is perhaps most apparent in our use of the term 'humane', as in 'acting humanely' (Fuller 2011). Whether this weak teleology of humanity is still coupled to man, as was the case in the late eighteenth, nineteenth, and twentieth centuries, is another question. The possibilities of human enhancement technology push us to answer in the negative. Paradoxically, I think that the weak sense of the term humanity (the self-recognition of man) is threatened by the potential fragmentation of man into something like subspecies with different natures described by Buchanan et al. in *From Chance to Choice: Genetics and Justice* (2000).[8] The strong sense of humanity might however be more amenable to reconstruction post-mankind: uncoupled from its identity with mankind. The fragmentation of mankind due to enhancement technology need not spell the end of universalist Enlightenment projects, only their reconception, perhaps with broader and/or more inclusive species horizons. However, I also think that resistance to this uncoupling is what is at stake in the deep-seated unease with enhancement in sport.

[8] "[W]e can no longer assume that there will be a single successor to what has been called human nature. We must consider the possibility that at some point in the future, different groups of human beings will follow divergent paths of development through the use of genetic technology. If this occurs, there will be different groups of beings, each with its own 'nature,' related to one another only through a common ancestor (the human race), just as there are now different species of animals who evolved from common ancestors through random mutation and natural selection." (Buchanan et al. 2000: 95).

This leaves human as a taxonomic classification that may or may not correspond with man and is in a different category than humanity, which carries with it an ethical, historical and teleological significance, making the ontology of humanity a different project than the ontology of the human which is normatively neutral. The taxonomic category of the human might well survive a fragmentation of mankind and endure as one of several or many post-man populations that may or may not attempt to reconstitute the universalist project of humanity.

I contend that a concern bordering on obsession with the purity or naturalness of sport is, if not yet a remnant of the age of man, then at least an attempt to hold on to it. The universal or global scope of the Olympics as both nationalist and cosmopolitan is broadly emblematic of this. Olympic sport, as the highest and most noble incarnation of sporting, aims at both demonstrating and pushing the potential of man in the sphere of physical accomplishments. The Olympic movement understands this ideal as a part of the broader project of humanity's or man's progress. But this ideal also relies on the coupling of man and humanity. Elite and especially Olympic sport is one of a myriad of ways in which the *nomos* of man is instituted.

The issue of enhancement in sport has become both symptomatic and emblematic of the end of the age of man, and a motivator for many opposing it. Enhancement and the trans-humanist movement that springs from it unbinds humanity from man and certainly from the human understood as a classification in biological taxonomy.

The intuitive reaction of the majority of the sporting public to enhancement is not only because it is cheating in the usual sense, that it gives an unfair advantage over other 'clean' athletes. It is somehow seen as a violation of the sacred but impossible to define naturalness of great sporting performance—the humanity of great sporting performance as being at the vanguard of human potential, and begging the question *Citius, Altius, Fortius*? The problem is that this Olympic question understood as an inflection of humanity's own weakly teleological question is, in the modern era of the Olympics, which began in Revolutionary France with *L'Olympiade de la République* held annually from 1796–1798, implicitly bound by the *nomos* of mankind, which in the Enlightenment mode of thinking included all humans; hence the triplet human-mankind-humanity.[9] Enhancement threatens to break apart the triplet. This is both the dream of the transhumanist project and the cause of the unease with enhancement in sport, which alongside military applications of enhancement technology will become the foremost testing ground for the transhumanist project.

This unease is not solely the domain of those dismissively referred to as 'bio-conservative luddites'. Rather, it is related to an experience of the normal that is, according to Husserl's phenomenology, originary and constitutive of the experience of the world. Experience of the world, understood in this sense, is coeval with an experience of normality or a horizon of normality that is experienced as shared with

[9] The instauration of *L'Olympiade de la République* is roughly contemporaneous with the *Déclaration des droits de l'Homme et du Citoyen.*

co-inhabitants or participants of the world. Put in another manner, I, as an individual subject, can only have a world insofar as it is shared by co-participants: those others with whom I do, or may, communicate with about the world and create meaningful sense structures, i.e. institutions understood in a very broad way, and projects. I share with these others a scope of normality that allows us to communicate and build up the sense-structures that constitute the world for me. Normality as an element of experience and as constituitive of the world is essential. It is the hinge by which individuals of a type relate to one another and to the world—thus constituting the species. It is unclear how or even that enhancement today is disrupting this normality, what is clear is that it raises the spectre of disruption.

The concrete content of the horizon of normality which constitutes a species is historically mediated insofar as it emerges in a lived-relation with others. These lived-relations are sedimented (to use Husserl's term) or concretised in meaningful worldly sense-structures or institutions. It is these historical institutions that give the species (e.g. mankind) its stability.[10] A scope of normality may change or expand as new manners of relating to the world are incorporated as normal. Husserl also opens the possibility for a shift in the normal scope of experience of a specific population group occurring in such a manner that we would have to speak of the emergence of a new species. I will more fully elaborate Husserl's conception of the normal and abnormal and even speciation as it relates the problem of enhancement. But I first want to briefly examine how this idea of normality relates to the two specific cases of Castor Semenya and Oscar Pistorius.

Castor Semenya was thought by some to be a 'gender freak'—naturally unnatural—and not deserving of the adulation that sporting institutions and the public give to champions.[11] This has to do with the perception of normality in the Husserlian sense. The horizon of normality that structures the European lifeworld, or certainly the lifeworld wherein the main sporting bodies (*qua* institutions) are situated, co-constitutes species-members in neat categories of the sedimented sense structures of male and female. As here we are dealing with a phenomenological distinction and not a physiological one, this binary distinction is ontologically prior to what we would call gender or sex distinctions. Both of those would be formalisations or idealisations of this initial binary distinction. This is not to essentialise this distinction. It remains historically constituted. What is more, the formalisations of this normal binary distinction (i.e. sex and gender made into real types or things) stream back into the lifeworld and structure our originary perceptual encounter with the world, without us having to be explicitly aware of it. Put simply, the normal

[10] This is an important point. Our species may have a greater stability than others precisely due to the fact that our horizon of normality is sedimented or concretised in sense-structures (culture, science) though we may not be the only species with stabilising sense-structures.

[11] In much of the literature surrounding the case of Semenya there seems to be an error surrounding the commonplace distinction between sex and gender. What is put into question by those demanding that Semenya undergo testing to verify her status as a female is her sex not her gender. As far as I know, Semenya's gender identification is unambiguous. She identifies herself as a woman (see, for example, Camporesi and Maugeri 2010).

scope of our species experience is such that all species-members that we encounter, if normal, should fall into this binary distinction (formalised as sexual difference).

Semenya was purported to violate this horizon of normality by not fitting onto this binary distinction. Whether this is indeed the case or not is somewhat irrelevant. What is relevant is that many in the sporting world reacted negatively to this perceived or reported abnormality. If there was not a question of technological enhancement then it was a case of the natural-unnatural. The suspected 'doping' in Semenya's case was not the addition of an impurity to the otherwise pure body, but an inborn impurity violating the normality (and purity) of the binary distinction between male and female. The normativity of the normal became readily apparent in this case. From the perspective of those demanding sex-testing, an athlete who did not fit clearly enough into a binary distinction of sexual difference violated the *nomos* of man—regardless of whether she had violated the *nomos* of the sport.

In addition to being naturally unnatural there was the additional transgression of being abnormally abnormal in a Husserlian sense. Her purported abnormality (being neither fully male nor female) was one that the majority of others, or at least the others involved in deciding that she should be tested etc., could not imagine or conceive of sharing. An abnormality that others could conceive of sharing, of experiencing the world as the abnormal person does, would be a normal abnormality. From the perspective of those interested in guarding the *nomos*, not only of the institution of the sport, which demands sexual difference on a binary scale, but also of man which demands the same thing, there was no question of declaring Semenya's abnormality a step forward in the project of humanity, or in its sporting inflection, understood in terms of taking the tradition of the human body to be a question.

The case of Oscar Pistorius—the so-called 'blade runner'—is also relevant here. Having had both his legs amputated below the knee, Pistorius now competes with 'full bodied' athletes with the help of bionic legs. Unsurprisingly, his entry into normal athletic competition has not been without some controversy. Following the decision to allow Pistorius to compete in the Olympics, Elio Locatelli, director of development of the International Association of Athletics Federation, complained that Pistorius diluted the purity of sport: another way of referring to the widespread belief in the normativity of an uncritical concept of nature and of man, both as related to a certain normal conception of the body.[12]

But Pistorius does not run as fast as the world's best 'full-bodied' athletes (i.e. those with two natural and 'whole' legs) and this may explain his rather more positive reception into the Olympic family. His case is generally looked upon as

[12] See, Camporesi (2008). Locatelli was quoted as saying "next will be another device where people can fly with something on their back".

inspirational.[13] I suspect that there would be far more complaints if a runner with Cheetah prosthetics actually won Olympic races and set world records. This would be the "paradigm shift in which disability becomes ability, disadvantage becomes advantage" (Camporesi 2008: 639); in other words, where the abnormal becomes the torchbearer for the progress of humanity. As it stands, Pistorius's blades fit relatively comfortably into the problematic, but well established, therapy-enhancement distinction. The use of therapeutic prosthetics has restored Pistorius to the normal scope of species functioning, albeit at the very far limits of this scope. Again in Husserlian language, I think that we could consider Pistorius's initial disability as a 'normal abnormality' it lays within a scope of experience that most could imaginatively envision as their own. The prosthetics he uses have restored him to normal function, meaning he is experienced as experiencing the world in a normal manner, in a fashion that is experienced and being shared by the majority of people. Of course when running, Pistorius's use of technology has made him an extreme outlier and one could easily argue that therapy would have involved fitting an artificial leg not a 'blade' and so Pistorius's case is not one of therapy.[14] It is this tension between therapy and enhancement in the Pistorius case that makes restored Pistorius (Pistorius with prosthetics) a possible instance of the abnormally abnormal. This is perhaps also why Locatelli's comments should not be dismissed outright as the remarks of an unreformed 'ableist'.

The admission of Pistorius into Olympic competition changes quite significantly what is considered to be not just a normal, but an optimal human body. Indeed, it brings into question just how human the optimal human body should be. And it also brings into question the role of sport in protecting or disconnecting the triplet: human-man-humanity. Without question, the Olympic movement continues to see itself as in the service of humanity. Pistorius's case is seen as inspirational in this sense because he has used technology to overcome his disability. But would the response be the same to an athlete who had her legs amputated for the express purpose of installing prosthetics and then proceeded to shatter all world records? I doubt it. Or imagine a future where animal tissues could be grafted onto human muscles to greatly improve performance. Yet, I do not see how these cases would greatly differ from Pistorius's.

---

[13] Pistorius's image recently graced billboards around the United Kingdom advertising super fast broadband. There is a certain amount of unnoticed irony here, as Pistorius does not run as fast as the fastest 'whole' and 'natural' athletes. On the other hand, the image of the technologically modified human is presented alongside superfast broadband as an analogy of humanity's progress—a progress which in its technological development may take humanity beyond man and beyond its traditional (human) form. I have yet to see Semenya on any billboards as the symbol of the next development in the teleological development and evolution of womanhood (and by extension humanity), one that overcomes the physical inferiority (in terms of strength and speed), which perhaps lies at the root of much sexual oppression and violence.

[14] I am grateful to Pieter Bonte for this point.

One of the reasons used to justify Pistorius's inclusion is that his prosthetics do not give him an advantage over natural legs. But the point of sport is to continuously push the scope of what the body is capable of doing. And very few top Olympic athletes have bodies that have not in some way been technologically mediated in their development. In the Husserlian language that I am using, Pistorius may fall close enough to the scope of the normal, on the right side of the therapy/enhancement distinction for doyens of Olympic sport to breath a sigh of relief and for the marketing departments of various multinational corporations to sign him up. But what of the next 'blade runner' whose prosthetics make her or him properly abnormally abnormal—makes him or her someone who experiences the world in such a radically different manner that I (a humble normally embodied subject) cannot imagine myself into her or his way of experiencing the world. Locatelli, in this sense, is perhaps interested in protecting the traditional body of sport and with it the traditional joining of human-man-humanity. His vision of serving humanity is to serve the *nomos* of man, to serve the *nomos* of the normal. We may decide that we do not share this vision of sport or of humanity, but we need to think seriously about what both alternatives entail before too quickly criticizing the conservationist of the human body.

7.6 Two Senses of the Normal Scope of Species Function: 'Natural' and 'Naturalistic'

There are two ways in which the idea of a normal scope of species function can be understood, one phenomenologically or naturally and the other naturalistically, meaning as understood by the natural sciences. Arguing from a phenomenological perspective, the scientific understanding of the normal scope of species functioning is founded on the more originary or natural understanding of the normal. The latter is built up as a fundamental character of experience and as a criterion through which we are able to communicate with others who are of the same type as us in terms of how they experience the world. The scope of normal experience is said to be proto or sub-scientific in exactly this sense. The concept of normality as constituted and refined by the biomedical sciences or sporting bodies is a formalisation and development of what is already constituted in the lifeworld of everyday social interactions. A creature outside that scope of normality would no longer be able to communicate with members of our species in a sufficiently complex manner so as to share a world in the phenomenological sense.

Envisioning how human enhancement technology would create a creature this far removed from us seems nearly impossible.[15] Somewhat less far fetched is the

[15] The position dubbed speculative post-humanism does postulate such a creature while admitting that we would be able to say almost nothing about how it would experience the world precisely because it would be so far outside of our scope of normality.

idea that beings might be created through enhancement that are outside our scope of normality in a limited manner, meaning that they are outside the scope of the normal in relation to certain capacities but not others. This might not rule out communication, but it might rule out shared participation in certain projects. There might be certain areas of endeavour that would become perhaps not fully incomprehensible to such creatures, but they could not imagine themselves experiencing the world in the requisite and relevant manner. This might be analogous to the way in which I can watch certain animals engage in activities and understand roughly what they are doing, but I cannot imagine having the experience that they are having. Husserl calls this emergence of a new type of being "primordial generation" or "speciation". This phenomenological notion of speciation, which Husserl says can never refer to one individual but must refer to a group for whom a new scope of normality manifests itself, would radically alter the conception of humanity that emerged from the Enlightenment and that I have described as the teleological project of mankind.

To refer back to an example I used earlier, when we say things like 'one should act humanely' this seems to suggest that humaneness is a goal not yet achieved universally. But it also presupposes that we might all be able to share in the experience of being humane. We may argue about what it is to act humanely, but the argument occurs within the scope of normal human experience. In this sense, there is a proto or sub-scientific manner of articulating what it would mean to act humanely—its recognition and formalization then becoming part of the project of humanity. The primordial generation of another species with another scope of normal experience could render the hope of a shared experience of what was humane, at the very least, much more difficult. It is this kind of fear that I think lies at the root of the deep unease caused by athletes like Pistorius and Semenya; they are emblematic of the possibility of a speciation that would sunder the joining of mankind, humanity, and the human. This does not in any way justify the treatment they have received, but I think it does help to explain one of the factors for the unease caused by a confrontation with the abnormal, and especially with the abnormally abnormal, or even its spectre.

What I contend and will try to work out in the remainder of this chapter, is that a normal scope of species functioning is constituted in social and interpersonal interactions between individuals, groups and institutions; what in the phenomenological register is referred to as intersubjectivity. This does not at all rule out that the bio-medical concept of the normal scope of species functioning—a concept which itself emerges out of the lifeworld—structures and orients the lifeworldly constitution of this scope. But it would entail that it does so through a sort of feedback loop wherein the natural or lifeworldly understanding of the normal is structured by the scientific idealisations that emerged out of it and developed beyond it. The intersubjective constitution of the scope of normal species function contains an often implicit ethical content. Everyday experience takes deviation from the normal scope of species functioning to have an ethical value. This has important consequences for our ethical understanding and evaluations of questions raised by enhancement in sport, and by the naturally enhanced individual such as Semenya.

7.7 The Constitution of Normality

The first thing to be reiterated is that the scope of normality being discussed here is in relation to constitution. It is in relation to the way the world appears to myself and to others, and the way the world appears to me as appearing to others. There is a scope of possible variation of how the world normally appears that is proper to what can be called a species community. This sharing of the world between species members means that it is something much different to say of an adversary (for example) 'she sees the world in a much different way than I do' than it is to say the same thing of the family dog. I cannot have a dog as my adversary precisely because our respective scopes of normality are so different that it does not make sense to talk about competing truth claims about the world that cross phenomenological species boundaries.

The normality of the world is also structured in terms of degrees that we can label as 'normally normal', 'normally abnormal', and 'abnormally abnormal'. In this section I will discuss the normally normal, in the next section the two variations of abnormality. The concept of normality for Husserl presupposes an intimate link with the operative functionality of the lived-body (the body as experienced as opposed to the body as an object) (Zahavi 2001: 88). The world is given perceptually to the body and the lived-body is taken by the subject to be the standard of normal functioning. This is the first and absolutely subject relative level of normality.[16] Our perception of the way other bodies, which appear to be of the same type as my own, negotiate the world around them provides an index of how the world is given to those bodies which either confirms the initial intuition of the normality of my own body or does not. Husserl argues that all experience of things in the world is accompanied by a co-experience of our own body as operatively functioning in a normal or abnormal manner (Husserl 1973: 57). Obviously this confirmation or disappointment occurs over time and multiple social interactions, but as all properly subjective experience is immediately intersubjective (in relation to others with whom I share the world) the confirmation of the normality of the body commences at a very early stage of life. I learn what is normal from those who are closest to me and the extension of my field or horizon of normality extends outwards in a centrifugal like motion as experience accumulates over time.

If the world appears to my body in a manner that is significantly different from the way that it appears to most other bodies that I perceive to be of the same general type as my own then my body also begins to appear to me as abnormal. To use an example provided by Havi Carel, for a body with certain forms of illness or disability, a mild incline that would for most be no trouble to traverse appears as insurmountable (Carel 2012). The constant experience of seeing other bodies doing with ease what my body cannot, verifies the self-perception of the body as abnormal.

---

[16] Although even at this most subject relative level Husserl maintains that normality is constituted in relation with others, the first other for most being the mother.

Abnormality is normally discussed in terms of various forms of deficiency, however an enhanced body (in the strong sense that I introduced above) would have a similar experience of abnormality.

The experience of the normality or abnormality of the body is only one side of the equation. It is the appearance of the world in a normal or abnormal style that provides the correlate to the body and the index against which the intersubjective verification of normality can take place. The immediate sensory givenness of the world to the body is completely subject relative. The subject relative givenness of the world is however always stabilised in interaction with others, despite it remaining subject relative. In other words, I perceive others as perceiving the world in a similar enough manner to myself that we are able to communicate about the objects and situations we encounter, potentially enter into disputes concerning truth claims about these objects or situations, and also embark upon common projects in the world. All of this further stabilises or objectifies the world for me. In the example above about walking up a mild incline, the world appears in an abnormal manner to an abnormal body, in most cases this abnormality in the world is confined to specific enough areas of praxis that the abnormal subject maintains the same ideal norm of world constitution as the normal subject(s). But in extreme instances of abnormality, this ceases to be the case.

There are of course varying degrees of deviation from the ideal norm within the scope of the normal conceived in the broadest sense. These are cases of abnormality, which can be divided into the normally abnormal and the abnormally abnormal. In relation to others, this distinction is between those species co-members whose worlds deviate significantly enough from the ideal norm that they can be considered abnormal, but with whom I can still engage in collective praxis; and those who are outside of my species, those with whom I cannot engage in a collective project.[17] This question of species belonging or membership relates to enhancement insofar as we ask specifically what kind of enhancements might disrupt our species norm and what kind of projects this might endanger. I think we can say that it would threaten the project of sport as we know it, as well as the social and political goals embodied by sport, and here I am thinking specifically of Olympic sport.

7.8 The Normally Abnormal and the Abnormally Abnormal: The Boundaries of the Human

As the species specific scope of normality is historically contingent, both the normally abnormal and the abnormally abnormal in relation to a certain species are likewise the products of historical contingency. Yet, only in the first can the abnormal be brought into the sphere of the normal within a species specific scope, without

[17] This is a similar point to the one that Nicholas Agar raises as an argument against "radical enhancement" in *Humanity's End, Why We Should Reject Radical Enhancement* (2010).

a new species emerging. In the first case, the surrounding world of the other is given to us as appearing to be in a manner that is substantially different from our own. There are of course varying degrees of this normal type of abnormality. A foreign culture might appear to us in this manner, or we might say that in order for it to appear to us as foreign it must appear to us in this manner. But the abnormality in this case is relative, meaning that it could be brought within our scope of normality, or at the very least we could imagine it being normal for us. In a foreign social situation, what appears to us as a meaningless gesture or a mundane object might appear for those who are at home in that context in an entirely different manner. And, as such, the perception of the gesture or object will motivate in the (relative) foreigner a set of experiences different from our own. Cultural difference, in this sense, provides a manner in which the varying degree of possible abnormalities becomes clear. However, increased exposure to what is relatively abnormal will lessen its degree of abnormality, to the extent that the foreign in this sense can be brought within my scope of normality while nonetheless remaining foreign. It could be that after enough exposure, the gesture or object that once was unintelligible or mundane to me, now takes on the same sense as it does for the locals—a sort of phenomenological rendering of 'when in Rome'. Or, it could be that the foreign gesture or object may never elicit or motivate in me the set of experiences that it does for the native of that culture, but I can imaginatively envision having the same experiences motivated. I can, put myself in the shoes (or eyes) of another through an imaginative act.

But there are limits to this ability to integrate abnormalities. It is only possible on the basis of sharing enough commonality with another that our structures of motivation are in the most general sense analogous. This question vis-à-vis the sports enhancement debate could be put as a question of whether we could imaginatively put ourselves into, for example, Pistorius's body. Another example may help to flesh out this sense of the normally abnormal. Phenomenologists of illness go to great descriptive and conceptual effort to explain the experience of being ill or disabled (Carel 2011). These are descriptions of abnormal conditions, however the possibility of comprehending such descriptions and understanding them as strong deviations from a norm of experience depends on there being both a ideal norm against which these experiences of illness appear as abnormal and enough of a shared dimension of experience that these phenomenological descriptions are comprehensible. For example, I have some idea of what it is like to come up against the limits of my body's capacity to scale tall structures, thus through an act of empathy I can understand—to some extent—what it would be for these type of experiences to become commonplace. Understanding, here, must be thought of in terms of motivation or the eliciting of one experience by another. Bringing someone into my phenomenological species community is a question of being able to imaginatively put myself into her or his flow of experience. These acts of empathy expand my horizon or scope of normality.

In these examples, it is clear that "the general style of the human body" (Husserl 1989) is the common element that makes an experience of normality and hence empathy possible. However, it would be a mistake to essentialize this general style.

Just as the cultural traditions that I experience as my own (and not foreign) are contingent on a specific history of development, the general style of the body is also contingent on the evolutionary development of the human body into its typical form. Phenomenologically speaking, nature and culture cannot be separated. In both a cultural sense and a natural sense (pertaining to the general style of the body), both the norm and its violations are built up historically in the contingent relations between living beings.

What falls irrevocably outside of my horizon of normality and thus my species horizon are those beings whose flows of experiences I cannot put myself into. What is irrevocably foreign to me must be considered abnormally abnormal. It cannot be brought into or even compared with my own sphere of normality.

How does this distinction between normally abnormal and abnormally abnormal relate to the question of enhancement? I think that the fear of the creation of enhanced beings is that at a certain point a divergence in the type of experience had by the enhanced and unenhanced or differently enhanced will be so great that there will no longer be a scope of normality that is shared. This is the breakup of the human and with it mankind couched in phenomenological terms—speciation or "primordial generation". The emergence of a new horizon of normality marks the creation of a new species (Husserl 1993: 319).

We can also think about this in less drastic terms. Instead of arriving at a situation where two horizons of normality are totally incommensurable ("speciation"), we might envision the emergence of sub-species that have something like overlapping but not fully commensurate horizons of normality. Take the examples of illness and disability. If a being was created via enhancement that experienced its own bodily vulnerability in a much different manner than we normally do now, such a being may have a very difficult time empathising with the ill or disabled. This does not only relate to cases of illness or disability. We can empathise with the experience of thirst or hunger and from that not only understand, but be able to put ourselves into various forms of comportment deriving from these experiences (Husserl 1989). A creature that did not experience bodily needs in a manner analogous to our (or my) own would most likely not be able to empathise in the same manner. Even the partial incommensurability of horizons of normality could have a serious impact on certain types of collective projects. This is not about the dangers of science fiction situations (although it is also about that); it is more a concern about exacerbating existing challenges to collective ethical endeavours. As Buchanan et al. remark, "history is replete with instances in which human beings have failed to empathise with their fellows simply because of quite superficial differences" (Buchanan et al. 2000).

We can also look at the cases of Pistorius and Semenya from the perspective of the normally abnormal/abnormally abnormal distinction, as both seem to present some challenges in this regard. Neither case is clear cut. Above I said Pistorius might be considered within the scope of the normal as he could be slotted onto the therapy side of the therapy/enhancement distinction (though not without some difficulty). Here I present reasons for considering Pistorius to be abnormally abnormal. I do not think that I could empathise with Pistorius's experience of running with Cheetah prosthetics, it is outside of my scope of normality because running in the manner of Pistorius is outside the general style of the species body as it has developed into its traditional form. I am more able to empathise with Usain Bolt, who runs faster than Pistorius, because we share in a general style of the species body that is constitutive of a scope or horizon of normality. In this aspect of experience, Pistorius is abnormally abnormal. This obviously is not enough for me to exclude Pistorius from what I would take to be my species-horizon. I am guessing that we share enough in common in other regards that communication and shared projects concerning most other dimensions of activity in the lifeworld would remain possible. I do not think that we can say that the Pistorius case is a crossing of any Rubicons, but I do think we can view it as a beginning of a fragmenting of the species. Pistorius remains my species-member, but he also begs the question of what technological modifications would push a being across the species boundary.

Semenya's case is perhaps more difficult. Her body is not technologically modified in the same manner as Pistorius's, and I imagine that the phenomenological account of what would set her apart is entirely hypothetical. It does not actually apply to her experience and can only serve as a kind of thought experiment and not a real concrete example. The horizon of normality, at least in Western European culture, is socially structured so as to be subject to a binary division according to sex (if not gender). The sporting world, by dividing nearly all sporting competition according to this binary opposition, functions to reinforce this division. It is normal to experience the world from one side of this binary divide and to experience others as belonging to one side or the other. In the case of Semenya, one version of the story about her was that she did not fit into this division. This violation of the norm was then ascribed a normative significance—one of the reasons why she, unlike Pistorius, does not adorn billboards around the world.[18] There was a negative value attached to not fitting neatly into this binary opposition of the sexes. Semenya's hypothetical situation would be that of the abnormally abnormal in relation to sex, as the species-norm entails both constitution and experience of the world from one side of this sexual binary.

But her case also demonstrates that the normal understood as a sort of species essence is socially and historically structured and mediated. I am speculating to say that there seems nothing about this person that would not allow for a full blown empathic relation as I would have with any other species co-member. It seems that in this kind of case the social structuring of the norm may threaten to override experience itself. Or if indeed there is something abnormal about the hypothetical Semenya, there seems no reason that it could not be integrated into the horizon of the normal, in a way that rather than violating the normal itself as experienced, instead acts upon certain elements of its social structuring (and their consequences) which have been given a negative valuation. The horizon of normality can be altered in some cases by an act of reason; it is not a necessary but rather a contingent structure.

[18] Thus the question here is more about 'sex diversity' or sexual polymorphism as a deviation from the normal than Semenya's eligibility to compete.

In a Marxist fashion, we might say that in some situations the horizon of the normal is something akin to a superstructure in experience that can outlast the practical conditions that created it in its present form and 'for a time hide their collapse' (Merleau-Ponty 2011). If the Pistorius case demonstrates why there may be some normativity in the norm, the Semenya case clearly demonstrates that this is not necessarily the case.

7.9 Looking Back Fondly or Running Ahead Blindly: The Role of Sport in the Age of Enhancement

In his essay, *Normative Aspects of the Human Body*, Ludwig Siep (2003), makes the case that the "traditional shape" of the human body, its normal scope of phenotypic variation, serves as an important source of values that are a significant determining factor in how we relate to one another, and especially how those whose own claims to justice are somehow muted or unarticulated (the poor, the unborn, future generations, the handicapped) are treated (Siep 2003: 174). This holds whether or not the link between the traditional shape and the values it helps found is conceived as essential in a strong sense or historically contingent, or some combination of the two, as I argue here. On this basis, Siep argues that the traditional shape of the human body should be regarded as a common heritage and should be protected and preserved in the manner that other cultural or natural aspects (e.g. biodiversity) of our heritage should be protected. The question that arises is whose common heritage are we talking about? The implicit response is, I think, mankind's. Siep argues further that if we are to endeavour to enhance human bodies to such an extent that their traditional shape is violated (or perhaps more radically to the extent that the enhanced body appears to non-enhanced bodies, or differently enhanced bodies as 'abnormally abnormal'), then there must be a public exploration regarding the possible consequences of this for established social norms.

I think that this argument is in many ways a persuasive one, or at least worth taking very seriously. I also think that it is closely interlinked with the debate over enhancement in sport. It is fair to say that the Olympic movement conceives of itself as part of a wider project of humanity as discussed above. The anti-enhancement argument within the Olympic movement would then manifest itself as a defence of the traditional shape of the human body and a celebration of its heritage and possibilities, which are bound by the horizons of its traditional shape: traditional human legs will allow us to run fast, but only so fast. Under Siep's argument, the limitations imposed by the traditional shape of the human body are a worthwhile price to pay for maintaining many of the Enlightenment values (most importantly the unity of mankind) that are underpinned by the commonality of the body.

This position is also one of the possibilities that emerges from the phenomenology of normality that I have tried to sketch. By this argument, the established scope of the normal and the phenomenological species community it establishes is worth trying to preserve because we are unaware of what consequences may arise from violating it. This also fits with a concern raised by Buchanan et al. when they note that "the effectiveness of people's motivation to act consistently on universal moral principles may depend significantly on whether they share a sense of membership in a single moral community [...] for all we know, it might turn out that if differences among groups in characteristics, other than a common rationality become pronounced enough, they would not treat each other as moral equals" (Buchanan et al. 2000: 95). According to the phenomenological approach, a moral community cannot be formed between beings that view one another as abnormally abnormal, who cannot imaginatively empathise with the flow of experiences of one another. This scope of empathy is limited by what Husserl calls the general style of the (human) body. By this model, the Olympics joins with the Universal Declaration of Human Rights and UNESCO as institutions dedicated to preserving and celebrating the moral community of man as a community of moral equals, and of promoting humanity as the ethical self-recognition of that community. In the case of the Olympics, this means celebrating and preserving the heritage of the traditional shape of the human body as normative: maintaining the normativity of the normal. I think that this is part of the "educational role in the modern world" that Pierre de Coubertin envisioned for the Olympic movement.[19]

Another vision is possible. Sports and the Olympics in particular might instead be envisioned as that sphere where humanity sheds the encumbering baggage of being tied to the traditional body of the human. This may well lead to a fragmenting of mankind and the splitting of the humanity-man-human triplet, but it need not put an end to the project of universal moral community. The degree to which someone like Pistorius has been taken up into the public imagination might suggest that the abnormally abnormal', that phenomenologically undomesticatible other, is a being much more radical than anyone dreams of creating or becoming. Rather than putting an end to Enlightenment dreams, enhancement might instead expand the scope of humanity while at the same time allowing individuals to shed the limitations and indeed some of the suffering that seems to very often accompany an attachment to our traditional—normal—form. This may seem like the riskier approach, but I am not sure that it is. The need for ever more stringent and renewed efforts to combat enhancement qua doping at the level of elite sports seem to show that the ethos of preserving the heritage of the traditional shape of the human body is not shared by all or even many in sport. Even those athletes who do not use banned substances or methods are very often technologically altering or mediating their bodies to the furthest extent allowed by both the rules and the means available to them. Perhaps most importantly, at the very least, this transhumanist rendering of the Olympic movement retains its humanist aspect, although no longer coupled to man or even the human. Paradoxically, the shift towards integrating the ethos of a transhumanist humanism into the Olympic movement may represent more of a return to its original

[19] See, the 'Olympic Movement Factsheet' http://www.olympic.org/Documents/Reference_documents_Factsheets/The_Olympic_Movement.pdf [last accessed 20/12/2011].

values than the continued cat and mouse game played by doping athletes and regulatory bodies. Indeed we may find that it is the transhumanist understanding which in the age of enhancement best expresses the original sentiment behind the modern Olympic Movement as conceived by Pierre de Coubertin: "Olympism is not a system it is a state of mind—It can permeate a wide variety of modes of expression and no single race or era can claim to have the monopoly of it."

References

Agar, N. 2010. *Humanity's end: Why we should reject radical enhancement*. Cambridge, MA: MIT Press.

Buchanan, A., D.W. Brock, N. Daniels, and D. Winkler. 2000. *From chance to choice: Genetics and justice*. Cambridge: Cambridge University Press.

Camporesi, S. 2008. Oscar Pistorius, enhancement and post-humans. *Journal of Medical Ethics* 34(9): 639.

Camporesi, S., and P. Maugeri. 2010. Castor Semenya: Sport, categories, and the creative role of ethics. *Journal of Medical Ethics* 36: 378–379.

Carel, H. 2011. Phenomenology and its application in medicine. *Theoretical Medicine and Bioethics* 32(1): 133–146.

Carel, H. 2012. Phenomenology as a resource for patients. *The Journal of Medicine and Philosophy* 37(2): 96–113.

Fuller, S. 2011. *Humanity 2.0, what it means to be human, past, present, and future*. London: Palgrave Macmillan.

Husserl, E. 1970. *Crisis of the European sciences and transcendental phenomenology*. Trans. D. Carr. Evanston: Northwestern University Press.

Husserl, E. 1973. In *Husserliana XIV Zur Phänomenologie der Intersubjektivität. Texte aus dem Nachlass. Zweiter Teil: 1921–1928*, ed. Kern Iso. Dordrecht: Kluwer Academic Publishers.

Husserl, E. 1989. *Ideas pertaining to a pure phenomenology and to a phenomenological philosophy. Second book: Studies in the phenomenology of constitution*. Trans. R. Rojcewicz, and A. Schuwer. Dordrecht: Kluwer.

Husserl, E. 1993. In *Die Krisis der europäischen Wissenschaften und die transzendentale Phänomenologie. Ergänzungsband. Texte aus dem Nachlass 1934–1937*, ed. R.M. Smid. Dordrecht: Kluwer.

Merleau-Ponty, M. 2011. *Le monde sensible et le monde de l'expression, notes de cours au Collège de France, 1953*. Geneva: Metis Presses.

Siep, L. 2003. Normative aspects of the human body. *The Journal of Medicine and Philosophy* 28(2): 171–185.

Wasserman, D. 2008. Performance-enhancing technologies and the values of athletic competition. *Philosophy and Public Policy Quarterly* 28(3–4): 22–27.

Zahavi, D. 2001. *Husserl and transcendental intersubjectivity*. Trans. E. Behnke. Athens: University of Ohio Press.

Part III
The Normative Value of Human Nature

Chapter 8
Doping Use As an Artistic Crime: On Natural Performances and Authentic Art

Andreas De Block

8.1 Introduction

Did Homerus actually exist, and if he existed, did he write the Iliad and the Odyssey? Did Shakespeare write the lion share of the works that are attributed to him, as most scholars believe? Or is there some truth in the claim of the so-called 'antistratfordians' that many of the poems and plays that are attributed to Shakespeare were in fact written by an anonymous, well-educated aristocrat? Questions like these keep literary historians awake, but they also seem to be of interest to the general public. Apparently, many of us care about the authorship of masterpieces, and not just about the masterpieces as such. Our caring about art authorship paves the way for art fraud. Art students and failed painters have made fortunes selling paintings they falsely attributed to old masters, and many of these forgers (among them David Stein, John Myatt, Han van Meegeren, Tom Keating and – more recently – Wolfgang Baltracchi) eventually ended up in prison for their fraudulent activities.

The issue of fakes and forgeries has legal and economic aspects, but it is also central to much theorizing in philosophical aesthetics. Although artistic fakes and forgeries share many or even all of the formal characteristics of original works, it would change our aesthetic appreciation if we were to discover, for instance, that a painting is not an original Hopper, but a perfect copy of one of Hopper's works, painted by an unknown Hungarian artist. I suspect that something quite similar explains the intuitive appeal of human nature arguments in the doping debate. This analogy entails that I agree with many athletes, sports fans and philosophers that doping use can violate the value of human nature. That said, I disagree with

I would like to thank Rafael De Clercq, Arnold Burms and the editors of this volume for their inspiring comments on earlier versions of this chapter.

A. De Block (✉)
Centre for Logic and Analytical Philosophy, KU Leuven University,
Kardinaal Mercierlaan 2 – bus 3200, 3000 Leuven, Belgium
e-mail: andreas.deblock@hiw.kuleuven.be

J. Tolleneer et al. (eds.), *Athletic Enhancement, Human Nature and Ethics*, International Library of Ethics, Law, and the New Medicine 52, DOI 10.1007/978-94-007-5101-9_8,

them about the nature of this value. The analogy with art forgery purports to show that the value that is at stake here is an aesthetic value rather than moral value.

The gist of my argument will be that much in the same way that an aesthetic judgment is always and in a fundamental way a judgment of origin and performance, the observable characteristics of an *athletic* performance are not the only things that matter to our aesthetic appreciation of this performance. If the observable characteristics were the only things that mattered, there wouldn't be an aesthetically relevant difference between authentic footage of an athletic performance and manipulated footage of an athletic performance. The information that distinguishes authentic footage of an athletic performance from visually identical manipulated footage influences both the aesthetic experience and the aesthetic judgment it engenders. I will argue that the difference between these judgments is grounded in the idea that an athletic performance is an aesthetically pleasing performance at least in part because it is perceived as both a human and a natural performance. The use of performance enhancing drugs (PEDs) often makes this performance simultaneously less natural and less human.

In this contribution, I will explore 'human nature' as an aesthetic value in sports, and I will sketch how this aesthetic value bears upon the doping debate. At the end of my contribution, I will say a few things on how aesthetic judgments relate to moral judgments in the doping debate.

8.2 Human Nature, Ethics and Aesthetics

Some philosophers believe that one can ethically anchor the ban on performance enhancing drugs and technologies in human nature. I believe this approach faces many problems, some of which are almost insurmountable.

A first series of problems is primarily conceptual. Two leading philosophers of biology, David Hull and John Dupre, have argued extensively that there is no such thing as human nature (Hull 1986; Dupre 2001). Today, it seems that their view is the consensus view, at least in most philosophical subdisciplines (but see Machery 2008). Likewise, there is a long tradition in philosophy (think of Lucretius and Spinoza) that straightforwardly rejects the distinction between natural and unnatural. Hence, philosophers who want to ground the doping ban in human nature have to make clear what human nature is, how it is to be distinguished from non-human nature, and how what is natural for humans is different from what is unnatural for humans. Moreover, it would be helpful if they could come up with a theoretically useful property to which the notion 'natural' can refer, so that the difference could be explained between 'natural' and other, more or less related notions such as 'innate' or 'normal' (Mameli 2008).

A second series of problems is moral in nature (pun intended). Suppose that we are able to delineate the natural from the unnatural, and human nature from non-human nature, then how exactly are these distinctions morally relevant? Even if one assumes that there is a theoretically useful property to which 'natural' refers, it is far from obvious what the relation is between this property and the normative claims that this property is supposed to play a role in (Levy 1980). Most importantly, philosophers who want to argue for the moral relevance of (human) nature need to show either that their account is untainted by the naturalistic fallacy or – even more challenging – that the naturalistic fallacy is no fallacy at all.

Third, if the distinction between natural and unnatural and/or between human nature and non-human nature is morally relevant, then to what extent are these distinctions helpful in tackling the issue of doping in sports? Can these distinctions really help to distinguish morally unproblematic performance enhancing substances and practices from morally problematic ones?

This short sketch of problems clarifies why the debunking of human nature arguments against the use of doping in sports is so successful. After all, the standard view in philosophy is (a) that there is no human nature (for instance, because there are no natural kinds in biology), and (b) that the alleged 'naturalness' of a behavior has both little meaning and little moral value. Finally, even those accepting the normative value of human nature, still have a hard time showing that human nature can be of any philosophical use in the doping debate, since the use of performance enhancing substances seems to be as much part of human nature as sports is. So why do some philosophers still try to construct a human nature argument against the use of doping in sports?

No doubt, one reason is that philosophers do not shy away from difficult tasks. It is no secret that quite a few philosophers defend unpopular theses simply because they are unpopular or part of the minority view. Yet, it would be incorrect and disrespectful to think that philosophers only use the human nature argument because of some sort of immature rebelliousness. The human nature argument really has some intuitive appeal. In my view, it derives its primary appeal from an intuition most of us share with regard to doping and sports. The problem is that most philosophers who build upon this intuition when they try to make a case against the use of certain performance enhancing substances (PEDs), mistake this intuition for a moral intuition, whereas the intuition is mostly aesthetical. The intuitive appeal of the human nature argument with regard to the doping ban is thus mainly derived from the aesthetic overtones of 'human nature' in sports. Hence, there is a normative value of (human) nature in sports, but it is primarily bound to an aesthetic normativity.

My claim – human nature has an aesthetic role to play in sports and can be used to argue against the use of PEDs – is not immune to many of the counterarguments I have sketched. The claim can only be defended, it seems, by showing that human nature exists, that it has normative value (aesthetic rather than moral in this case), and that its aesthetic value is relevant to the debate on PEDs in sports. Before I tackle these important issues, however, I will analyze the intuition and its aesthetic appeal, by developing an analogy between doping use in sports and artistic forgery.

8.3 Doped Athletes As Artistic Frauds

Some art forgers like Tom Keating and Han van Meegeren claim to have started their forging because they felt that there was something rotten with the artistic (sub) culture in general and the art critics in particular. The focus on big names, rather than on true craftsmanship and genuine beauty, was a thorn in their flesh. Hence, the critical acclaim their forgeries received was the proof they needed: their artistic skills hadn't changed, but now that they painted under the name of a famous artist, the critics that had ignored them when they were painting under their own name, now applauded the work they produced.

Stories like these might be dismissed as attempts by the forger to rationalize his crimes. Nonetheless, the sentiment they express is also found among some art theorists. In *The act of creation*, Arthur Koestler argues that the critical acclaim for van Meegeren's forgeries of Vermeer reveals the snobbery in art circles (Koestler 1964). In his view, only the formal characteristics of a painting should have an impact on our aesthetic judgment. Who had painted the work, or when it was painted, might matter to art historians, but it should not matter aesthetically. Still, most of us think that authorship makes an aesthetic difference for most, if not all works of art:

> Galleries display paintings which are, we hope and expect, authentic in the sense of being painted by the artists to whom they are attributed. They do so in order that the public may appreciate the aesthetic merits of those works. We think it important, even essential, to see the painter's canvas, rather than any copy thereof, in order fully to appreciate the work. (Currie 1985: 153)

Dennis Dutton has argued that this concern is not – or not always – born out of snobbery, but is in fact a genuinely aesthetic concern. According to Dutton, the meaning of a work of art affects the aesthetic appreciation, and rightly so. This meaning is not only determined by the observable characteristics of a painting or a poem, but also by different sorts of information. For example, whether or not the formal characteristics were meant to be part of a work of art makes an aesthetically important difference. A pattern in the sand that was created by the wind will be valued differently than a pattern in the sand that was created by a skilled artist, even though the two patterns are perceptually indistinguishable. Similarly, most people would agree that knowledge about the tradition it belongs to, and knowledge about the major influences on the artist, tend to help us to appreciate the work. Sometimes this knowledge helps us to discover certain observable properties: it tells us where and how to look. Yet, even if it does not lead to such discoveries, this knowledge is relevant. This is nicely illustrated by Walton who speaks of "the tremendous aesthetic difference we are likely to experience between a film of a dancer moving very slowly and a still picture of him, even if 'objectively' the two images are very nearly identical. We might find the former studied, calm, deliberate, laborious, and the latter dynamic, energetic, flowing, or frenzied." (Walton 1970: 346) Walton's example focuses on knowledge about the medium, but knowledge about the artist (who?, when?, why?) is no less important. After all, our aesthetic appreciation is always in part an appreciation of the artist's performance. Discovering that a work of art is a forgery alters the meaning of the work of art, because the performance that has led to the work of art is crucial for its meaning (Dutton 1979).

Most philosophers of art agree that the meaning of a work of art is important for our appreciation – a meaning that is constituted by circumstances connected to the work's origin-, but they disagree whether this importance should be seen as primarily artistic or as primarily aesthetic (Lessing 1965).[1] I consider this last debate as mostly semantic, but insofar as the disagreement is not based on semantics, I am on Dutton's side, who has always defended the view that authenticity matters artistically. Yet, for the purpose of this paper it does not really matter whether the appreciation that bears upon this information is artistic or aesthetic, as long as one accepts that the appreciation of a work of art is in part influenced by information on the artist and the artist's performance.

There are quite a few similarities between art forgeries and doping use in sports. Both the forger and the doped athlete are frauds – at least within a culture like ours that prohibits the use of doping and the forging of art. The art forger deceives the art lovers and the art critics much in the same way as the doped cyclist deceives sports fans and sports writers. Both types of deception can have serious legal consequences as is illustrated by the trials of Han van Meegeren and Floyd Landis. Furthermore, the art forger and the doped athlete often have strong economic incentives for their malicious deceit. That said, these similarities are not very illuminating for my purpose here. The sketched similarities have too much to do with the misrepresenting of the true nature of a particular work of art or the misrepresenting of the true nature of a particular athletic performance. The issue that I want to tackle here is more fundamental: whereas the deception is usually the consequence of the ban on PEDs, I want to inquire whether there are good (aesthetic or other) reasons to ban PEDs. There is one similarity between art forgery and doping use that seems particularly relevant here: much like our appreciation of a work of art is always an appreciation of the artist's performance, our appreciation of an athletic performance is always an appreciation of what an athlete has achieved in his performance. Or, more to the point, just like our appreciation of a painting is influenced by the knowledge that this painting is not painted by the artist to whom it has been previously attributed, our appreciation of the athlete's performance is altered when we stop seeing it as a human performance because the athlete used PEDs.

Before I focus on PEDs and how their use might undermine the aesthetic qualities of sports, I will elaborate another example of how information affects our aesthetic appreciation of sports. If you watch television and see a female athlete jump over 2m10, this performance grasps your attention. You might like to see the replay of that jump over and over again to admire the sheer perfection of its execution. Suppose that you learn that it was in fact a man – disguised as a female athlete – who delivered the performance. Most of us would be less interested in seeing the same performance in slow motion. This jump would simply be much less extraordinary. The reason is that the standards for male high jumping differ from the standards of female high jumping. These standards have an effect on how the perceptually indistinguishable

[1] But see, e.g. Beardsley (1965: 301):"I think it is interesting in itself to know, if we can, how the artist's mind works, but I do not see that this has any bearing upon the value of what he produces. For that value is independent of the manner of production, even of whether the work was produced by an animal or by a computer or by a volcano or by a falling slop-bucket."

performances strike us aesthetically. The outstanding high jump is aesthetically and athletically overwhelming not because of its absolute properties, but because it is an overwhelming performance *for a woman* (Walton 1970). Suppose now that you are told that it was not a man or a woman who jumped over 2m10, but that what you saw was a really realistic computer generated animation. In that case, our aesthetic appreciation would change more drastically. Even though some spectators might continue to watch the jumps, and even though some spectators might still be interested in seeing the computer generated jump in slow motion, my guess is that their interest would shift, as would the aesthetic criteria that are applied. What makes for a beautiful and aesthetically interesting jump in the first case (the female high jumper), is different from what makes for a beautiful and aesthetically interesting jump in the third case (the computer generated female high jumper). For example, the life-likeness of the computer generated jump can influence our aesthetic appreciation in the third case, but it obviously is of no relevance at all to our judgment about the performance of the female athlete. The vivid movement of the female high jumper would be aesthetically relevant in the case of the computer generated jump, but rather trivial and aesthetically uninteresting if the jump was performed by a human athlete. Similarly, the basis for our admiration would shift completely, once we learn that we had witnessed a computer generated animation. We would then be admiring the impressive quality of the technique of 3D-visualization and the impressive programming that visualizes the drama of an excellent female high jump.

Information regarding (the origin) of an athletic performance influences our appreciation of the observable characteristic of that performance. Such information need not be related to enhancement, as the example above shows, but it almost goes without saying that knowing that some kind of enhancement was involved, shapes our aesthetic appreciation to a large extent. Consider for instance mechanical enhancement: some people's interest in watching a mountain stage in the *Tour de France* would wane if we knew that everyone was riding a motorized bike, and nobody's aesthetic experience would remain the same, even if the observable characteristics would not change at all. Pharmacological enhancement, the kind of enhancement that is most targeted by sports officials and philosophers, has very similar effects. Some of us might still aesthetically enjoy Ben Johnson's run in the 100 m final at the 1984 Summer Olympics, but the aesthetic pleasure surely has been altered due to the information we now have on how the performance was actually achieved.

Perhaps the analogy I have developed suggests a few things that I do not wish to suggest. First, I am not defending the view that sports is a form of art. I only argue that sports is in part an aesthetic activity, and that sports can be beautiful. Whether or not this beauty is artistic beauty is not really relevant for the point I make. Second, I am not claiming that the beauty of an athletic performance disappears as soon as we know that it was partially driven by the use of PEDs. Many forgeries can still be aesthetically pleasing for those who know that they are forgeries. It is even conceivable – though not always very realistic – that works that are known to be forgeries are considered to be aesthetically superior to some of the works of the artist to whom they had been falsely attributed. Likewise, athletic performances on PEDs might sometimes be more aesthetically pleasing than 'natural' athletic performances. The analogy I make only builds upon the idea that information regarding the origin of a performance matters from aesthetic point of view. Third, people might focus on the deceit that is involved in the use of PEDs, and conclude that this morally despicable origin influences our aesthetic appreciation. In a recent paper, Lacerda and Mumford (2010: 191) make a statement along these lines:

> The beauty of a winning 100 meter sprint can be flawed for us, for instance, if it then emerges that the winner had taken performance enhancing drugs. To be truly beautiful, the win must be fairly achieved and so, the moral quality of the performance adds to its aesthetic value.

I agree with Lacerda and Mumford that the moral quality of the performance can add to its aesthetic value. Yet, even if we assume for a moment that there is nothing morally wrong about the use of PEDs, and even if no deceit whatsoever is involved in the use of it, I would argue that it would still make a difference for the aesthetic value of the performance. I suppose that no-one considers it to be morally wrong to perfectly mimic a Tour de France mountain stage with motorized bicycles, but that performance would nonetheless have a different aesthetic value than the original mountain stage.

8.4 Athletic Performance and Human Nature

Artistic performance is more important for some 'arts' than for others. For instance, our aesthetic judgment of firework resembles our aesthetic judgment of natural beauty in the sense that the origins do not really change our appreciation. Some would even argue that firework and circus acts can never constitute real art because the aesthetic experience they engender is engendered almost exclusively by the observable characteristics (McFee 2005). Sports seems to be at the other end of the spectrum: observable characteristics play a more limited role, and the emphasis is mostly – though not entirely – on the performance. In between those two extremes are the visual, the literary and the performance arts. Interestingly, both extremes on the continuum are usually not plagued by forgeries. In the case of fireworks, this is so because it does not really matter who has conceived the firework. In the case of sports, people usually watch the performance, which makes it almost impossible to misattribute the athletic performance.[2] Conversely, an important difference between forgeries and athletic performance is that the forgery is not considered a forgery because it is unnatural, while I claim that the unnatural does play a role in our aesthetic appreciation of sports. If we would learn that Jim Morrison never used

[2] There are, however, a few cases in which this did happen. In 1991, Abbes Tehami won the Brussels Marathon. Yet he was soon disqualified after one of his rivals noticed that he started the race with a mustache and finished it without moustache. It turned out that his coach had run the first seven and a half miles with mustache, and then handed the race number on to Tehami (Zimnuich 2009).

heroine, that would probably only have a minimal impact on our appreciation of his songs, and I am not certain that it would positively influence our aesthetic appreciation of his compositions. This difference is important because it highlights a difficulty with the very notion of the natural. Our aesthetic appreciation of a work of art depends to some extent on the identity of the painter. This issue is of little importance to the aesthetics of sports, since there is usually no doubt about the identity of the athlete who has played the game or who has run the race.

In sports, the origin of an athletic performance has to be natural and human in origin, in order to deserve our fullest aesthetic appreciation. To some extent, one can argue that the same holds for art: chimp art (if art at all) has not the same aesthetic value as human art, in part because we think that chimp art was not meant as art. But are painting chimpansees and athletes on PEDs really on a par? In *Naming and Necessity* (1980), Saul Kripke has given a very influential account of how (proper) names refer. According to Kripke, the referent of a proper name can be determined by reconstructing a causal-historical chain. His causal theory of reference implies that a name 'Oswald' refers to the individual that received this name when his parents pointed at him and baptized him 'Oswald'. When other people use the name 'Oswald' they refer to Oswald because there is a causal chain that connects the parents (or the other baptizers) to everyone else who uses the name 'Oswald'. Likewise, a Vermeer painting can be determined as a Vermeer painting because there is a historical-causal chain between the initial baptism of the painting and the people who now want to refer to the Vermeer-painting. Obviously, many things can – and often do – go wrong in the reconstruction of this causal chain, but at least we know what we are looking for when we want to be sure whether we are dealing with a Vermeer: we are less interested with description of the formal characteristics of a painting than with the historical and contingent connection of the painting with the painter Vermeer. Kripke has extended his causal theory of reference to natural kinds. The details of his theory are not relevant here, but his conclusion is that the referent of a biological taxon – which is a natural kind term according to Kripke – can be determined by essential properties. So it is possible to determine precisely whether X is a human being, based on a number of essential properties of *Homo sapiens*. That said, even if Kripke's account of biological species would be convincing, it is not what we are interested in the case of PED-driven athletic performances. We usually do not doubt whether the athlete belongs biologically to our species, we doubt whether his performance is human, and it is not sufficient to determine – e.g. through DNA-testing – that the athlete is human in order to make sure that his performance is properly human. What we are actually interested in, it seems, is a (precise) determination of the extension of the terms 'natural' and 'natural performance'.

Here, we witness the re-emergence of the conceptual problem with 'human nature' that was already mentioned at the beginning of this contribution. How should we conceive the 'naturalness' of the athletic performance that is supposedly undermined by the use of PEDs, but less so by exercising and by following a strict diet? It seems that this naturalness is constituted – at least in part – by what is understood as some sort of mismatch between a human being and the substance that this human being uses. The performance that results from this mismatch is then termed 'unnatural'.[3] However, we are still pretty much in the dark about why the very notion of 'unnatural' is chosen to characterize this mismatch, not in the least because what constitutes such a mismatch varies over time. In the remainder of this section, I sketch and explore a solution for this problem, a solution that builds upon the notion of family resemblances.

If unnatural performances are seen as a result of a mismatch between a human athlete and a particular substance, what then are the prototypical matches between these two parameters? In my view, prototypical matches tend to involve substances that are either healthy or that belong to the set of substances that most (healthy) people tend to use.[4] Someone who only eats and drinks what the ordinary healthy individual eats and drinks is considered an athlete whose performances are natural. The more substances differ from these prototypically natural items (for humans), the more these substances and the performances they result in, are considered to be unnatural. If the substance is only slightly unhealthy, but a regular part of the average person's diet, the effect on our aesthetic appreciation of the performance can still be weak, assuming that the effects are additive.[5] If the substance is both extremely toxic and – probably as a consequence – absent from other people's diet, the resulting performance will generally be less aesthetically appreciated.

Historical variation in judgments about naturalness can then be accounted for (a) by historical changes in diet and historical changes in what we consider to be healthy, and (b) by changing conventions on what counts as too different from the prototypical natural substance. Hence, the term 'natural' is not a scientific term, but scientific information undoubtedly has a bearing on what we consider to be natural substances and performances. In other words, culture can have some influence, both on the borders and on the prototypical cases of what counts as a natural performance. In the early years of the Tour de France, for instance, multiple gears were not allowed, while there was no doping control whatsoever, and it was known that most riders used strychnine to ease the pain in their legs (Augendre 1996). A very similar cultural influence on what is seen as aesthetically relevant authorship can be witnessed in art. As Dennis Dutton put it (in 1979):

> There might come a time (…) when electronically produced accelerandos will become accepted procedure in piano recording. But note that this will alter our conception of what will count as achievement in a recorded piano performance. We will no longer say things like, "Didn't she play that run beautifully?" but rather "Don't they do marvelous tempo

[3] Hence, the 'naturalness' or 'unnaturalness' of a substance depends on this triangular structure that consists of the product, the athlete and the performance. It would be meaningless (or at least confusing) to ask whether the product is unnatural without taking the other two components into account.

[4] The plasticity of the term 'health' does not undermine my position. Yet, it does explain part of the cultural and historical variation in what is considered a natural performance.

[5] My assumption is that the effects are usually additive. Hence, I do not think that as soon as a newly discovered substance becomes undeniably healthy – whatever that may mean-, it would also become 'natural'. If this substance differs too significantly from the things that are part of our day-to-day diet, it would probably still be seen as unnatural.

> engineering at Decca?" We may expect that engineers will be given credit on record albums, not for having (presumably) faithfully reproduced the sounds the artist has produced, but for having altered those sounds in ways previously left to the performer (Dutton 1979: 310).

I realize that underscoring the influence of cultural factors on what is seen as 'natural' in sports makes it difficult to judge in border cases whether or not a particular product should be banned. The utility of my proposal is further jeopardized by the difficulty of assessing the relative weight that each of the variables (health and familiarity) has on our evaluation of a performance's naturalness. That said, this approach might nonetheless elucidate why the border cases are border cases. I think that in this debate, not much more can be reasonably expected from any philosophical account.

8.5 Aesthetic Value and Virtuosity

So far, I have sketched an analogy between doping and forgery in art. This analogy builds upon the idea that our appreciation of an athletic performance is partially affected by information about its origins, much in the same way as our appreciation of a work of art is influenced by such information. Yet, the information that matters differs between the two cases. Very few art critics would be eager to learn what Titian ate the morning he started working on his Venus of Urbino. Someone who is fond of archery, however, really wants to know whether the archer has taken valium just before the target archery competition. So, there is a point where the analogy breaks down. But where exactly does it break down? Is it valid to assume that the analogy can substantiate the claim that 'human nature' is an aesthetic value in sports, rather than a moral value? Can we really assert that not only in the arts, but also in sports, the origins of a performance affect our *aesthetic* appreciation? This issue is particularly pressing because the origins of behavior/performance tend to be relevant for moral judgments, too. We all know that intention, cultural background and medical history of a person are often essential for the moral evaluation of his or her actions. So the question becomes why information about the (natural) origins of an athletic performance influences our aesthetic appreciation of this performance rather than its moral appreciation.

One answer to this question would be that it is difficult to see how the 'naturalness' of a performance can be a morally relevant property. Yet, in a sense this answer begs the question. On top of that, it does not provide a positive argument why the naturalness of an athletic performance would be an aesthetic value. Another, slightly more satisfactory answer would be that naturalness sometimes influences our aesthetic appreciation. For example, we are sometimes moved by something in a natural environment, because we think of that something as a part of nature. If it turns out that the natural scene is actually man-made, the aesthetic appreciation changes dramatically (Carroll 1993). Although this is a correct observation, the observation itself does not entail that naturalness is an aesthetic value *in sports*, especially not since the naturalness of our environment differs quite substantially from the naturalness of athletic performances.

I think that the aesthetic virtue of virtuosity harbors a more fruitful take on this issue. In a seminal paper on artistic virtuosity, Thomas Carson Mark defines works of virtuosity as works in which the display of skill is made a central feature. The paintings of the Dutch masters of the seventeenth Century (Vermeer, Hals, Gillig, …) are such works of virtuosity, according to Mark. "[C]onsider the still life paintings by various Dutch artists of the seventeenth century, in which such objects as crystal goblets, jewelry, glass spheres, dewdrops, and oysters on the half shell are consummately represented. Whatever else is true of them, these pictures show a deliberate collecting of hard- to-paint items whose convincing representation counts as a display of the skill of the painter." (Mark 1980: 28–29). Now suppose for a moment that David Hockney and Charles Falco are right and that that Vermeer and many other Dutch masters used a *camera obscura* to paint those 'convincing representations'. Wouldn't that alter our aesthetic appreciation of these paintings? Of course, few critics endorse the view that the paintings of Vermeer are artistically worthless if he really used a *camera obscura*. Yet, we would be tempted to think of his work as less virtuoso than we thought it to be. The paintings would simply not be the displays of skill that we thought they were before we knew that Vermeer used a *camera obscura*. This shows how virtuosity depends on constraints. Levinson gives another example that underscores this point. According to Levinson, our interest in Perec's novel *La Disparition* – the novel in which the letter *e* was not used – is to a large extent an interest in how the author has managed to surmount the challenges he has set himself (Levinson 2003). Maybe in sports the successful manoeuvering within constraints is even more important than in the arts (see for example Suits 1978). Of course, most of the skills that are needed to perform in sports are not the skills that we associate with art. But the virtuosity that we admire in sports depends on the constraints that the athlete has to deal with, much in the same way that the virtuosity in art depends on artistic constraints. Hence, changing these constraints tends to affect our appreciation of the athlete's/artist's virtuosity. In sports, the naturalness of the performance is (the term for) one of those constraints.

I know that virtuosity comes in different types. Even an academic paper or presentation can be a display of virtuosity. So the point here is not that sports is an art or that the surmounting of challenges gives an artistic aura to any activity. The point is rather that virtuosity is an aesthetic value that can be found in quite different activities. It might not be that difficult to conceive of a moral virtuosity – not to be confused with moral virtuousness -, but even then virtuosity would still be a primarily aesthetic value, and not a moral one.

8.6 Disentangling Ethics and Aesthetics

I have argued that the use of some PEDs affects and even violates the aesthetic value of an athletic performance because our aesthetic appreciation of such performance partially depends upon the origin of this performance. In my view, the meaning (and the beauty) of an athletic performance is constituted by its being human and natural, and the use of (some) PEDs undermines the natural or human character of an athletic performance. In this final section, I want to explore why many people still think that the 'human nature'-argument against doping is (primarily) an ethical argument, rather than an aesthetic one.

Moral and aesthetic values sometimes overlap. In the *Tractatus*, Wittgenstein famously wrote that "ethics and aesthetics are one and the same", and Kant noted in his *Critique of Judgment* that "the Beautiful is the symbol of the morally Good". The overlap may be particularly important when our aesthetic judgment is about human actions or performances, as is the case here. In the case of PED use in sports, I think that it is perfectly legitimate if our aesthetic appreciation of an athletic performance also involves an ethical appreciation of the moral strength or weakness that has led to the athletic performance. Again, the link with art forgery is obvious. Art forgers are liars and deceivers, who exploit other people's genuine interest in good and authentic art. This lying and cheating is obviously immoral, and should influence our aesthetic judgment, at least to some degree.

Although I endorse the view that immoral behavior can aesthetically compromise a work of art or an athletic performance, I do hold that moral values are neither necessary nor sufficient for our aesthetic depreciation of PED-driven athletic performances and art forgeries. After all, the immorality that is involved in these behaviors, is (logically) secondary: the cheating and deceiving only becomes necessary because a copy is aesthetically or artistically – and thus economically – less valuable than the original work. If people would enjoy a van Meegeren as much as a Vermeer, the incentive to manipulate others into believing that van Meegeren's paintings should be attributed to Vermeer would disappear. To some extent, the same line of reasoning can be applied to doping use in sports. If people would find an athletic performance on steroids equally attractive and beautiful as an athletic performance by someone who has not taken steroids, the steroid user would be less motivated to hide his drug use. Moreover, because people do not want to be fooled by doping users, sport-internal and legal rules have been created that make these behaviors illegal.

The dependence of moral and legal issues on the aesthetic appreciation of an athletic performance entails that most of our aesthetic judgments about art forgeries and drug enhanced performances are influenced by moral information. Yet, it is also plausible that this entanglement gives rise to some kind of confusion with regard to the aesthetic and the moral valuing at stake here. This confusion might be further aggravated because health is not only one of the 'markers' of the natural, but also has moral overtones. Psychological research has shown that healthy people are seen as ethically clean, whereas unhealthy people tend to be seen as morally flawed (Stein and Nemeroff 1995). This should come as no surprise. After all, many moral philosophers consider self-inflicted harm to be morally problematic, for instance because it indirectly harms other people (children, spouse, parents, friends) as well. Yet, non-philosophical factors/reasons probably are more important in shaping people's moral attitudes towards health and illness. In many popular stories (fairy tales, urban legends, novels), immoral behavior tends to be linked to all kinds of illnesses, while health and beauty are often presented as an indicator of virtue (Sontag 1978).

Additionally, many of the strictest rules we were subjected to as a child had to do with health and food, which makes it psychologically understandable that so many people regard health as a moral duty.

According to many psychologists and neuroscientists, disgust is the emotional link between the unhealthy, the ugly and the immoral. Although disgust evolved as a defense against the intake of toxic food, it was later evolutionarily co-opted for moral purposes (Kelly 2011). Members of an out-group often elicit disgust, much in the same way as rotten meat or rotten vegetables. Hence, one can speculate that the oral intake of certain substances elicits disgust, thereby emotionally strengthening the moral condemnation of the doped athlete. If you are what you eat, you might become as disgusting (or as dirty) as the substances you use to enhance your performance.

More in general, disgust might play an important role in the rejection of the transhumanist endeavor. The blurring of the distinction between the human and the non-human is what Freud would have dubbed an uncanny experience, an experience that unsettles us in such a way that it produces a mixture of fear and disgust (Miller 1998). Objects such as robots and statues which seem 'almost human', elicit this negative emotional response. In that sense, the comparison of doped athletes with robots and machines in a human disguise is quite telling, because it suggests that we might experience such athletes as uncanny. The uncanny is generally taken to be an aesthetic experience (or feeling, or property), but because the uncanny involves disgust, it often gives rise to moral attitudes as well. The moral condemnation of uncanny -'almost human'- athletes is then a byproduct of the Janus-headed – aesthetic as well as moral – emotion(s) that these athletes elicit.

8.7 Conclusion

In this chapter, I have emphasized how the use of PEDs threatens the aesthetic value of sports. This is not to deny that there are (important) moral problems with doping, as I have tried to make clear as well. That said, I have paid little attention to the question whether 'human nature' can be a moral value outside the athletic realm. Arguably, the answer to this question partly depends on the weight one wants to give to disgust in the project of normative ethics (Kelly 2011). If one considers disgust to be a viable basis for moral judgments against, for instance, human cloning and cognitive enhancement, one would probably also view 'unnatural performances' as violations of a moral value because they are unnatural. Since I have shown myself to be skeptical with regard to human nature as a moral value in sports, it must be clear that my sympathies are not with those who claim that 'purity' – understood as the absence of animal traces or as the absence of pharmacological traces – is a morally relevant category. At the very least, I would maintain that purity is not primarily a moral category, while it is undoubtedly an aesthetic category. I think the issue of performance enhancing drugs in sports underscores that point.

References

Augendre, J. 1996. *Le Tour de France: Panorama d'un siècle*. Paris: Société du Tour de France.
Beardsley, M. 1965. On the creation of art. *Journal of Aesthetics and Art Criticism* 23: 291–302.
Carroll, N. 1993. On being moved by nature: Between religion and natural history. In *Landscape, natural beauty and the arts*, ed. S. Kemal and I. Gaskell. Cambridge: Cambridge University Press.
Currie, G. 1985. The authentic and the aesthetic. *American Philosophical Quarterly* 22: 153–160.
Dupre, J. 2001. *Human nature and the limits of science*. Oxford: Oxford University Press.
Dutton, D. 1979. Artistic crimes: The problem of forgery in the arts. *British Journal of Aesthetics* 19: 304–314.
Hull, D. 1986. On human nature. *Proceedings of the Biennial Meeting of the Philosophy of Science Association* 2: 3–13.
Kelly, D. 2011. *Yuck! The nature and moral significance of disgust*. Cambridge: MIT Press.
Koestler, A. 1964. *The act of creation*. New York: Penguin.
Kripke, S. 1980. *Naming and necessity*. Cambridge: Harvard University Press.
Lacerda, T., and S. Mumford. 2010. Genius in art and sport. *Journal of the Philosophy of Sport* 37: 182–193.
Lessing, A. 1965. What is wrong with a forgery? *Journal of Aesthetics and Art Criticism* 23: 461–471.
Levinson, J. 2003. Elster on artistic creativity. In *The creation of art*, ed. B. Gaut and P. Livingston, 235–256. Cambridge: Cambridge University Press.
Levy, D. 1980. Perversions and the unnatural as moral categories. *Ethics* 90: 191–202.
Machery, E. 2008. A plea for human nature. *Philosophical Psychology* 21: 321–329.
Mameli, M. 2008. On innateness: The clutter hypothesis and the cluster hypothesis. *The Journal of Philosophy* 105: 719–736.
Mark, T.C. 1980. On works of virtuosity. *The Journal of Philosophy* 77: 28–45.
McFee, G. 2005. The artistic and the aesthetic. *British Journal of Aesthetics* 45: 368–387.
Miller, I. 1998. *The anatomy of disgust*. Cambridge: Harvard University Press.
Sontag, S. 1978. *Illness as metaphor*. London: Penguin.
Stein, R., and C. Nemeroff. 1995. Moral overtones of food: Judgments of others based on what they eat. *Personality & Social Psychology Bulletin* 21: 480–490.
Suits, B. 1978. *The grasshopper: Games, life and utopia*. London: University of Toronto Press.
Walton, K.L. 1970. Categories of art. *Philosophical Review* 79: 334–367.
Zimniuch, F. 2009. *Crooked: A history of cheating in sport*. Lanham: Taylor Trade Publishing.

Chapter 9
Something from Nothing or Nothing from Something? Performance-Enhancing Drugs, Risk, and the Natures of Contest and of Humans

M. Andrew Holowchak

9.1 Introduction

Over the decades, there have been many different scholarly objections to the use of performance-enhancing drugs (PEDs) in competitive sports—PEDs are unnatural, PEDs give users an unfair advantage, PEDs are harmful (or potentially so), and PEDs give advantage without requiring additional skill. The issue generally tends to center on harm done to sport through sanction of the use of PEDs or harm done to the individuals using them. The two are often treated as if they are independent issues.

In this undertaking, I examine enhanced performance in athletic competitions from an Aretic perspective—a philosophical view of competitive sport, which draws heavily from the virtue-based accounts of Aristotle and of the Stoics. Focusing on PEDs, I analyze potential both for harm done to sport (part one) and for harm done to athletes (part two).

In the first part, I look at PEDs and their potential for harm to sport, independently of the issue of potential harm to individuals. Examining the nature of sport, I argue that sanction of the use of PEDs would not cause harm to sport. Though their use would seem to give athletes using them something for nothing, there seems to be nothing philosophically objectionable to something for nothing.

In the second part, I look at use of PEDs and their potential for harm to individuals. Examining human nature from the Aretic viewpoint I commend, I argue that PEDs ought not to be sanctioned. On the one hand, given our current state of knowledge pertaining to their potential for harm to individuals, they are significantly dangerous. On the other hand, PEDs offer athletes who take them not a something-for-nothing, but a nothing-for-something exchange.

M.A. Holowchak (✉)
Department of Philosophy, Rider University, 2083 Lawrenceville Road
Lawrenceville, NJ 08648, USA
e-mail: mholowchak@hotmail.com

J. Tolleneer et al. (eds.), *Athletic Enhancement, Human Nature and Ethics*, International Library of Ethics, Law, and the New Medicine 52, DOI 10.1007/978-94-007-5101-9_9,

In a word, PEDs do not seem to be inconsistent with the nature of sport, but they seem to be inconsistent with human nature. This conclusion, however, is hasty. Harm done to sport and harm done to individuals are not separable issues. Humans are social creatures and any harm done to individuals, as participants in any sanctioned social activity, is in a manner of speaking harm done to that activity as well. Thus, PEDs are inconsistent with human nature and the nature of sport.

9.2 PEDs and the Nature of Sport

9.2.1 Something for Nothing

Imagine a scenario in which the harder, more intelligently, and more persistently a person works over time at some task, like architecture, the worse he becomes. An architect wins acclaim for an avant-garde bridge he designs, but thereafter seems to have diminishing returns on his architectural creations. What is worse, his architectural creations seem to diminish in value in proportion to the effort and study he puts into his craft. The scenario, if seriously entertained, seems diabolical, for whatever anyone might say about the vacuity of life, no one can say that persistent and intelligent application to some task, whether worth doing or not, does not in general bear some fruit over time. There seems to be at least some sense of cosmic justice in that sentiment.

What applies to designing bridges applies to all other human activities. Intelligent application toward some task in the main always leads to enhanced performance. To a large extent, competitive sport is about performance-enhancement, for performance-enhancement is a simple way to measure some of the things we value in sports as well as everyday life—viz., persistence, hard work, patience, planning, and diligence.

Occasionally, new ergogenic aids—such as drag-minimizing swimming suits, bench-pressing shirts, and aluminum baseball bats, *inter alia*—are introduced to certain sports in a manner that asks nothing of athletes but promises enhanced performance. They offer athletes something for nothing. Performance-enhancing drugs are among those aids.

In this section, I examine the issue of use of something-for-nothing ergogenic aids in competitive sports by analysis of the nature of sport. Are something-for-nothing ergogenic aids, like PEDs, incompatible with the nature of sport? The question first invites us to examine carefully "ergogenicity".

There are roughly five sorts of ergogenic aids apropos of competitive sport: mechanical, psychological, physiological, nutritional, and pharmacological (Holowchak 2002, 76–7). Mechanical aids are technical improvements that allow for increased efficiency and, thus, improved performance within the rules of a game. Psychological aids are mental techniques or tactics—some laudable, like distraction-control techniques; some dubious, like trash-talking—used by competitors to gain an advantage. Physiological aids are practices, like body-fat reduction to enhance neuromuscular efficiency, which are designed to improve physiological functioning. Nutritional aids, like optimal protein and carbohydrate consumption, are the advantages of improved nutrition. Finally, pharmacological aids are performance-enhancing drugs, such as caffeine and amphetamines for enhanced alertness, EPO for enhanced red-blood-cell activity, and anabolic drugs for enlarged musculature.

Not all ergogenic aids are equally problematic.

On the one hand, some ergogenic aids promise enhancement of performance, but do so only by enhancement of skill or expenditure of effort. For instance, certain rubbers, added to table-tennis bats, promise improved play by allowing the ball to come more speedily off one's bat on contact. The gain of speed, however, comes with the price of loss of control. Players opting for such speedier rubbers have to learn how to compensate for such loss of control. In short, it takes considerable effort and skill to use speedy rubbers, which are, it comes as no surprise, commonly used mostly by elite players and by novices.

On the other hand, other ergogenic aids allow for enhanced performance without expenditure of athletic skill or effort. In the sport of Powerlifting, use of knee wraps enables powerlifters to squat some 25 kilos more without expenditure of effort or enhancement of skill. That allows those athletes to enhance performance without requiring enhancement of skill or expenditure of effort. As I have argued elsewhere, that might suggest to many adoption of the following principle, called the principle of enhancement of skill (PES), apropos of ergogenic aids: *All ergogenic aids that do not enhance athletes' skills or require expenditure of effort should be disallowed* (Holowchak 2002, 82). With 109 characters, the problem of something for nothing seemed to be solved.

Nonetheless, upon what grounds could the principle be justified? Something for nothing, like knee wraps for powerlifters that enable them to squat some 25 kilos more without expenditure of effort or enhancement of skill, seems a rooted part of competitive sports. Resting on intuitions that seemed sound, I could find nothing to ground PES, which seemed intuitively correct, though not intuitively unshakable—hence the need for grounding. Moreover, the principle said nothing about ergogenic aids that enhanced skills or required expenditure of effort, without harm.

I entertained a weaker principle, PES*: All ergogenic aids that enhance athletes' skills or require expenditure of effort without significant harm to athletes should be encouraged. PES*, I noted, encouraged use of those aids that did enhance athletes' skills, while it did not shut the door on other sorts of ergogenic aids, like ergogenic knee wraps, that provided little or no enhancement of skill and required no additional effort (Holowchak 2002, 82–3). This solution, however, skirted the issue of something for nothing.

Here we have a genuine imbroglio. There seems to be no persuasive reason for disallowing ergogenic aids that do not enhance skills or require expenditure of energy. Nonetheless, something seems most unsettling about their inclusion.

9.2.2 Suits and the Nature of 'Games'

Why it should be problematic that an athlete should gain something for nothing might seem bewildering. Perhaps this is the sort of problem that only a real sports purist or philosopher of sport entertains.

Something for nothing seems to be a problem, because it seems inconsistent with the reasons why we agree to participate in athletic competitions in the first place. Stated baldly, something for nothing seems to be at odds with the nature of sport, a species of game.

The notion that sport has a nature might seem fatuous, as much of the talk of the essence or nature of something is often reviled today by philosophers as mawkish Platonism. Moreover, has not Wittgenstein showed that that there is nothing in common, other than family resemblance, to the things we call "games"? (1973, §66)

In *The Grasshopper: Games, Life and Utopia,* philosopher Bernard Suits has taken Wittgenstein's family-resemblance thesis to task vis-à-vis "games". Suits has argued that there is something in common to almost all the things we call games and that that commonality is the best explanation of why games are called "games". "[T] he commonality I detect is a good explanation of why they are all, or very nearly all, called games, and that family resemblance is not a good reason. For if family resemblance were the reason they are nearly all called games, then it would be puzzling…" (2005, 173). Suits defines game-playing thus:

> To play a game is to attempt to achieve a specific state of affairs [prelusory goal], using only means permitted by rules [lusory means], where the rules prohibit use of more efficient in favor of less efficient means [constitutive rules], and where the rules are accepted just because they make possible such activity [lusory attitude] (2005, 48–9)

The prelusory goal is the end that is describable independently of the game. It is a "*specific achievable state of* affairs" that "can be described before, or independently of, any game of which it may be, or come to be, a part". The constitutive rules function to disallow the most efficient lusory means to the prelusory goal. "Lusory means are means which are permitted (are legal or legitimate) in the attempt to achieve prelusory goals" (2005, 50–1). Finally, the lusory attitude is the voluntary participation of the constitutive rules, because they make possible the game. "In games I obey the rules just because such obedience is a necessary condition for my engaging in the activity such obedience makes possible" (2005, 45). Put succinctly, "[P]laying a game is the voluntary attempt to overcome unnecessary obstacles" (2005, 54–5).

Suits notes that the lusory attitude is what cements together all other elements of game-playing. A participant in a game, such as an athletic contest, agrees to contest for the prelusory goal pursuant to the constitutive rules. In the Yoke-Walk in Strongman, a contestant agrees to yoke himself to a Π–shaped apparatus that weighs some 450 kilos (roughly 1,000 lb) by setting the top of the apparatus on his traps and upper back, holding on to the "legs" of the apparatus with his hands, and walking it past a finish line many meters away. Note that the constitutive rules forbid more efficient means of attaining the prelusory goal of carrying the apparatus across the finish line. One could, for instance, carry it with a forklift, but that would no longer be a test of strength, but one of forklift-use efficiency.

Nonetheless competitive sport, Suits emphasizes and rightly so, is about inefficiency, not efficiency. The constitutive rules are about stipulating, in a manner that is as precise as possible, just how the prelusory goal is to be attained. In general, it specifies only a narrow range of means of attaining the prelusory goal, thereby disallowing numerous other, generally more efficient, ways of achieving the prelusory goal. In sum, inefficiency is the nature of competitive sport.

A competitive athlete's task, then, is about achieving the prelusory goal by adherence to the inefficient means specified by the constitutive rules. In effect, the winner is the one who works most effectively and efficiently within the framework of the constitutive rules. Winning, thus, is about using inefficient means efficiently.

9.2.3 Harm to Sport

Therein lies the imbroglio. It is laudable for any athlete to gain advantage or victory through improved skill or heightened expenditure of effort. No one ever denounced a competitor for having won because of skill or effort. Yet what of a competitor who gains advantage or victory without skill or effort? A professional golfer, Rodney Ruff, designs a new set of clubs, *CorrectAll*, to minify the ill effects of mishits, while increasing distance. Ruff's clubs, consistent with the guidelines of the PGA, improve his game, without any changes in his preparation, so much that he goes from being a middle-of-the-pack professional to the top-ranked professional in the world.

Perhaps the scenario of *CorrectAll* does not seem so objectionable, only because Ruff himself took the added effort to think through the problem of utmost mechanical efficiency and design his own clubs. He has, one might reasonably argue, earned the success through years of mulling over mechanical design for improved efficiency. He compensated for his limits of physical talent by inordinate mental efforts.

Yet what if an ultraist fan of Ruff, instead of Ruff, designs *CorrectAll* just so Ruff can defeat the world's best golfers of his day? The fan gives the set of clubs, designed to look much like his old set, to Ruff and keeps his discovery away from all other golfers. Ruff thereafter becomes the world's number-one golfer—something that would not have been possible otherwise, given his limited capacities as a golfer. He now has a decided advantage over all other golfers due to his clubs and has done nothing to earn that advantage.

Overall, it is not so much that Ruff has gotten something for nothing that is objectionable, but that he has gotten something for nothing and none of the other professionals have had access to the invention of his ultraist fan. Had everyone been allowed access to the *CorrectAll* clubs, then, it seems, there would not have been any "peccancy" or impropriety. Everyone's game would have improved by the same or roughly the same amount and Ruff would have been left without advantage. At this point, to disallow the clubs would be to embrace PES and we found no reason to disallow PES.

What I have just described is merely improved mechanical efficiency—literally, clubs compensating for lack of mechanical efficiency—through technological advances, designed to help golfers. There is a huge market for mechanical ergogenic aids in all sports. Athletes want them, manufacturers want to supply athletes with them, and mavens of various professional sports do not seem to mind them.

Imagine now *SuperStraight*—viz., a set of clubs so designed that one has only to line up the head perpendicular to the flag for the ball to go straight to the flag. The clubs force their user to have perfect mechanical efficiency of swing prior to the ball being struck so that a perfectly straight set up of any club to the ball, prior to the swing, results in a perfectly straight striking of the ball. In effect, it is impossible to hook or slice with these clubs. Moreover, the club is so evenly weighted that it demands an even velocity in the backswing and the same acceleration of every golfer in the forward-swing. Thus, if any two golfers using the same club on the same hole set up precisely the same way prior to swinging, their balls will wind up in precisely the same spot on that hole. Golfers need only to account for set up, club selection, and sloping of the fairways and shifts in wind. Variously styled putters also have been designed for different distances—e.g., a 25-ft putter guarantees its user that a golf ball struck by it will travel roughly 25 ft on a straight surface. Putting now becomes solely a matter of lining up straight the putter head and reading the slopes of greens.

Nonetheless, something has gone wrong with *CorrectAll* and incredibly wrong with *SuperStraight.* With *CorrectAll,* golfers get enhanced performance without enhancement of skill or effort. What is troublesome, however, is that the clubs are designed to compensate for defects in a golfer's swing. Because of that, it will be of greater benefit to golfers with improper mechanics than it will for golfers with proper mechanics. Thus, it will narrow the gap between elite-level golfers and all others—an unwelcome result, especially for elite golfers, who spend the most time and effort in improving mechanical efficiency.

With *SuperStraight,* the gap between elite-level golfers and all others is narrowed considerably more. Elite golfers are now recognizable not by their mechanical superiority, but by their ability to read distances and select the right club, line up their clubs perfectly prior to shots, and read slopes on fairways and greens. The most difficult aspect of the game—mechanical efficiency of golfers' swings—has been eliminated. The greatest skill of elite golfers now is having a keen eye. The nature of golf itself has been changed.

In sum, both scenarios compensate for lack of skill and of effort. There seems unquestionably to be some harm done to sport by disadvantaging those golfers that work hardest and most intelligently—the elite-level competitors. Because of that, there is some reason to think that neither set of clubs would ever be sanctioned by the PGA, even if their use would gain currency among non-professionals. Therefore, there is good reason to sanction what we might call the principle of diminution of skill or PDS: *All ergogenic aids that offer improved performance by diminishing the role of skill or effort ought to be disallowed.*

9.2.4 The Fragility and Difficulty of Creating Contests

Are PEDs any different from mechanical ergogenic aids, like *SuperStraight* clubs? John Gleaves has us consider them to be similar by taking away the harm done to users through social condonation of dangerous or potentially dangerous drugs. He has us consider non-harmful PEDs that are manufactured and available to all competitive athletes. In such a way, he takes himself to be addressing fundamental philosophical issues that those persons, who focus on harm to users, seem to ignore (2010, 71–2). The difficulties he envisages are three—suitability, durability, and continuity—and each of these difficulties is harm done to the nature of sport.

Gleaves begins with the assumption that there would be a great reluctance of the majority of competitors to take PEDs. First, in time, "physiological performance [of enhanced athletes would be] improved drastically enough to harm the suitability of a sport's test". PEDs would make a sport's initial test of athletes' skills unsuitable, thereby distancing enhanced athletes from unenhanced athletes and creating a need for two competitive tiers: one for enhanced athletes and one for unenhanced athletes. Few communities, Gleaves thinks, could support two tiers of venues (2010, 275).

Second, there is the problem of durability of suitable contests. Even if communities could accommodate the uneven bifurcation due to tiering, the separation into tiers would likely lead to decay of certain sports through loss of empathy. Gleaves has us consider moving back the walls of baseball stadia in an effort to compensate for enhanced ballplayers. Such compensation by "enhancing" stadia would have the effect of distancing the majority of other, unenhanced ballplayers, who would then seldom hit homeruns. Distancing might result in the demise of baseball through loss of interest. The majority of ballplayers, unenhanced, just would not be able to empathize with enhanced ballplayers. The game for them would change, for, as he is won't to note, "well-designed tests are both fragile and difficult to create" (2010, 275).

Thus far, there is a glaring problem with Gleaves's account. Drawing from personal observation, he states that if PEDs were allowed, "only a small percentage of athletes would (or could) ever use them" (2010, 275). Harm aside, what is the reason for thinking that the majority of athletes would not use PEDs, if made available and if unharmful? In a footnote, he writes: "I will admit that this belief is a bias on my part and that I have found no empirical data to either confirm or deny this position. It does however seem the most believable given the various factors surrounding their use" (2010, 281). What are those various factors? If he is referring to their potential for physiological harm, that is something he rules out at his paper's start. If he is referring to the illegality of PEDs, that point is uninteresting, if only because illegality is perhaps exclusively driven by the perceived harm of PEDs to users. Overall, Gleaves's caveat is unpersuasive. Experience shows plainly that athletes will generally do whatever it takes to gain any advantage, especially if they can do so without risk of harm to themselves. Consequently, if not harmful, there is every reason to believe all or nearly all athletes would take PEDs.

Gleaves offers still another argument, based on distancing—one immune to my objection above. That is the argument from continuity. The noticeable gains of athletes on PEDs would create a wall of separation between achievements of earlier athletes in some sport and athletes on PEDs. In time, there would be loss of "continuity with previous tests". Loss of continuity would then result in increasing the challenges for enhanced athletes, which would in effect be creating new sports for enhanced athletes.

This argument of Gleaves is given succor by the conundrum in which Major League Baseball (MLB) officials find themselves apropos of the home-run record. Barry Bonds's use of PEDs enabled him to shatter Babe Ruth's record of 714 and then Hank Aaron's record of 755 and become the holder of what is generally judged to be the most prized record in baseball. Bonds, not Ruth or Aaron, is the true Sultan of Swat. Yet MLB has come down hard on the use of PEDs and has in place today a strict testing policy. Now that ballplayers are off-the-sauce, so to speak, and the 50-home-run mark is once again sacred, no one quite knows what to do with the statistical data of Bonds and all other ballplayers, known to have taken PEDs.

Yet why does Gleaves believe loss of continuity is such a problem? Time and technological advances are distancers. Many quarterbacks in the National Football League (NFL) today are the size of offensive linemen decades ago and it is unusual to find an offensive lineman that does not exceed 300 lb. Players in all positions are larger, faster, and stronger and their skills have improved due to enhanced, football-specific training methods. Such loss of continuity has nowise diminished the interest in American professional football by its mavens. Moreover, when enhanced athletes were popping baseballs out of their parks during the "steroid years", baseball was at its peak in popularity—no more so than when McGwire and Sosa battled each other in 1998 for the 70-homerun mark. Loss of continuity through heightened spectacle might be a problem for sports purists, but it is unlikely much of a problem for the general public, which wishes to be entertained, and enhanced athletes makes for enhanced entertainment.

There is an additional difficulty. Gleaves takes himself to be offering three, presumably independent, arguments that are philosophically based, against use of PEDs. Yet all of Gleaves's arguments concern distancing as a premise and there is a fundamental problem with distancing. It presumes that there will be a general loss of interest on the part of fans of sports as the athletes' achievements surpass the *in potentia* achievements of fans. That seems dubious, to say the least. Many fans, I suspect, are attracted to specific athletes or teams in proportion to their deeds being perceived as superhuman or bewildering. Fans love to see Jake Long of the Miami Dolphins pancake block a 290-lb defensive end just because they could never envisage themselves doing such a thing. It is the same with teams. Fans love to see the New England Patriots win their division year after year and make their run for another Super-Bowl victory, as such consistency of excellence seems far removed from everyday living.

In sum, use of PEDs does not create threats to the suitability and durability of contests, only to their continuity, and loss of continuity is just not the problem Gleaves makes it out to be. Thus, none of Gleaves's arguments have shown that use of PEDs create intractable problems for competitive sports and their athletes. It might well be that well-designed games are difficult to create, but it is unlikely that they are as fragile as Gleaves thinks them to be.

In the second main section of his paper, Gleaves takes a different tack. Following Alisdair MacIntyre's account of internal goods in *After Virtue*—viz., that the goods of some practice like a competitive sport are only recognizable within the context of that practice—Gleaves argues that "using performance-enhancing drugs may alter the way athletes train, compete and even view their sport". For athletes whose sport is their life, this alteration is a change of life, which might affect the internal goods of their sport. He adds, "It seems equally likely that the drugs may provide short cuts, make training less relevant or change the nature of training in negative ways". He sums, "Without overwhelmingly good reasons to introduce pharmaceutical performance enhancers, there is no reason to risk losing the intrinsically valuable internal goods found within living the life of the athlete" (2010, 277).

Gleaves's internal-goods argument is unavailing for, at least, two reasons. First, the premises of his argument are possibility claims and possibility claims, qua claims of possibility and not fact, do not give any good reasons for thinking that the risk of loss of internal goods by sanction of PEDs in competitive sports is real. The claim *P is possible* is generally only of interest when it is largely conceded that *P is impossible* (or nearly so) is true. Thus, Gleaves's argument does nowise provide adequate support for the conclusion *PEDs are harmful to sport,* only the conclusion *PEDs might be harmful to sport*—e.g., "using safe performance-enhancing drugs potentially risks certain internal goods of many sports" (2010, 272)—and that conclusion is too weak to be used as a reason for disallowing PEDs in competitive sports.

Gleaves concedes that use of PEDs in competitive sports might also make training more rewarding, but adds, in an effort to close the door on further argument, "this is far from certain" (2010, 277). It is not so far from certain as he thinks it is. As a former user of PEDs—I took, for instance, the steroid Dianabol—I can cite my own experience as some evidence that use of PEDs is rewarding. I was able to handle considerably more weight while lifting weights. That itself gave a considerable jolt to my attitude toward training. I also noted quicker recovery from intense training sessions. Quick recovery allowed for more intense training sessions and more numerous training sessions, which allowed for speedier gains. The main drawback of their use was that my manner of living was slightly upset by the nagging concern that adding Dianabol to my system was something that might have unforeseen negative consequences at some future date. Yet that is a concern that Gleaves disallows, as he wishes readers to consider only PEDs that are harmless.

Second, there is the troublesome qualifying clause (italicized) in Gleaves's conclusion, "*Without overwhelmingly good reasons to introduce pharmaceutical performance enhancers*, there is no reason to risk losing the intrinsically valuable internal goods found within living the life of the athlete" (2010, 277). I suspect that considerable enhancement of athletic performance with no possibility of harm to one's health is an overwhelmingly good reason.

Overall, nothing Gleaves has stated shows that PEDs should be banned, because he has not shown that they pose an above-board threat to sport. He has shown merely that they pose a possible threat to sport. Yet in conceding that they might also add to the internal goods of sport, he concedes that PEDs might also be a boon to sport. Thus, he has shown nothing. PEDs offer athletes competitive advantages without, it seems, affecting their skills. If made available to all competitors, it seems likely that the advantages too would be had by all, or nearly so, with no handicap for elite athletes. Thus, PEDs are not inconsistent with PES* or PDS and Gleaves has given us no reason for banning them.

Overall, Gleaves has intentionally skirted the issue of harm to individuals through his scenario of envisaging harmless PEDs, but in arguing for the harm done to sport through inclusion of such harmless PEDs, he is not "primarily addressing philosophical issues" (2010, 271) any more than those writers, such as myself, who have focused attention to the potential for great harm of PEDs to individuals. Overall, Gleaves gives a spirited attempt at a fresh look at the issue of PEDs in competitive sport, but one that is gainless. PEDs do not seem to be inconsistent with the nature of sport and, thus, their use by athletes poses no harm to sport.

We are left, at this part's end, with no reason to think PEDs are inconsistent with sport and, thus, no reason for their disallowance.

9.3 PEDs and Human Nature

It is not clear that the tack of intentionally skirting the issue of the potential harm of PEDs to individuals is availing. One might argue, as I have done elsewhere, that PEDs are harmful to sport, just because they pose a considerable threat to the athletes who choose to take them (Holowchak 2001a) and social sanction of drugs that are potentially very dangerous—and here the soft epithet "potentially" is followed by "very dangerous"—is harrowing. Allowing athletes to expose themselves to undue risks just for enhanced performance, especially in spectator sports, is social sanction of savagery.

9.3.1 Liberalism in Defense of PEDs

Yet there is the liberal argument, in effect, that the risks persons wish to take are their own risks and social policies put into place that tell them what they can and cannot do with their own body—disallowance of PEDs being one such policy—have the potential for greater harm than the harms individuals might willingly impose on themselves. Before proceeding with this argument, I have something to say about how liberalism, which is generally taken only as a political theory, as Rawls insists its aim is his account of social justice, is also and pivotally an ethical theory.

9.3.1.1 Value Liberalism

As a political theory, liberalism is not only parturient with ethical implications, but also rests on an axiological assumption—namely that liberty—in the guise of freedom from, freedom to, or both—is what drives rights-aimed political theories. As an ethical theory, liberalism is the view that individuals are substratally autonomous, choosing beings that are happiest when they are capable of acting as each sees fit, so long as each agent's actions do not impede others' capacities to act as they see fit.

There is doubtless no monolithic theory of liberalism as an ethical view, but at least three variants are extant: subjectivist value liberalism, progressivist value liberalism, and pluralist value liberalism. Subjectivist value liberalism goes back at least as far as John Locke, and perhaps as far as Thomas Hobbes (1994, I.xiii–xv). Locke maintains that values are reducible to tastes.

> The mind has a different relish, as well as the Palate; and you will as fruitlessly endeavour to delight all Man with Riches or Glory, (which yet some Men place their Happiness in,) as you would satisfy all men's Hunger with Cheese or Lobsters; which, though very agreeable and delicious fare to some, are to others extremely nauseous and offensive: And many People would with reason preferr [*sic*] the griping of an hungry Belly, to those Dishes, which are a Feast to others. Hence it was, I think, that the Philosophers of old did in vain enquire, whether the *Summum bonum* consisted in Riches, or bodily Delights, or Virtue, or Contemplation: And they might have as reasonably disputed, whether the best Relish were to be found in Apples, Plumbs or Nuts.... For ... pleasant Tastes depend not on the things themselves, but their agreeableness to this or that particulare [*sic*] Palate, wherein there is great variety... (1975, II.xxii).

Ethical disputes, thus, are as desultory as disputes concerning the tastiest food-type.

Individualist value liberalism is the view of John Stuart Mill in his watershed work *On Liberty*. Mill writes: "[I]n things which do not primarily concern others, individuality should assert itself. Where not the person's own character but the traditions or customs of other people are the rule of conduct, there is wanting one of the principal ingredients of human happiness, and quite the chief ingredient of individual and social progress" (1985, 120). Individuality makes a person more valuable to himself and, consequently, to others. Individuality enriches, diversifies, animates, and elevates life.

When a society is so structured that genius can flourish—when, following Wilhelm von Humboldt, there is freedom and variety of situations for persons to flourish—the benefits are incalculable. The undeveloped learn and are enlivened and humanity progresses. "[A]ll the good things which exist are the fruits of originality". Moreover, one's own mode of laying out one's life is best not because it is best, but because it is one's own mode. Finally, compelling others to think or act one way corrupts the compeller (1985, 129–33). In short, individuality allows for fullest human flourishing through individual and social progress. Much of that progress comes through behavior in keeping with the maxims of justice, which are collective moral requirements that are of utmost social utility (1979, 58–62).

Pluralist value liberalism is the view advocated by Isaiah Berlin. Berlin states that values are many and they compete for our attention in varied ways. Aiming at one might mean neglecting another.

> [E]qually it seems to me that the belief that some single formula can in principle be found whereby all the diverse ends of men can be harmoniously realized is demonstrably false. If, as I believe, the ends of man are many, and not all of them are in principle compatible with each other, then the possibility of conflict—and of tragedy—can never wholly be eliminated from human life, either personal or social. The necessity of choosing between absolute claims is then an inescapable characteristic of the human condition. This gives its value to freedom as Acton had conceived of it—as an end in itself, and not as a temporary need, arising out of our confused notions and irrational and disordered lives, a predicament which a panacea could one day put right (1984, 31).

Berlin's pluralism is not necessarily reducible to an inimical form of subjectivism, as it is consistent with and he takes it as objectivist. What is critical here is Berlin's statement that "the necessity between choosing between absolute claims" is itself what "gives its value to freedom … as end in itself". The human condition, then, is inescapably one of freedom. Yet to be free is to be constrained to choose continually between competing values—e.g., equality, justice, happiness, security, and public order—which have some empirical or *a priori* claim to objectivity (1984, 32).

Berlin is committed to value-pluralism and the negative liberty it entails. Freedom is a matter of choosing ends, "without claiming eternal validity for them". He adds, "Principles are no less sacred because their duration cannot be guaranteed" (1984, 33). Berlin unfortunately fails to amplify on how the objectivity of the values is secured empirically or *a priori*.

9.3.1.2 Liberalism in Philosophy of Sport

Perhaps the most ardent and grandiloquent spokesperson over the years for the liberal defense of PEDs in sport has been W. Miller Brown. Brown's views seem to be in the subjectivist camp.

In "Ethics, Drugs, and Sport", Brown asserts that arguments against the use of PEDs ignore one substantial human value—autonomy.

> What is in question here is a powerful moral vision which may be taken to guide our judgments about both the games we play and who may play them together. It is a vision we all share to some extent, but one which I think is increasingly open to doubt. For it embodies, I believe, an essentially conservative attitude toward the human condition which neglects to give sufficient weight to *a primary human value*: freedom. It opts for relative stability, predictability, and control in human affairs, rather than novelty, change, surprise, and creativity. It presupposes a relative fixity in the human condition as opposed to an evolving transformation of what we are…. If the question is, as I believe it is, "who or what am I?", the answer can be sought either in a race or on long solitary runs. But the answer will be relative. It will be: in these shoes, on this day, following those workouts and with the aid of whatever substances, given the current state of my knowledge, have become part of my diet, medication, and preparation. Such self-knowledge is as changing as our knowledge in general, whose sources, too, lie in human freedom. Indeed, I suspect that if we survive as a

> species for another century, our understanding not only of what we are and how we have become what we are, but of what we can make of ourselves, will be so transformed as to render athletics as we know them now obsolete (1980, 22).

For Brown, it is autonomy versus control, stability and predictability versus novelty, change, surprise, and creativity. Five years later in "Paternalism, Drugs, and Sport," Brown writes:

> The values, perhaps even a conception of what is good for human life, are associated with sports, not because of their nature, but due to the way we choose to play them. We can indeed forbid the use of drugs in athletics in general, just as we do in the case of children. But ironically, in adopting such a paternalistic stance of insisting that we know better than the athletes themselves how to achieve some more general good which they myopically ignore, we must deny them the very attributes we claim to value: self-reliance, personal achievement, and autonomy (1985, 21).

The issues are autonomy, rationality, and paternalism. In forbidding PEDs, we take it upon ourselves to decide for all athletes what is in their best interest. In doing so, we disallow them self-reliance, achievement, and autonomy.

In "Practices and Prudence," Brown insists that the objectivist view of goods has been impracticable.

> Where it was possible to consider the goods to be achieved by sports practices as agent-independent, little headway seemed possible even in a society where it could be assumed that the basic goods of life were distributed in a just manner. The problem is that although risk appraisal is a statistical matter, risk taking is a personal one. Risk taking is not agent-independent, but, rather, relative to the overall circumstances and goals of individuals (1990, 246).

Even if one should be able to objectify risk-appraisal, and Brown seems to concede that risk-appraisal can be objectified, there are still the issues of autonomy and rationality. Because individuals are free and rational, statistical appraisals of risk-taking vis-à-vis actions that concern them and no others—viz., the taking of PEDs—might be objective, but risk-taking, given the human condition, is a personal affair.

Mill's notion of individualist value liberalism is perhaps assumed by William J. Morgan in a postscript in *Leftist Theories of Sport*. For Morgan, there are two elements of any social structure that promote human flourishing. First, societies flourish when they are structured in a decentralized and democratic manner—viz., in such manner as to promote individuation and autonomy. Second, the political element of any society must remain entirely neutral apropos any notion of a particular good—viz., it can reference particular goods, but it cannot promote them, though it might reference and promote more general goods. "[T]o say that the state must remain neutral is not to say that it is disallowed from making reference to *any* conception of the good, but only reference to any *particular* conception of the good. So it can perfectly well justify its intervention in the social world by appealing to a general, common conception of the good" (2002, 481).

Borrowing from Karl Marx and John Rawls, in idealized decentralized democratic societies, there is a realm of necessity and a realm of social perfectionism.

The realm of necessity, following Marx, concerns material production. The realm of social perfection, following Rawls, concerns excellences, of which sport is one. "[T]he realm of social perfectionism … comprises all the practices of a society that contribute to the advancement of human excellence in its various forms". Morgan does not assert that sport will play a significant role in a well-ordered, democratically diverse society, only that it will play some role. He says, "[I]f it is plausible to count sport among those practices dedicated to excellence and its appropriate advancement, then any society that fails to make some provision for sport is a deficient society…" (2002, 479–80). It is uncontroversial, as his postscript shows, that Morgan thinks the antecedent is true.

Bob Simon in *Fair Play* advocates what seems to be a Berlinian approach to value pluralist liberalism. The values of sport, Simon argues, are not merely a reflection of or reducible to those of society. "These values are internal in that they arise from the core character of competitive sport and are not mere reflections of wider social values, which can in principle and sometimes in practice diverge significantly from those found in competitive sport". These unique values are excellence, discipline, dedication, respect for the rules of any game, and excellences of particular sports that guide evaluation of the play of oneself and others. Most importantly, there is sport as the mutual quest for excellence itself (1985, 199–203).

Simon does not say that all of the values that characterize competitive sport at a given time are unique to sport, nor does he deem it necessary that there is any one value that is unique to sport through all time. What he seems to be stating is that sport is in one sense defined at a particular time by a cluster of values that is itself unique within a given society. Sport, thus, has an inner normative logic. Yet in defining sport as a mutual quest for excellence, his view of sport has a normative slant that is not unlike my own Aretism.

9.3.2 *Animadverting Value Liberalism*

Central to any version of value liberalism is the notion of liberty, given human rationality and autonomy. Subjectivist value liberalism of the Lockean sort presumes human autonomy and the subjectivity of values. Individualist value liberalism of the Millean sort presumes autonomy with the aim of facilitating individuation for the sake of social progress. Pluralist value relativism assumes human autonomy and the choice between several values, presumed "objective" in some sense, though not infallibly so, that compete for humans' attention. Each, in effect, is a morality in which autonomy is the substratal moral value or the fundamental axiological concept, which drives inquiry into other values.

The argument from risk, to which I referred at the beginning of the section of liberalism as a defense of PEDs and which is captured quite cleanly by W. Miller Brown, is tantalizing and ought to be reexamined thoroughly. Not allowing individuals to choose their own manner of living, risks notwithstanding, is a greater harm than social sanction of potentially very dangerous social practices, like taking certain PEDs. I draw from several of Brown's premises.

1. One of the primary human values is autonomy.
2. Autonomy presupposes not a fixed, but an evolving, notion of "human nature".
3. There is neither any *telos* to which human nature is evolving nor any one value that trumps any other.
4. The values that different persons embrace can only be judged malefic, if they are malefic to others.
5. So, imposition of some form of the good life on another is malefic (1–4).
6. PEDs, if malefic, are malefic only to the agent taking them.
7. So, disallowing any athlete the use of PEDs is malefic (4–6).

The argument is subtle. Nonetheless, I think each of the first four premises is challengeable. First, I grant that autonomy is an important human value. Yet, for value liberalism, autonomy has to be the substratal value or the key axiological concept that makes the notion of values possible. That is arguable, at least, and perhaps false. Second, I challenge the presumed assumption of Brown that autonomy is an atomic ideal. Humans, I have elsewhere argued (Holowchak 2008, 2009), are social creatures and, thus, autonomy must not be viewed independently of humans' social nature. To grasp fully the proper sense of "autonomy", one should think of "autonomy within" and not "autonomy from".

Second, the evolving notion suggested by premise two seems arbitrary and outside of human control. Though humans evolve pursuant to the changes in the global milieu, they are in large part responsible for numerous changes to that global milieu and numerous changes to their own species. Technological advances will have much to say shortly about changes to the human species over time. Thus, the sort of autonomy suggested by premise two seems too arbitrary to be aidful. Moreover, granting that human nature, following evolutionary biology, is not fixed but evolving, it does not follow that that evolving nature is not perfectible, which premise two seems to suggest, due to its arbitrariness. Selective breeding proves the possibility of perfectibility, given some notion that the characteristics, being selected, are perfections or improvements. Rationality seems to be one such perfection.

Third, it is not clear that nature does not point humans in the direction of any one *telos* and that there is no *archē* among values. I have argued in numerous publications that there is considerable merit to virtue-ethics and eudaemonist approaches to ethics (Holowchak 2004, 2008). The arguments put forth are not irrefragable, but they at least give reasons to call to account the truth of premise three.

Finally, I maintain that premise four is baldly false. *Pace* many consequentialist accounts of the good life, there is no neat dividing-line between action and thought, between harm to oneself and harm to others. One indulgent of malicious thoughts, but who can refrain from acting on them, is no more virtuous than one who acts on such thoughts. As the Stoics were fond of saying, one can drown in 2 in. of water as well as in an ocean of water. Moreover, as is acknowledged commonly today, harm seen to be exclusive to oneself is seldom, if ever, so. Such harms insidiously always seem to have unforeseen effects on others.

9.3.3 Chryssipus's Footrace

The chief defects of the value-liberal argument from risk-taking are, to coin a neologism, "axiologizing" autonomy and privileging individuals. The two defects cannot be grasped independently of each other, for the meaning of "autonomy" is dependent on the meaning of "individual". When individuals are construed as metaphysical atoms, viz., entities whose good is graspable independently of social embeddedness, autonomy tends to get cashed out egoistically—generally in terms of individuals' wants and desires. Reason, then, in the words of David Hume, is truly a slave to the passions. As Bob Brecher states, the position is morally unavailing, tenebrous, and paradoxical.

> Just as the alleged incorrigibility of what we want serves finally to undermine moral justification, so its allegedly value-free conception of harm in the end obscures and even makes paradoxical the liberal insistence that interference with individual autonomy is justified only 'to prevent harm to others'. For in failing to see individuals as embedded within the society and various sub-groups in which they live, it cannot deal adequately with those harms that are morality-dependent, harms which cannot be recognized from a value-neutral vantage-point (1998, 155).

"Harm" too is a concept whose meaning is not independent of "individual". Individuals, as Aristotle often says, are inescapably social animals.

Given humans' social nature, the key to happiness is balancing self- and other-concern. This is best illustrated by what the ancient Greek and Roman Stoic philosophers, in effect, call *oikeiosis*. *Oikeiosis* is a matter of familiarizing oneself with all things—more precisely and literally, making all persons (and all things), to a greater and lesser degree, one's own. Infants begin with an attachment to themselves and their own constitution. They recognize themselves as embodied. In time, they grasp themselves as cosmic entities that have responsibilities not only to themselves, but to others.

Oikeiosis is best captured by the Stoic Hierocles (Holowchak 2008, 75), who writes analogically of ten concentric circles to illustrate each person's ethical duties (Holowchak 2008, I.8). They are, in order of significance, the circles of:

1. one's mind and body;
2. parents, siblings, wife, and children;
3. one's uncles and aunts, grandparents, nephews, nieces, and cousins;
4. other relatives;
5. local residents (i.e., neighbors);
6. fellow demesmen (a unit of a city-state);
7. fellow citizens of a city-state;
8. people of neighboring city-states;
9. fellow countrymen; and
10. the entire human race.

What Hierocles's circles show is that human actions are cosmos-embedded and that any one action can have unforeseen implications for the entire human race.

The Stoic Chrysippus illustrates that notion of cosmic-embeddedness of all human actions by use of a spider's web. All things in the physical cosmos are held together by an all-pervasive matter, *pneuma,* which has a certain cosmic tension (*tonos*), which might be understood as comprising the several, intricate threads of a spider's web. Following the web analogy, an action at one part of the web is felt by the spider at another part of it (Holowchak 2008, 20). It is the same with human actions. Actions which seem to involve no one else but an agent have unseen implications for persons other than the agent.

Actions in keeping with *oikeiosis* do not require completely sacrificing self-interest to help others. If anything, following Hierocles, there is a privileging of oneself and those persons most intimate to oneself, though such privileging ought never to come at the expense of others, for that would be an action contrary to nature as well as human nature and, thus, inconsistent with virtue.

> For one person to increase his welfare at the cost of the other person's welfare is more contrary to Nature than death, poverty, pain, or any other thing that can happen to one's body or one's external possessions. It destroys human communal living and human society. If we are each about to plunder and carry off another's goods for the sake of our own, that will necessarily destroy what is in fact most according to Nature—namely the social life of human beings (2001a, b, III.21).

Consequently, an essential part of *oikeiosis* is knowing not only what to appropriate, but what not to appropriate.

Chryssipus, Cicero reports, says striking a balance between self- and other-concern does not rule out healthy competition among apprentices, so long as "competitors" play by the rules. Chrysippus says:

> When a man enters the footrace, ... it is his duty to put forth all his strength and strive with all his might to win, but he ought never with his foot to trip or with his hand to foul a competitor. Thus, in the stadium of life, it is not unfair for anyone to seek to obtain what is needful for his own advantage, but he has no right to wrest it from his neighbor (2001, III.42).

For Stoics, what is conducive to one's advantage are the sorts of things that lead to virtue and those things do not vary from person to person. The sort of footrace that best describes the balance at which a Stoic aims between self- and other-concern is cooperative competition toward the same goal and the prize of virtue, unlike victory in an athletic competition, can go to more than one victor. In the footrace of life, the contest will be better the more winners it produces—a point that Seneca makes: "Wisdom has this reward, among other things: No one can be vanquished by another, except when he is ascending. When one has arrived at the top, however, it is a draw. There is no place for further ascent; the contest is decided". Hence, mutual aid is ethically desirable in the contest of life.

Overall, the aim here is not merely to do one's duty, dictated by reason. Unlike Kant, duty, for the Stoics, is bounded by capabilities, which vary from person to person. Each of us (or nearly so) has a capacity to do certain things, like tell the truth or form genuine friendships, but it is not the case that each of us can do other things, like bail out a friend during financial hardship or help one who is excessively maudlin in dealing with reality. Thus, it is not because of rationality alone that one is duty-bound to do some action; it is because of rationality *and* a certain set of capabilities that enable one to do that action that one is bound to do it. Remove these capacities and one may no longer be bound to do that action.

9.3.4 Victory, Not Virtue

What has Stoicism to do with use or disuse of PEDs in competitive sports? Following Suits, when individuals consent to compete in competitions, they agree to play by the constitutive rules in order to make the competition possible. They adopt a lusory attitude. Yet no violation of lusory attitude occurs, if the lusory means one uses—that includes use of PEDs—are in keeping with the rules of a competition. Thus grasped, there is nothing about use of PEDs that is inconsistent with the nature of sport.

Unfortunately, Suits is silent vis-à-vis the ethical dimension of game-playing. Bob Simon, whose views on sport I cashed out as being similar to or a form of value-pluralist liberalism, and I are not. Simon regards competitive sports as "a mutual quest for excellence through challenge". His notion "quest" entails fair play and that entails that not all means of winning are legitimate, because not all means are moral (2002, 167–9). In developing my view "Aretism", I have argued that competitive contests themselves are not goods, but might be conducive toward what is good, if they are contested in such a way that virtue and integration with others are the real aims (Holowchak and Reid 2011, 203–11). If not, they are inconsistent with human nature.

Here I proffer two arguments to show that use of PEDs is inconsistent with human nature. The first I have put forth in several publications and, given our current state of ignorance apropos of the harm done to users by PEDS, it is still cogent—i.e., there is no additional information via an additional premise to show its conclusion unlikely to be true. The second is a new argument—an argument from sufficiency.

First, PEDs like steroids and HGH ought to be disallowed, because there is good reason to think that their use is potentially very dangerous for them who take them. A society that sanctions use of such drugs places the spectacle of enhanced performance ahead of safety of the agents of such performance. Thus, it is social sanction of brutality for the sake of spectacle—a sign of a morally degenerative society.

My argument applies not only to use of potentially very dangerous drugs, but also to dangerous sports, like Ultimate Fighting Championship, and dangerous practices in sports, like checking from behind in ice hockey, that are given social sanction. Societies ought always to err on the side of restraint, when it comes to sanctioning potentially harmful practices. One cannot reply that risk-taking is a personal decision. That begs the very question of human rationality and human sanity. One can readily argue that persons fixed on practices that have significant risks are, at least subconsciously, guilt-driven or perhaps even have a death wish. Disallowance of such dangerous practices is not promotion of any one conception of the good life; it is instead promotion of a salubrious social milieu in which all persons can flourish in life-enhancing activities.

This argument, vis-à-vis PEDs, is based on certain contingencies—viz., whether such potentially harmful drugs to users turn out to be harmful drugs to users. Thus, my argument too is based on a possibility claim as a premise, but, in doing so, I am qualifying not merely harm, but significant harm, hence my premise, though based on possibility, is substantive and not vapid. Nonetheless, if it should turn out that use of PEDs, used in moderation, are not very dangerous, this argument becomes effete, because the new evidence, added to the premises, will not strongly support the disallowance of PEDs.

Second, there is the argument from sufficiency. PEDs ought to be disallowed, because they are *de trop*. It is not so much that they offer something for nothing, but that, strictly speaking, they offer nothing from something.

Pursuant to Aretism, athletic challenges are praiseworthy only when they promote virtue and social cohesion. Within the confines of a competitive athletic challenge, it is not winning that counts, but the aretic commitment to playing to win within the confines of the constituent rules, which make the game possible. Authenticity and integrity are critical. Not giving one's full effort to win a game throughout the game, even in lopsided competitions, might be considered disrespectful to one's opponent, in individual sports, or teammates and opponents, in team sports. It is also disrespectful of spectators. Treating others with less than full respect is inconsistent with virtuous activity. Thus, victory gained without consideration for virtue is a gain of nothing from something—more specifically, a gain of something insubstantial by losing something substantial.

Moreover, undue emphasis on or preoccupation with new means to obtain the ends of athletic games betrays a preoccupation with winning, not with fair competing. There is nothing essentially wrong with devising new lusory means to obtain the ends of a contest, as long as a competitor remains within the constitutive rules of that contest. Still, one obsessed with devising new means to gain any edge—and use of PEDs is one such mean—is very probably one who places the end of winning before the aretic commitment to play to win. That again is a gain of nothing from something.

Finally, use of PEDs is symptomatic of what I have elsewhere called "panagonism"—the tendency to view any quantification of athletic achievement, independent of its function in procuring victory in a competition, as a noteworthy or praiseworthy athletic achievement.

Too often in high-profile competitive sports today, athletes are judged not by their contribution to their team's commitment to playing for victory, but by individual displays of talent that seem to stand on their own—e.g., a remarkable one-handed catch in American football, a prodigious 475-ft homerun in baseball, or a showy slam-dunk in basketball. The wide receiver in football is then evaluated by certain quantified displays of talent—e.g., number of catches, number of catches versus drops, and number of one-hand catches; the hitter, by total homeruns, homeruns per plate appearance, and number of dingers over, say, 450 ft; the basketball player, by points scored and number of slams; etc. Such data do not show the overall worth of a player to his team. They fail to show whether the wide receiver is an effective decoy or blocker, when the ball is not thrown to him; whether the hitter does anything other than hit dingers; whether the basketball player prevents opponents from scoring points, while he scores points.

It is unquestionable that use of PEDs makes the wide receiver, the hitter, and the basketball player more powerful and more efficient. Each can perform more spectacularly and each can recover more effectively after physical exertion. In short, use of PEDs will very likely dilate the problem of panagonism by allowing athletes who use them—e.g., Clemens, Sosa, and Bonds in baseball—to stand apart from them who do not. Athletes who stand apart from others seem to rise above sport itself. When that occurs, competitions become meaningful only insofar as the superior athletes distance themselves through superior displays of athleticism. The gain is nugatory; the loss, substantial. Thus, PEDs allow a gain of nothing from something.

9.4 Upshot

In this undertaking, I have examined the issue of social sanction of PEDs from harm done to sport as well as harm to individuals. Using a Suitsian definition of game-playing, I concluded in part one that there is nothing about the nature of sport, as a species of game, that rules out inclusion of PEDs. Using my Aretic perspective of competitive sport, it was found in part two that PEDs ought not to be socially sanctioned, because of the potential for significant harm to individuals using them and because the promise of enhanced performance is not so much a something-for-nothing exchange, but rather a nothing-for-something exchange. If virtuous living is the proper end of right-intended human activity, as I suggest it is, then it is likely that inclusion of PEDs in competitive sports will make that end more difficult to attain.

In showing that PEDs, because they are very likely inconsistent with pursuit of virtue, are very likely inconsistent with human nature, I have shown as well that they are inconsistent with sport. Because humans are social creatures, what is inconsistent with human nature is inconsistent with any right-aimed social activity, of which sport is one. Thus, harm done to sport and harm done to individuals are not separable issues.

References

Berlin, I. 1984. Two concepts of liberty. In *Liberalism and its critics*, ed. Sandel Michael. New York: New York University Press.

Bob Brecher, B. 1998. *Getting what you want: A critique of liberal morality*. New York: Routledge.

Brown, W.M. 1980. Ethics, drugs, and sport. *Journal of the Philosophy of Sport* VII: 15–23.

Brown, W.M. 1985. Paternalism, drugs, and the nature of sports. *Journal of the Philosophy of Sport* XII: 14–22.

Brown, W.M. 1990. Practices and prudence. In *Philosophic inquiry in sport*, ed. W.J. Morgan and K.V. Meier, 71–84. Champaign: Human Kinetics.

Cicero. 2001 [1913]. *On duties*. Trans. Walter Miller. Cambridge: Harvard University Press.

Gleaves, J. 2010. No harm, no foul? Justifying bans on safe performance-enhancing drugs. *Sport, Ethics and Philosophy* 4(3): 269–283.

Hobbes, T. 1994. *Leviathan*. Indianapolis: Hackett Publishing Company, Inc.

Holowchak, M.A. 2002. Ergogenic aids and the limits of human performance in sport: Ethical issues, aesthetic considerations. *Journal of the Philosophy of Sport* 29(1): 75–86.

Holowchak, M.A. 2004. *Happiness and Greek ethical thought*. London: Continuum Books.
Holowchak, M.A. 2008. *The Stoics: A guide for the perplexed*. London: Continuum Books.
Holowchak, M.A. 2009. Happiness and justice in liberal society: Autonomy as political integration. In *Fundamentals of philosophy*, ed. D. Stewart, J. Petrick, and H.G. Blocker. Upper Saddle River: Prentice Hall.
Holowchak, M.A., and H. Reid. 2011. *Aretism: An ancient sports philosophy for the modern sports world*. Lanham: Lexington Books.
Locke, J. 1975 [1706]. *An essay concerning human understanding*, ed. Peter H. Nidditch. Oxford: Oxford University Press.
Mill, J.S. 1979. *Utilitarianism*. Indianapolis: Hackett Publishing Company, Inc.
Mill, J.S. 1985. *On liberty*. New York: Penguin.
Morgan, W.J. 2002. Sport in the larger scheme of things. In *Philosophy of sport: Critical readings, crucial issues*, ed. M.Andrew Holowchak. Upper Saddle River: Prentice-Hall.
Simon, B. 1985. *Fair play: Sports and social values*. Englewood Cliffs: Prentice-Hall.
Simon, S. 2002. Sportsmanship and fairness in the pursuit of victory. In *Philosophy of sport: Critical readings, crucial issues*, ed. M. Andrew Holowchak. Upper Saddle River: Prentice-Hall.
Suits, B. 2005. *The Grasshopper: Games, life and utopia*. Orchard Park: Broadview Press Ltd.
Witgenstein, L. 1973. *Philosophical investigations*. Upper Saddle River: Prentice-Hall.

Chapter 10
Transhuman Athletes and Pathological Perfectionism: Recognising Limits in Sports and Human Nature

Michael J. McNamee

10.1 Introduction

One traditional way of conceiving sports is to think of them as perfectionist practices; sports challenge practitioners to extend themselves, to enhance the limits of their capacities and performances. What enhancements are possible and which are desirable are much disputed. The discussions and negotiations arise among and between various interested parties: athletes, coaches, sports scientists, sports medics, and of course national and international governing bodies of sports. Philosophers, and those who specialise in the ethics of sports too have a role to play in articulating, criticising, defending, and justifying what is good and right in sports. Of course, scholars follow and change fashion; some are conservative, others avant garde.

In recent discussions in sports and bioethics there have appeared essays where it has been suggested that sports may become a site where the modification of human nature itself might be played out (cf. McNamee 2007; Miah 2003). In this essay I want first to probe the plausibility of entertaining seriously this issue by rehearsing an argument by Daniels (2009) to the effect that anyone making such a claim has failed to consider properly the nature of human nature. Secondly, were such transformation to be possible, we might think of the idea of a transhuman athlete as a realistic possibility. But what, and how, should we think of this possibility? I suggest that we would be forced to consider the normative status of human nature itself and the (potential) desirability of biotechnological modifications and the resultant transformations of our understandings of elite sports. To that end I argue here, tentatively, that such a desire is a failure to consider the goodness of human and athletic vulnerability. I argue against the idea of the transhuman athlete, and reject it as a form of pathological perfectionism.

M.J. McNamee (✉)
Department of Philosophy, Humanities and Law in Healthcare,
College of Human and Health Sciences, Swansea University,
Swansea SA2 8PP, UK
e-mail: m.j.mcnamee@swansea.ac.uk

J. Tolleneer et al. (eds.), *Athletic Enhancement, Human Nature and Ethics*, International Library of Ethics, Law, and the New Medicine 52, DOI 10.1007/978-94-007-5101-9_10,

10.2 "Will London 2012 Be the First Transhuman Olympic Games?"

There is nothing like a good headline to capture the imagination. The question that forms the title for this section is adopted from a web-based article by the British scholar, Andy Miah, who has written much about the fusions of technology, medicine and sports. In promoting a public engagement event between various interested groups, Miah writes of the – apparently- first Transhuman games:

> Today, elite sports find themselves in increasingly unchartered waters. More than ever before, athletes are using technology to optimize their biology for performance and many of their methods are not even tested for by the authorities. From genetic tests for sport performance to the use of superhuman prosthetic enhancements, this subject reaches parts that present-day anti-doping rules cannot reach. These technologies have changed elite sports, as we know them, but the next decade promises even more of an overhaul to what we think being good at sport means. As we approach the London 2012 Games, this debate will consider the ethical implications of new technology in sport, asking what distinguishes the cheat from the innovator. We will ask whether the debate about the ethics of athletic performance is all but over, as the winners' podium makes space for the transhuman athlete. (Miah 2011)

To be fair, his remarks are not made in the context of a sober philosophical examination of the issues. His remarks are to be understood as arising from the context of promoting a public event. Nevertheless, it is worth noting how he moves from uncontentious openings to a highly provocative concluding question. It is undoubtedly true that athletes more than ever before are using technology to optimise athletic performances. Of course their use of technology is not in itself new,; only the widespread use and the level of sophistication is. And people are now using genetic tests for talent identification (as one specific example take the test for the presence of ACTN3 which is strongly associated with power production and fast twitch muscle fibre production (Genetic Technologies 2009)) though these are discredited by scientists (Pitsiladis and Wang 2011) and ethics scholars alike (McNamee et al. 2009). Miah moves to ask whether we are about to see the transhuman athlete? But what kind of figure might she, or much likely he, cut?

10.3 What Is Transhumanism?

Transhumanism is a loosely defined movement that has developed gradually over the past two decades. Indeed , since 2008, it uses the label H+ as a (marketing-friendly, and perhaps less threatening) shorthand. The modern 'philosophy' of transhumanism was authored by Max More, PhD in 1990 in the essay "TRANSHUMANISM Toward a Futurist Philosophy". According to More.

> Transhumanism is a class of philosophies of life that seek the continuation and acceleration of the evolution of intelligent life beyond its currently human form and human limitations by means of science and technology, guided by life-promoting principles and values. Transhumanism promotes an interdisciplinary approach to understanding and evaluating the opportunities for enhancing the human condition and the human organism opened up by

> the advancement of technology. Attention is given both to present technologies, like genetic engineering and information technology, and anticipated future ones, such as molecular nanotechnology and artificial intelligence. (http://humanityplus.org/philosophy/, last accessed 17/2/2012)

One may read this, as I do, somewhat sceptically. There is no threat here, not danger, just the promise of a rosy-tinted technologically enhanced future of human happiness. That technologies, as Heidegger (1993) observed long ago, can be used for good or bad ends is suppressed. Later in their promotional literature we find it refined thus:

> Transhumanism can be viewed as an extension of humanism, from which it is partially derived. Humanists believe that humans matter, that individuals matter. We might not be perfect, but we can make things better by promoting rational thinking, freedom, tolerance, democracy, and concern for our fellow human beings. Transhumanists agree with this but also emphasize what we have the potential to become. Just as we use rational means to improve the human condition and the external world, we can also use such means to *improve ourselves, the human organism*. In doing so, we are not limited to traditional humanistic methods, such as education and cultural development. We can also use technological means that will eventually enable us to move beyond what some would think of as "human". (http://www.transhumanism.org/resources/FAQv21.pdf, last accessed 17/2/2012, emphasis added)

When we go to Miah's web site, http://www.andymiah.net/, that boasts the eye-catching headline about the first transhuman games, we find not a robot or cyborg, but an all too human figure: specifically the finely sculpted body of champion female body builder Francesca Steele-Edwards, fake tan and all. For males – and it certainly seems that it is males who are more interested in human enhancement if we look at those writing in the literature (or indeed those athletes who are caught doping) – the figure of the female bodybuilder might be sexually attractive. But what exactly is "transhuman" about female bodybuilding? Well, nothing, in short, must be the answer to that specific question. The morphology achieved over years of disciplined training and diet may be viewed either as an impressive body transformation or, less sympathetically, as an example of someone with a body identity disorder according to your preferences or norms. Consider the italicised remark in the above quotation: notice the slide between "improving ourselves" and altering the "human organism" *per se*. Whichever way we wish to view Ms Steele Edwards, we must agree that her near unique muscular and fat-free appearance cannot be understood as an example of someone who has transcended or indeed transformed human nature. Why not? Well, it is worth asking a sceptical question to draw this out: "When does modifying one's appearance constitute modifying one's nature?" Posed this way we can see readily the answer: no amount of modifying *oneself* amounts to modifying *human nature* because, as Daniels (2009) argues, human nature is a population concept. It follows, therefore, that no amount of modifications to any *individual* can constitute a change in human nature.

In order to head off one possible rejoinder, it is worth noting that this conceptual objection is not based on some archaic notion of human nature as fixed or unchanging. As Juengst (2009) notes in the contexts of genetic modification, we should not think of human nature as static; the gene pool is changing all the time, but in miniscule ways.

On reason we might suppose that *talk* of transcending human nature finds an all too welcoming home in the contexts of (elite) sports, is because we readily think of elite athletes pushing the boundaries of human potentials (Miah 2003). Clearly, though, working on aspects of one's embodiment and embodied capacities will count as (m) enhancement, the notion of shedding our nature is another thing altogether. The two can all too lazily be brought together. But that is just conceptual sloppiness. Thus if the idea of transhuman athletes showing up at the 2012 Olympics is eye- catching but utterly improbable, we may still ask questions about the modifications athletes consider in their attempts to achieve or fulfil ambitions of sporting perfectionism. Some words about the nature of sports and the ideas of perfectionism are in order.

10.4 Sports As Perfectionist Practices

We understand sports as social practices in which a range of embodied capacities are developed in order to achieve an artificially constructed goal, using only means that are permitted (see variously Brown 1992; McNamee 1995; Morgan 1993; Suits 1978). Pursuit of the ends of sports requires and promotes various human (though not uniquely human) capacities of co-ordination, concentration, effort, flexibility, skill, strength, speed, and so on. What is uniquely human about sports, however, is their institutionalised rule governed nature. The rules give structure and meaning to play, which is otherwise observed throughout the whole animal kingdom. And even if ethnologists were to object that in animal play we may observe a form of ritualised rule following that is the genesis of sports (Sansome 1992) it is the formal rules that set apart sports as human practices.

Mention of the defining rules of sports, begs questions as to their nature. It is a commonplace to observe that there are both constitutive and regulative rules in sports.[1] While the former constitute or define the nature of the activity (the pitch must be thus long, the ball must be pressurised to this degree, the net must be so high, the playing time may not exceed so many hours or days, and so on) the regulative rules prescribe and proscribe what can be understood as play within that sport (one may not harm an opponent, abuse the officials, use doping substances and so on). This proper recognition of the place of regulative rules is pregnant with ethical possibilities. If we reject technical reductionism – the view that sports are no more than exercise of technical prowess where morality (so to speak) goes on holiday,[2] we must by virtue of the presence of regulative rules recognise that there is a logical

[1] While the distinction goes back to Kant, the modern locus is usually thought to be John Searle's work in the philosophy of language where the distinction is developed. It should be noted, however, that the constitutive-regulative distinction was explored thoroughly before Searle's work in an essay by Geoffrey Midgley (1959) in a paper given to the Aristotelian Society in the UK.

[2] This memorable phrase was first coined by Reddiford (1981).

demand for ethical rules within sports. This is not – of course – to suggest that sport is and must be ethically positive. But formalists in the philosophy of sport rightly point out that there is a sense in which players who do not conform to the regulative rules of a given sport are logically not playing the game. Accepting this point we must recognise the limited nature of the conception of sports as merely technical enterprises. The cultivation of sporting excellence thus requires a range of personal characteristics (virtues) that both suffuse our attempts to secure victory in sports *and* are the basis of public admiration and self-worth.

In striving for success in sports we come to understand ourselves, in some ways, as athletes who are at various stages of learning, improving, maintaining and, perhaps for those lucky and gifted few, perfecting abilities and indeed ourselves. As we are initiated into the practice we become better, and we recognise objective standards of performance and appreciation; we come to learn 'where we stand' in relation to historical standards of excellence.

What brings the ideas of changing human nature and understanding sports as rule governed activities together? For my own part, at least one interesting answer to this question lies in the idea of how understanding and accepting, or breaking and rejecting rules demands a consideration of the roles that the concepts of limit or limitations play in our understanding of perfectionism. And that in turn begs questions as to the means by which perfectionism may admirably or ignobly be pursued.

10.5 The Idea of Perfection: Some Conceptual Remarks

It is more typical in the modern west to think of a life well lived in terms of the well-being experienced by the subjects themselves. An ancient way of looking at matters, which finds its home in some Greek philosophy, is to think of it in terms of the achievement of and striving for excellence or success. Even for a life understood as other than the achievement of subjective experiences of pleasure, satisfaction, or wellbeing there are different kinds of objectivists. The kind of objectivist account that I will touch upon here follows Aristotle classically, and authors like Phillipa Foot (2001) or Thomas Hurka (1993) more contemporaneously.[3] What unites these figures, separated in time by thousands of years, is the idea that in seeking a life of perfection, what is being perfected (what excellences are being sought) is to be understood against an understanding of what it is to be human, or what nature is to be found in human being. I will not delve in detail into their theoretically subtle positions; it is rather the general approach that will interest me here. Nevertheless, it is necessary to do at least some ground-clearing. I will do this partly by stipulation and partly by some preliminary and basic conceptual analysis.

[3] See Breivik (2010) for an application of Hurka's perfectionist thesis to sports.

In the philosophical literature a distinction is usually drawn between relative and absolute perfection. Relative perfection relates to some task or performance. One is to be evaluated as im/perfect in relation to some specified aspect, thing, role, performance, and so on. By contrast, absolute perfection relates to the whole of one's being. This latter understanding, particularly in Christian theology or ethics, is thought to be beyond the province of humankind, whose nature is fallen after the Original Sin of Adam and Eve. Only God, as they understand it, can be perfect in absolute terms. Let me make it clear that I am not interested in the idea of sporting gods, unless this is taken in a metaphorical way to explain the presence or absence of luck or chance and their role in the determination of winning and losing. Of course one might talk of perfect societies, such as Plato's Republic or Marx's socialist state, perhaps even the Kingdom of Heaven, but it is individual perfection that is the subject of my present concerns.

While there are many ways of thinking about the concept of perfection I wish to set out a crude dichotomy. As with many social, political and ethical concepts, one may attempt to define them negatively or positively. Consider, for example, the concept of freedom. We may define freedom either as the *absence* of external constraints (freedom from….) or the presence of conditions or abilities (freedom to….) (Berlin 1969). Perhaps a more widely utilised social concept that admits of negative and positive conceptualisation is health. When most students are asked to give a definition of health (a task I have undertaken with many cohorts over the years) the first and most typical retort is to negative definition. Health, they say, is being in a condition without illness, infirmity or disease. As a definition it is incomplete. It focuses on what health is not but does not specify precisely what it is. That is why we understand it as a negative definition. We should note that this definitional strategy is far from useless: health systems around the world went a long way to cultivating that personal and social good predicated precisely on this negative conceptualisation or definition of health.

We may readily ask: Is perfection the absence of flaws, or the complete presence of characteristics, without betterment. Let us agree, then, that we may too make some profit by considering perfection in negative terms. What does this amount to? Well, we may speak of perfection in conceptually negative terms in two ways. First, one may speak of it in terms of its being the absence of features that render it incomplete or, secondly, in terms of features that tarnish what is present. In either case perfection is understood as the absence of its opposite: imperfection(s). Again, while not entirely fruitful this mode of definition is not useless either.

I have said that we may speak of human perfection in terms of a whole or of parts. And I have rejected discussion of absolute human perfectibility by fiat. I will later reject it by rehearsing an argument of Aristotle's concerning human potentials. Less problematically, we may consider perfection in relation to a task or a role, or in relation to a person's entire being. Elision between these two aspects is, I suggest, one source of conceptual and ethical problems for the *telos* of sportsmen and women: their ultimate goal. Athletic perfection is to be understood, then, in relative terms.

10.6 Athletic Perfection

To talk of athletic perfection is already to situate perfection under a partial aspect. One concerns oneself with the perfection of the human *qua* athlete. Typically this might be thought of as "technical perfection" as Passmore (1970) described it. There are helpful and problematic aspects to this label that are worth attending to under this aspect. By calling such perfection "technical" Passmore draws attention to the scope of application: one is perfect in the execution of a task or a role. We speak easily and naturally of the perfect delivery in cricket; the perfect serve in tennis; the perfect 10.0 in gymnastics; the perfect 180 in darts; the perfect penalty in football [soccer] and so on. We distinguish between the good and the im/perfect with equal ease: if only the penalty had been placed 5 cm further to the left; if only the weightlifter could have held their balance for a split second longer with the bar above their head it would have been a world record; if only the ball had not hit the net cord it would have been a championship winning ace and so on. These are all technical failings. We can recognise their relative excellence if not their perfection. One may still win the gymnastics gold without scoring perfect 10s in every discipline. One can win the world cup without breaking world records or perfect performances. It is often said that some athletes, though not the fastest, are championship runners. They turn in their best performances when it counts: at the Olympics or the World Championships beating competitors who have better personal best times.

So perfection is not necessarily demanded in one's attempt to secure victory. Of course, not all failings are technical in kind. One problem, witnessed in sports ethics discussions, is to think of failings such as cheating or doping *as if it were* a merely technical failing. One has merely broken a rule. As we have seen not all rules are of a technical kind. And those regarding doping have an explicit ethical component since one of the three criteria used to ban certain means of performance enhancement is that they offend the normative criterion: the spirit of sport (World Anti Doping Code 2012). Independent, then, of the idea that an athlete may produce an apparently perfect performance, we understand easily enough the idea that the observed behaviour can be sullied or tarnished by the mode of preparation for their performance or the conditions under which they have performed.

The avoidance (or at least minimisation) of error or imperfection is very often a goal that coaches have. This idea of an error free performance is something that must be understood in a contextually sensitive way. It makes more sense in some sports than others. Best (1978) drew a distinction that is now a commonplace in the philosophy of sports between aesthetic and purposive sports. All sports are instrumental activities: the ends (what Bernard Suits (1978) has called the "pre-lusory goal") can be achieved by any means that do not contravene the rules. In aesthetic sports one cannot specify what is to be achieved, however, without specifying the means of achievement. In such sports the manner of achievement is fused with the criteria for success in those activities. The distinction is a good and useful one from an analytical perspective. When one considers more precisely the modes of praise and blame in sports talk we realise how the boundaries are not quite as neat when placed in the service of a discussion of perfectionism.

Take for example an international diver who is not world class. Nevertheless in the world championships s/he may elect (more likely their coach will tell them) to perform a dive with a low tariff, one which they can perform excellently but which will attract a lower score than one who performs a more complex higher tariff dive to the same degree of proficiency. Equally one can imagine a perfect defensive display from a football team that was overcome by a superlative goal, or a piece of outrageous luck, in which no single defender was at fault. Such performances may, while error- free, nevertheless be lacking in many aspects that we both admire and might expect in a perfect performance. Perhaps better, consider a golfer who scores a hole-in-one on every one of the 18 holes on the course. But in each stroke the ball either bounced or deflected off a tree or spectator or buggy before reaching its destination. We might think of the golfer proclaiming that they had scored the perfect round: 18. Most people, I suggest, would rather say that although the score could not be bettered, the round was less than perfect. What other things do these intuitions trade on over and above an error-free performance?

For my part, it seems that the best of all possible performances – the most perfect, if that makes any sense – refers to the optimisation or actualisation of some fuller specification of the activity. That is to say, one may give an error-free performance, or a performance that – through factors other than merit – is an excellent one but that conforms only to our negative conceptualisation. Conceptual space still remains open for performances that are better in aesthetic or ethical terms than mere error-free ones.[4]

10.7 The Transhuman Athlete As Perfect

If what motivates at least some transhumanists is the idea of transcending human vulnerability, ought we to think of the transhuman athlete as the perfect athlete? Might we not think of them (negatively) as the athlete who would not tire nor break down with injury, or (positively) who could concentrate more fully, who could strategise better by imagining/remembering all [possible] moves by their opponents in order to counter them, or who was faster or stronger, or who – in short – could outperform all (their merely human) competitors?

What kind of transformation is entailed in our thought experiment? Is it merely one where there is an incremental advance on our previous performances? Or, invincibility has our transhuman athlete assumed a new identity?[5] Can the transformation be that total? What motives might drive the sports engineer or biotechnologist in producing or preparing such a creature? Is it the case that with such athletic perfection they must

[4] While I will not expand upon it here, it seems that this problem is particularly apt for those sports which measure relative performances, especially those where there is a theoretically maximal/ minimal achievable limit as *inter alia* the case with golf or darts or snooker. More might be said of the structure of these games and the role that luck can play in them, but I shall not attempt it here.

[5] I assume here that the point of such transformation is a competitive advantage: after all, why go to the bother of radical transformations if they are available to all?

have transcended humanity itself? Or, to put it the other way round, must we transcend humanity in order to achieve sporting perfection? There is certainly something very odd about this idea that to become the greatest ever athlete- footballer, golfer, or whatever- we must leave behind the very nature-bound species that we wish to be seen as the apex of.[6]

Something like that seems to be the goal of the would-be transhuman athlete. Perhaps it could put in the following schematic way. What the transhumanist athlete seeks is a shift from relative perfection (understood negatively) to a positive and absolute conception.

Why should we accept or admire such a shift? First, note that the idea is far from new. Long before anyone invented the word 'transhumanism' or its apparent successor 'posthumanism', Francis Bacon and Rene Descartes had both spoken of relieving man of his estate; a condition of vulnerability to disease, misery and suffering (McKenny 1997). Secondly, is this, or at least something very like it, the idea that would drive the transhuman athlete? As Holm notes, transhumanism is not simply a *laissez faire* ideology in which a 1,000 flowers might bloom (despite the protestation of its libertarian spokesmen). Though it recognises and values highly individual freedoms to choose what Anders Sandberg (2001) calls 'morphological freedom', it embodies an implicit philosophical anthropology (Holm 2007).

Transhumanism incorporates a perfectionist dialogue, though its *telos* is unspecified. The idealised transhuman is characterised by: increase in intelligence; increase in certain perceptual qualities; new transhuman experiences; control of emotion; control of the body (including perfect health); and, last but not least, infinite life (Holm 2006). The Olympic motto, "citius, altius, fortius" (faster, higher, stronger) would taken on a whole new meaning with athletes biotechnologically engineered to remove all stains of human imperfection.

Of course for mere mortals, the idea of unlimited enhancement is unthinkable. As Aristotle argued, one cannot simply actualise all of one's potentials.[7] In his Aristotelian discussion of human potential, Scheffler (1991) notes that there are three myths of human potentials: human potentialities are neither fixed, nor harmonious, nor uniformly valuable. In keeping with these thoughts we should not expect enhancements to be equally valuable in the life of all or any particular agents. Nor should we expect that we may combine any or all enhancements that are desired by any particular agent, a point commonly made of the heterogeneity of value as well as desire. Finally, given our developmental nature we ought to expect that certain goods are time dependent (Slote 1983) and therefore that any enhancements are likely to differ in their desirability according to the role that the capacity plays in the lives of the enhanced at various points during the life course. The slogan one often hears then, that one is exhorted to "be all that you can be" is mission impossible. What makes it impossible to perfect oneself in this absolute way is our nature. Of course it is precisely this that the transhuman (athlete) seeks to transcend.

---

[6] Preface by "See McNamee and Edwards (2006)".

[7] See Metaphysics Book 5, chapter 4 and Book 7, chapter 4 (Aristotle (1953); Holm and McNamee (2011)).

10.8 Stories of Transcendence

Imagine that you are an elite athlete. First, I want you to consider the possibility of enhancing your performance considerably beyond your current level. Think of it as a magic bullet, that which anabolic steroids are surely not. Now I know there are problems with considering performance enhancing substances in this way; it is not as if they typically remove the need for effort. Often they are taken so as to enable greater effort, to train harder and recover quicker. But indulge me. What I am asking you to think is far short of transhuman transformation. Secondly, I don not want to consider whether you will make the decision in reference to the possibility of taking risks or of being caught by any doping control offices. Let us agree, hypothetically, that you cannot be caught. This then is not a strategic or instrumental risk: benefit question about whether it is rational to pursue your goal or ends through this means. Rather, thirdly, what I invite you to consider is the meaning of the victory/ies that are subsequent upon the transformation that you are given. In particular one must consider the extent to which the subsequent, Olympic gold winning performances are meaningful, all things considered.

There might be any number of ways of considering this problem. The sources of moral reflection and argumentation are heterogeneous. Nevertheless, for the purposes of this essay, since there are more analytic arguments to be found elsewhere (Brown 1992; McNamee 2008), I want to explore the extent to which rhetoric – in the form of a moral fable or myth – can inform wise judgements about the nature and limitations of our aspirations in and for sports.

One place to start is in that great repository of western wisdom, the Greek myths. We have something of an analogous idea in Homer's Odyssey. Returning home from the victorious siege and sacking of Troy, Odysseus's ship and all his men find themselves at the mercy of Neptune, shipwrecked in a storm. Not all of the Gods are against Odysseus though, and he finds himself sought out and rescued by the ageless and beautiful goddess Kalypso who has taken more than a shine to the cunning hero. Returning to an unspecified island haven she offers Odysseus a life, in contrast to the vagaries of the mortal one he has laboured in thus far, one of endless pleasure, without suffering or fatigue. This life of course stands in marked contrast to the vulnerabilities, struggles and eventual death that define the shape of a human life. Odysseus, as we all know, declines Kalypso's offer.[8] He chooses to continue his Odyssey home to return to Ithaca and his wife Penelope and their mutual and mortal life there.

Martha Nussbaum, in her essay "Transcending humanity", explores the idea that a mortal, if offered, might reasonably reject the life with a Goddess such as Kalypso. What he opts for is his quest for Penelope: "he is choosing, quite simply, what is his: his own history, the form of a human life and the possibilities of accidents, love, and achievement

[8] Not immediately, it must be noted: he enjoys several years of mortal pleasures before making his mind up.

that inhabit that form." (1990: 366). Nussbaum says that there is something elusive about the speech Homer puts in the mouth of the hero Odysseus:

> we don't quite know what it would be for this hero, known for his courage, craft, resourcefulness, and loyal love to enter in to a life in which coverage would atrophy, in which cunning and resourcefulness would have little point, since the risks with which they grapple would be removed, and in which love, insofar as it appears a toll, would be very different in shape from the love that connects man to wife and child in the human world of the poem. The very possibility makes one uneasy: for where, and who, in such a life, would our hero actually be? Do we wish for him a good result that involves a transformation so total that he might not remain himself? (1990: 366)

In a way what Odysseus chooses ultimately is his identity as a mortal. One might in his situation think as an utilitarian, weighing up pros and cons, risks and benefits. But this is not the type of reasoning he employs nor the kind of decision that he makes. His is not one about efficiency or effectiveness of means rather than ends; to the contrary it is about the nature of ends. What kind of life is the life of our mortal hero: well, mortal is the only answer he can bring himself to. He discards the opportunity to leave his own mortal frame. Does he have immortal longings?[9] Homer does not tell us. To consider a modern analogue: were one to be transformed to a robot who operated as an exceptional firefighter, I suspect that we would not ethically admire their performance. We would not consider their actions heroic. Why not? Well, what is it that the robot risks? Nothing. He cannot be said to risk anything since the very employment of that word in this context makes no sense for a machine without agency. What is missing are the possibilities of choice, fear, an understanding of life that might be saved, or risked, and so on. The differences are myriad. And just as they are myriad when we compare human with machine, is there any reason to suspect that we ought properly to adopt a different disposition towards transhuman athletes? Would any kind of adverbial vocabulary, such as we employ of virtuous and vicious, be apt? The nature and limits of human excellence shape the kinds of admiration we have for those who aim at perfection, humanly conceived. The framework gives meaning to the action and its evaluation.

I have taken Nussbaum's words that "Human limits structure the human excellences, and give excellent action its significance." (1990: 378) with considerable seriousness. I am neither unaware nor unsympathetic to the claim that it is all too easy to essentialise from this position: that we use some biomedically validated and positivistic notion of humanity to rule the roost. And that would be problematic, not least of all for disability sportsmen and women and their prosthetically influenced present and future. One need look no further than the unfortunate case of the challenged sexual identity of the African athlete Caster Semenya, to see how nature does not obey to our categories male: female in a simple dichotomous way. Equally, I take seriously the idea that human nature admits of considerable variations and

[9] It will be clear to readers that I have not taken the attractions of the human transcendence of humanity with an honest equanimity: *mea culpa*. For discussion that is more balanced and more nuanced in relation to Nussbaum see Kerr (1997: 1–22).

that species typical norms may help regarding health and classification of sporting events. I understand that functional classifications of competition classes predicated on species typical functioning are not without problems. Athletes with disabilities may offer interesting case studies to test our intuitions about the limits of sporting contests as shared by human athletes. Moreover, possibilities for human/animal chimeras – most readily through xenotransplantation – represent more radical and more problematic issues than do sports transformations. It is not beyond the realms of ethicists imagination, however, that we conceive of such tests in athletic milieus. After all, the ancient Romans developed a penchant for it, but whether such contests and contexts are to be thought of as sporting is another matter.[10] Nevertheless, I want to re-assert here: it is the framework that supplies the context for our evaluations of perfection. And the framework we have, and will always have, is a human one.

10.9 Concluding Remarks

Bernard Williams (1995) reminds us that 'humanity' is a name not merely for species but for a quality, a way of being in the world. That humanity, our nature and modes of being, structure what is both possible and admirable for us. Of course we know that the last few hundred years of western philosophy and culture – understood loosely as the enlightenment – have brought us to realise all too keenly that the terms "human", "natural" and "good" belong to conceptually distinct categories. Nevertheless, we can still look for the kinds of things that define the best of human being and living, however loose and contested as that may be. What we are left with as philosophers of sport, or biotechnology, or medical engineering, is the task of articulating human excellences, athletic, moral, technical, theoretical and so on. And their pathologies of course.

Where perfectionism in sport is understood in negative terms within a relative framework, it is *ceteris paribus* unproblematic. Under such a description it will typically be seen as the eradication of technical error or human failing. Where it is viewed as transcending human nature I take it to be pathological; removing human limits from sports is tantamount to removing the framework that gives it sense and meaning. While some may be attracted to watching robots or cyborgs – or indeed transhumans whatever they may come to look like – in contests, I am not. Nor shall I honour that movement with the appellation "athletes". Theirs will be transformation of humanity, not a transcendence of it. And such a desire, as I have tentatively argued here, embodies a pathological conception of perfectionism.

[10] I say this, mindful of the fact that the South African rugby player, Brian Habana, has already performed in a race with a cheetah. See http://www.youtube.com/watch?v=7_iAQXyDJns. Last accessed 5.3.12.

References

Aristotle. 1953. Metaphysics, Book 5 Ch 4; Book 7 Ch 4 (Trans W.D. Ross). Oxford: Clarendon Press

Berlin, I. 1969. *Four essays on liberty*. Oxford: Oxford University Press.

Best, D. 1978. *Philosophy and human movement*. London: George Allen & Unwin.

Breivik, G. 2010. Philosophical perfectionism – Consequences and implications for sport. *Sport, Ethics and Philosophy* 4(1): 87–105.

Brown, W.M. 1992. Practices and prudence. *Journal of the Philosophy of Sport* XVII: 71–84.

Daniels, N. 2009. Can anyone really be talking about ethically modifying human nature. In *Enhancing human capacities*, ed. N. Bostrom and J. Savulescu, 25–43. Oxford: Oxford University Press.

Foot, P. 2001. *Natural goodness*. Oxford: Clarendon.

Genetic Technologies. 2009. *Sports performance ACTN3 Sports Gene Test_2006* [online]. Available from URL: http://www.gtg.com.au/archives/migration/2/110/383/ACTN3%20web%20brochure.pdf. Last accessed 27 Feb 2009.

Heidegger, M. 1993. *The question concerning technology*. New York: Harper Collins.

Holm, S. 2006. The nature of human welfare. In *Future perfect?* ed. P.M. Scott and C. Deane-Drummond. London: T & T Clark.

Holm, S. 2007. Naturalness and anthropology in modern bioethics, with a special view to trans- and post-humanism. In *Theology and science – Issues for future dialogue*, ed. Kragh Helge. Aarhus: University of Aarhus Press.

Holm, S., and M.J. McNamee. 2011. Physical enhancement: What baseline, whose judgement? In *Enhancing human capabilities*, ed. R.J.H. ter Meulen, G. Kahane, and J. Savulescu, 291–303. Oxford: Oxford University Press.

Hurka, T. 1993. *Perfectionism*. Oxford: Oxford University Press.

Juengst, E., et al. 2009. What's taxonomy got to do with it? 'Species integrity', human rights and science policy. In *Human enhancement*, ed. J. Savulescu and N. Bostrom, 43–58. Oxford: Oxford University Press.

Kerr, F. 1997. *Immortal longings*. London: SPCK.

McKenny, G. 1997. *To relieve the human condition*. Brockport: SUNY.

McNamee, M.J. 1995. Sporting practices, institutions and virtues: A critique and a restatement. *Journal of the Philosophy of Sport* XXII: 61–83.

McNamee, M.J. 2007. Whose Prometheus?: Transhumanism, biotechnology and the moral topography of sports medicine. *Sport, Ethics and Philosophy* 1(2): 181–194.

McNamee, M.J. 2008. *Sports, virtues and vices*. Abingdon: Routledge.

McNamee, M.J., and S.D. Edwards. 2006. Transhumanism, medical technology, and slippery slopes. *Journal of Medical Ethics* 32: 513–518.

Miah, A. 2003. *Genetically modified athletes*. Abingdon: Routledge.

Miah, A. 2011. *Will London be the first transhuman games?* http://www.andymiah.net/2011/01/09/2012-the-first-transhuman-games/. Accessed 18 Mar 2011.

Midgley, G.C.J. 1959. Linguistic rules. *Proceedings of the Aristotelian Society* 1960: 271–290.

Morgan, W.P. 1993. *Leftist theories of sports*. Chicago: University of Chicago Press.

Nussbaum, M.C. 1990. *Love's knowledge*. Oxford: Oxford University Press.

Passmore, J.A. 1970. *The perfectibility of man*. London: Duckworth.

Pitsiladis, Y., and G. Wang. 2011. Necessary advances in exercise genomics and likely pitfalls. *Journal of Applied Physiology* 110: 1150–1151.

Reddiford, G. 1981. Morality and the games player. *Physical Education Review* 4(1): 8–16.

Sandberg, A. 2001. *Morphological freedom: Why we not just want it, but need it*. http://www.nada.kth.se/~asa/Texts/MorphologicalFreedom.htm. Last accessed 17 Feb 2012.

Sansome, D. 1992. *Greek athletics and the genesis of sports*. Berkeley: University of California Press.
Scheffler, I. 1991. *In praise of the cognitive emotions*. London: Routledge.
Slote, M.D. 1983. *Goods and virtues*. Oxford: Clarendon.
Suits, B. 1978. *The grasshopper: Games, life and utopia*. Toronto: University of Toronto Press.
van McNamee, M.J., I. Hilvoorde, A. Mueller, and S. Holm. 2009. Genetic testing and sports medicine ethics. *Sports Medicine* 39(5): 339–344.
Williams, B.A.O. 1995. *Making sense of humanity*. Cambridge: Cambridge University Press.
World Anti Doping Code. 2012. www.wada-ama.org/en/world-anti-doping-program/sports-and-anti-doping-organiszations/the-code/. Accessed 5 Mar 2012.

Part IV
Socio-Cultural and Empirical Approaches

Chapter 11
"Definitely Not for Women": An Online Community's Reflections on Women's Use of Performance Enhancing Drugs in Recreational Sports

Marianne Raakilde Jespersen

11.1 Introduction

On the face of it, doping seems to be the most objectionable version of athletic enhancement. It evokes intense reactions and the consequences, if use is detected, seem far more severe than for other means of improving performances, such as for example Fastskin swimsuits (Magdalinski 2009). Enhancement from the inside not only triggers a sense of injustice and moral panic, but also raises fundamental concerns about human nature.

When considering the use of performance enhancing drugs in recreational sports, there is, for good reasons, a tendency to focus on young men's use of androgenic anabolic steroids (AAS). We tend, however, to forget that other segments of the population might use drugs to enhance their performance when exercising too. We especially tend to forget that women might use drugs.

Very few women, primarily elite body-builders (Evans-Brown and McVeigh 2009), use AAS and for obvious reasons: the use of AAS has an irreversible masculizing effect on the female body. Numerous tragic stories from former East Germany show that this is not a desirable effect (Franke and Berendok 1997). But since women, at least Danish women (Nielsen et al. 2002), are known to be more likely than men to medicate on an everyday basis, it would be surprising if the situation was completely different when it comes to performance enhancement. Some surveys suggest that men in general are more willing to use (performance) enhancing drugs than women (Møldrup and Hansen 2006; Breivik et al. 2009). Another possible explanation could

The article is based upon a study of Danish women's use of performance enhancing drugs. The study is made possible by a grant from The Danish Ministry of Culture Committee on Sports Research (J.nr. TKIF 2010–045)

M. Raakilde Jespersen (✉)
Sport Science Section, Department of Public Health, Aarhus University,
Dalgas Avenue 4, 8000 Aarhus C, Denmark
e-mail: marianne@sport.au.dk

J. Tolleneer et al. (eds.), *Athletic Enhancement, Human Nature and Ethics*, International Library of Ethics, Law, and the New Medicine 52, DOI 10.1007/978-94-007-5101-9_11,

be that women might tend to use other types of drugs (for example ephedrine, human growth hormone, human chorionic gonadotropin (HCG), clenbuterol) that less often figure in surveys, or are ignored due to the focus on AAS. Searching one of the oldest and largest Danish internet fora for exercise and training (www.bodyhouse.nu) reveals that women do in fact have an interest in and do use performance enhancing drugs. Furthermore it becomes apparent that women's use of such drugs in particular ignites discussions and reflections upon human nature itself. It seems, therefore, that paradoxically women's use of performance enhancing drugs, although often neglected, is regarded as much more serious when discussed in relation to human nature, than men's use of such substances.

11.2 Culture of Optimization

The use of substances to obtain a desirable effect is neither new nor just an issue for sport, it is a universal phenomenon: "Drug cultures are endlessly varied, but drugs in general are more or less ubiquitous among our species." (Jay 2011: 10). At any time in history and in every geographical location people have been using different kinds of drugs, prepared in various ways, to get a specific experience, be it stimulating, strengthening, calming, healing, euphoric, spiritual, or enhancing. Drugs are used in rituals and in everyday life, medically as well as recreationally, and they figure on Donald E. Brown's list of human universals as "medicine" and "mood- and consciousness-altering techniques and/or substances" (Brown 1991; quoted from Pinker 2002: 437). The pursuit of elixirs and remedies which might secure youth, eternal life etc. is a recurring theme in human cultural history, as is the search for aphrodisiacs, and the even more pragmatic use of herbs, leaves and roots to get through daily life. There are multiple examples of the ancient Greeks and Egyptians, the Roman gladiators, Vikings, and medieval knights, taking stimulating drugs before feats or battles – or in order to be able to fight at all, when injured (Verroken 2003). Using drugs in order to do something exceptional is old news.

It could be argued, however, that in twenty-first century Western society the exceptional situation has in many ways become the norm. Modern Western life is, in general, characterised by an imperative to perform. *Citius, altius, fortius* is no longer limited to the world of sports but is, at least metaphorically, characteristic of modern life from cradle to grave, professionally as well as personally. This imperative to perform and the explicit, articulated, as well as the internalized, urge to evolve, expand and enhance has, since the beginning of the eighteenth century, been encouraged politically and culturally. In this *culture of optimization* we celebrate initiatives meant to increase the outcome and to make processes and productions more effective.

In some ways this is just another example of the industrial mentality that has characterized the modern period, but in its contemporary form the emphasis on optimization and the use of buzzwords like *lean* obfuscate the actuality – the retrenchments and reductions. The initiatives have an air of being for the better and the end results are often presented through a positive bodily metaphor such as fitter or trimmed. Optimization has become a goal on the personal level as well: We strive to be better persons, husbands, mothers, families, friends, colleagues, etc., to be healthier, smarter, more loving, caring, giving, to be our true selves and fulfil our potential, etc. The eighteenth century project of *Bildung* has permeated every level of our society and our lives. And all through this period the body has been central to the definition and articulation of the individual. While the process of becoming a subject is increasingly turned into a project, as Anthony Giddens (1991) has phrased, the body is playing an ever-greater part in this project, as, among others, Chris Shilling (2005, 2006, 2008) has demonstrated. We stand however, Frank Furedi argues, at a point of "radical redefinition of what a human being is" (Furedi 2011). Developments in biomedicine and technology point the way to the day we might redefine what it is to be human. Human enhancement gives individuals a very concrete possibility of creating significant parts of themselves. The body stops being a given that you have to live with. Instead you are free to change its shape and abilities – according to your wishes and ambitions.

One road to such enhancement is via the use of drugs. But even though drug use is universal (as discussed above), and even though we are encouraged to optimize and maybe even feel an urge to develop continually, not all kinds of biotechniques and drugs are equally acceptable socially, culturally or ethically (Jay 2011: 34). Some drugs, for instance, tend to generate panic and initiate moral crusades while others are used without anyone noticing. This difference can partly be ascribed to medicalization (Conrad 2007), partly to cultural norms defining legitimate use of drugs. Furthermore, as Coomber has pointed out, when looking at drug use in general, it is notable that the users differ immensely (Coomber 2007). Take, for instance, the population of performance enhancing drug users in sport. On the one hand there are the elite athletes who, despite differing sporting abilities, are relatively homogeneous in their desire to win at any cost – at least theoretically. On the other hand there is the large group of recreational athletes who come from diverse backgrounds and have a wide range of desires and motivations. While the use of drugs to enhance might be seen as an example of a universal element of human culture, and as, in John Hoberman's words, "a basic characteristic of modern civilization" (Hoberman 2005: 21), the acceptance of a specific kind of use changes according to history and geography. We must bear in mind that even human universals such as drug use and what Savulescu, Foddy and Clayton describe as 'mankind's' inherent need to get better ("To choose to be better, is to be human" – Savulescu et al. 2004: 670), may be significantly influenced by tradition and culture, as are decisions about what is considered acceptable in terms of enhancement and what is viewed as 'unnatural'.

11.3 Performance Enhancing Drugs in Recreational Sport

Sport offers numerous examples of this inseparable relationship between nature and culture and often presents these issues in a radicalized way, ahead of the rest of society. While elite athletes in general keep quiet about using performance enhancing drugs, recreational athletes seem more willing to share experiences, good advices and reflections, for instance in online communities on the Internet. These fora provide us with an insight into contemporary reflections on body, gender, sexual relations, performance, self-image etc., and of special interest here: reflections upon human nature in relation to athletic enhancement.

Furthermore, when reading these internet fora, and especially the reactions to instances of performance enhancing drug use, two minorities stand out: Women and young people. In Western cultural history women have traditionally been associated symbolically with body and nature and do as a category provide a privileged site for reflections upon and negotiations of human nature. Women's use of performance enhancing drugs has however, to date, often evaded such reflection.

Performance enhancing drugs such as androgenic anabolic steroids (AAS), growth hormone, clenbuterol and (human) chorionic gonadotrophin have been used in elite sport for the past 60 years in order to enhance athletic performance and give the athlete a competitive advantage.[1] Today the use of these drugs seems, however, to have diffused into the general population, where they are used in order to optimize the physical appearance as well as to enhance performance, especially in those sports where strength is important (Christiansen 2009; Mogensen 2004; Bach 2005). There have been several surveys in Denmark that have looked at the use of AAS among recreational athletes, and the general conclusion is that these drugs are used primarily by young men. (Kulturministeriet 1999; Mogensen 2004; Bach 2005; Singhammer and Ibsen 2010; Anti Doping Danmark 2004, 2009).

The use of AAS is associated with a variety of side effects (Porta 2008). Due to methodological, ethical and logistical complications in administering doses corresponding to the large doses taken by users, it is, however, not possible to use randomised clinical trials to investigate the (side) effects of steroids used for enhancement. A lot of the data on side effects are therefore based on case reports, observations and self-reporting. Self reported side effects include acne, loss of hair and, for men, gynecomastia, i.e. abnormal development of large mammary glands resulting in breast enlargement. More serious side effects are irreversible masculinization of the body (for women), permanent testicular atrophy (for men), and damage to the liver, heart, circulation, muscles and tendons. Furthermore, the use of AAS has been associated with depression, aggression and suicide (after use has ended) (Evans-Brown and McVeigh 2009; Hartgens and Kuipers 2004; Thiblin and Petersson 2004).

11.4 Internet Fora as Indicators

In Denmark at least, the focus of legislation, doping control, research and media attention in relation to the use of drugs in recreational sports has so far primarily been on young men and AAS. Searching the Danish Internet forum www.bodyhouse.nu,

[1] In the following section on androgenic anabolic steroids I am greatly indebted to Andreas Kimergård who has kindly shared his knowledge.

however, reveals that it is not only young men who have an interest in these drugs – even if this profile does predominate. In the following I want to use this forum as a gateway to knowledge about recreational athletes' use of performance enhancing drugs, which in this forum is discussed openly. This makes it an important source of information, in spite of the limitation that most members are male and are into fitness or bodybuilding, and thus not representative of the total group of recreational athletes in Denmark.

Using Internet fora as a source for research has some methodological implications. In relation to the present study it is important to be aware that (1) 'Answers' are limited by the wording of the search – either on key words or in topics; (2) Posts may be placed in an unexpected thread, meaning that important posts may be missed; (3) Due to the amount of and continuous piling up of posts, it is not feasible to read all posts. As in qualitative interviews it is necessary to judge when a point of theoretical saturation has been reached and new posts do not bring new, additional information (Gillham 2005); (4) Posts and profiles are written representations that might be fictional. Though this is problematic in relation to the individual information on gender, height, weight, use of hormones etc. (demographical information), the representations still offer an insight into reflections, values, norms etc.; (5) Due to missing and unreliable demographic information posts on these fora cannot be regarded as representative but only indicative.

This study is, in other words, not representative, but it can give an insight into a particular group of people's reflections on the use of performance enhancing drugs in relation to the body and to human nature. In my close reading of posts, I do not deal with the correctness of stated facts, or with real risks etc., in order not to be censorious in my approach and due to uncertainty concerning what the drugs taken actually contain. I will concentrate exclusively on the notions, conceptions and ideas expressed by the members in their posts. I have, furthermore, chosen not to become a member of the community myself in order to remain personally detached. Although the information on www.bodyhouse.nu is publicly accessible, I have, in line with Wilkinson and Thelwall (2011), chosen to maintain the informants' anonymity, since my interpretations of their comments could be experienced as a violation of their privacy.

I have limited my research on www.bodyhouse.nu to posts concerning women in the subfora "Doping and hormones" and "Training and strength". In the first subforum I have limited my search to the following topics "Weightloss (fatburners)" "HGH, IGF-1, Insulin" and "Anabolic steroids". In the second subforum my search has been limited to the topic "Body log". I have been focusing especially on posts with subjects containing gender-specific words like "Women + efedrin – side effects?" or "Clen for girls?? Stupid??", or those written by members with a name signifying gender like "Fitness queen", even though these names might be misleading. For each selected post I have traced the members' profiles and looked at their earlier posts. I have generally read through posts 1 year back (to 2010), but have followed threads back to earlier than 2010 when they concerned women's use of drugs. My material consists of 59 threads and 11 profiles.

The performance enhancing drugs mentioned in this material in relation to women are a combination of stimulants (often used as fatburners), e.g. ephedrine, caffeine,

ECA (=Ephedrine+Caffeine+Aspirin); regular weight loss products, e.g. Letigen,[2] Slimex; hormones, e.g. human growth hormone (HGH), human chorionic gonadotrophin (HCG), thyroid hormones (T3: Triiodothyronine; and T4: Thyroxin), Levitoron, Clenbuterol; and AAS, e.g. brand names like Rexobol, Anavar, Winstrol, Deca.[3]

11.5 Asking for Advice

The website forum is structured in different themes, called subfora, and topics. When you mail a comment, question, etc., you choose the topic under which you wish to place your post. The website codex forbids the sale of drugs and the advertising of specific products. The members are primarily male. Female members often specify their sex in their username, e.g. "Fitness Queen" or "The hcg girl", in the subject, e.g. "T3 (Thyro 3) help to dosing for a woman on 25" (F, 24.12.10; # 6), or in the posts, e.g. "I am a girl of 22, have been training for 4 years and would like a little more definition" (F, 06.10.11; # 8). Being female is clearly not common in this community.

Several men place posts on behalf of their girlfriends. In their accounts the women are dissatisfied with their bodies, they want to lose 5–10 kilos, be trimmed and gain more defined muscles, and they want drugs to speed up the process and keep up motivation. The men often emphasize that their girlfriends do eat proper food and do exercise, as if to state, almost automatically, that they (the men) know that these are the factors it all comes down to in the end. In general members ask for advice on drugs, often in combination with diets, or if taken along with for example contraceptive pills. Some of them ask out of sheer lack of knowledge or lack of personal experience with these drugs, others because they presuppose that the drugs affect men and women differently in terms of doses, side effects etc. Their notion of gender difference is, in other words, based on the biological body:

> Is it [ephedrine] more harmful for girls than for boys? It is for my girlfriend…! (M, 04.01.10; # 24)[4]

---

[2] Letigen was a medical weight loss product made by the Danish pharmaceutical firm Nycomed. It contained *inter alia* ephedrine and caffeine and was withdrawn in 2002 due to suspicion that it could cause thrombosis in the brain and heart, manic psychosis, cerebral haemorrhage and sudden death (Nielsen 2002; Carlsen 2002).

[3] There are references to Melanotan (also known as the "Barbie drug" due to its supposed effect: Slimming, tanning and increasing sexual libido) as well, but specific posts about Melanotan are in a separate thread. I have chosen to focus on topics on drugs known and used for their performance enhancing effect and have, therefore, not included melanotan, even though it could be argued that it is just as enhancing as for example fatburners, at least cosmetically.

[4] All quotations concern women's use of performance enhancing drugs. Quotations are made anonymous. Each thread is given a number (#x), accompanied by the period of time, the posts span over. In every quotation the contributors are numbered, if more than one (*x). If they reply more than once in a single quotation, they appear with the same number. Numbering the contributors blurs any sign of gender. I have marked the assumed gender (M – F) based on an interpretation of

> T3 is definitely not for women!!! :0 :0 a woman's pancreas [sic] is much more sensitive than a man's, and T3 may cause great harm (*1, M)
> [...] But take it easy at the start, women are much more sensitive than we men ☺ (*2, M) (30.01.03; # 42)

The repeated assertions of the biological differences between male and female bodies ensure, according to Tara Magdalinski, "that gender appears to be rooted in the organism and is not merely constructed through culture." (Magdalinski 2009: 94). The reflections and statements in the quotations above are based on a fundamental understanding of female biology as being more sensitive than male biology. The interpretation of biological differences is clearly gendered as the word "sensitive" shows. The emphasis on sensitivity and frailty is, however, not new in relation to women and drugs. In his book *Dope Girls*, Marek Kohn notes that in the 1920s "the idea of frailty recurred again and again in representations of the 'dope girls'." (Kohn 2001: 8–9). Kohn, furthermore, suggests that concern for drug use might be a way of articulating an otherwise not legitimate concern for a particular group (Kohn 2001: 8). Hence, stressing sensitivity and frailty in association with drugs and female biology in the postings may be a way of articulating a concern for women that in today's Western gender discourse is not legitimate, and as such it may be a strategy of confirming traditional gender positions.

Diets and drugs are, in the Bodyhouse community, used as a short cut to obtain an ideal body that would otherwise "be close to unobtainable with an ordinary healthy and varied diet" (F, 28.01.10; # 45) – "no matter how much we're training" (F, 14.01.11; # 32). The female members express a wish to sculpt the body in a certain way and their motivation seems largely cosmetic, rather than health related or linked to a need for a competitive athletic advantage, even though they report training several times a week. Some of the male members ask for supportive knowledge in order to prevent their girlfriends from starting drug diets:

> my girlfriend is about to start on T3 ☹ don't ask me what a fucked up idea she has got... [...] as her boyfriend I would like to be the one to tell her about it (M, 28.09.06; # 2)

It is not unusual for the male members to refer to their girlfriends' earlier experience with drugs like ephedrine, ECA, clenbuterol, HCG or T3, or to their own experience with these of drugs. This indicates that performance enhancing drugs are not unusual in Danish fitness milieus and that this acceptance of drugs diffuses from men who are

their Danish username – this entails, as mentioned, a risk of misinterpretation. Furthermore, translations of quotations are mine. The quotations are translated culturally from Danish to English, leaving out misspellings, abbreviations, slang etc. in order to communicate their semantic content and present them in a readable way. This, however, removes important markers of style. The case thus might read: (*3[number of respondent], M[ale], 10.11.11 [date of post/period of correspondence: DD.MM.YY]; # 65 [number of thread]). In quotations I have used (), when the brackets apperar in the original text, and [] when I have added the brackets to explain something. References following the quotations are in ().

training to their wives, girlfriends etc. who might not be training. This diffusion seems to be well known in the Bodyhouse-community and is mentioned explicitly as the responsibility of the man:

> As her boyfriend you should keep her away from any drugs at all when it is just a teenage idea. She just wants to lose 10 kilos and then she will be happy again? (*1, M)
> [...] Stop her before it is too late (*1, M)
> [...] T3 for girls is simply shit. It is not at all comparable to men using it! (*2, M)
> [...] She might have started...That is the first mistake since she probably got the idea of taking pills because you are into it...Second mistake is that you have accepted her starting on T3...Especially when it is as 'dangerous' for women as it is...Third mistake is that you haven't searched properly enough in here in threads on T3 and women...[...] I know what I would do.. (*3, M)
> (02.10.06; # 2)

11.6 Semi-science, Slang and Locker Room Rhetorics

The culture of the community is stereotypically masculine. The focus is on performance, strength and power in the traditional shape of muscles, the ability to lift heavier weights and to endure tough diets or cycles, i.e. taking drugs for a specific period of time. The pivotal point is strength of character and will power. Whimpering is definitely not accepted, neither for men nor women:

> Listen sweetie, if you were a guy you would have been dragged through the mud already and the thread would have been deleted. There are no easy solutions in this universe and if you are not at least dedicated you just have to live with being fat/thin/whatever (M, 12.01.11, # 33)

What counts is, as one of the male members phrases it: "will, determination, engagement and a strong psyche" (M, 28.01.10, # 45). Members are encouraged to fight for their goals, to make an effort, sacrifice themselves, and not just rely on drugs:

> I just want to get rid of it! I really don't care how (*1, F)
> What kind of an attitude is that, lazy bitch, sorry about my language, but come on. No drugs will give you any results unless you sacrifice yourself for it (*2, M)
> (13.01.11, # 33)

Giving up, whining etc. is explicitly defined by some members as not masculine, or rather as typically feminine:

> is your problem as for a lot of other bitches a lack of will power and no backbone ☺ If I am not mistaken it is just an excuse for using illicit subtances to lose weight [...] (M, 28.01.10; # 45)

The idea that women, in general, are weaker and have no strength of character, functions as a confirmation of a stereotypical masculinity. Female members, however, reply that men give up as well:

> [...] how hard can it be to lose weight?? You don't even have to think about losing muscle mass...jesus.. ☹ (*1, M)
> Well but it is obviously pretty hard as you can see...and if you read some of the bloke's blogs in here it is obviously hard for you guys as well, I have by now read about many who after a few days had to stop because it was getting too hard for them... (*2, F)
> (21.02.10; # 45)

As this quotation shows, the rhetoric has a tendency to be harsh and direct. Furthermore, it often contains sexual connotations and comments, which is enforced by the small pictures that accompany each member's username on the posts. When not cartoonish, these images are in general pictures of body builders, or undressed women in sexual positions (for men). Many members have signatures following their posts. These signatures are statements that confirm the identity of the member and signal attitude, character, and personality.

The Bodyhouse community is reinforced by specific slang and technical jargon related to body-building, fitness and drug use. The posts are rhetorically a combination of locker room small talk and semi scientific discourse. Many of the posts contain questions about and advice on what to do in order to reach a specific goal related to for example weight loss, muscle building and definition etc., which food supplies or drugs to use, and how to use them. Descriptions of diets, instructions on training, on the use of drugs, doses etc. are heavily loaded with technical jargon, numbers (e.g. mg, percentage of fat) and often also references to presumably valid sources of information. The information is, in general, of a variable quality, and the validity of it is often discussed, but in spite of rhetorical shortcomings the semi-scientific discourse gives it an authoritative and reliable air, which then influences advice and 'truths' of a more hearsay character. The community is ascribed an authority which is constantly confirmed by new as well as old members (both male and female):

> Made a profile in here, since I am brand new in this world! I have read many of the topics and am quite sure that I can get the best advice from you! (F, 05.01.08; # 55)

> BH [BodyHouse] [is] still a huge store of knowledge about these subjects, based on self studies, education and experience. In my opinion invaluable at times, even though we all hate each other. ☺ (M, 13.01.11; # 33)

A lot of the members, men as well as women, share their subjective and often very private experiences either in answers to others' questions, in discussions, or in logs used to minutely register their advances, back steps, doses, diets, reflections etc. The posts often have a confessional character:

> YES!!! Finally on the third day things got going – haven't been to the toilet for 3 days, well sorry about my frankness.... (F, 03.02.10; # 45)

> Friday night I slipped...two pieces of bread were suddenly the most delicious I had ever seen...stopped at that...until Saturday...when an evil fight inside of me started full time... and I hadn't the willpower to stand up against it...on that day I ate all the yucky stuff. Almost like 'el loading day' once more...(though not that bad)...But I got really sick and was THAT close to puke in the bus...Lesson learned...? Well you could say that... Next day my weight clearly showed it...gaining 1.4 kilos!!!! But then I went cold turkey on Sunday and made ready to start afresh again... (F, 09.02.02, # 45)

The confessional mode of telling is a private genre dedicated to sharing personal experiences and what is regarded as private and secret – in short *everything*. Since the eighteenth century, the most private has in general been associated with the body – often in moments of sexual arousal, sickness, pain, desire etc. (Brooks 1993: 33). In the Bodyhouse community bodily experiences are regarded as individual (e.g. it is "different how we react to things [drugs]", F, 29.11.09; #1) and personal side effects are reported:

> But I have never felt so shitty, as I did the first three days [on a clenbuterol diet]:-(ugh – nausea, headache, stomach ache, burning hot. No cramps – luckily (: I keep up the spirits and do whatever I can to make it work out well (F, 08.05.11; #13)

Articulating these experiences verbally – translating them into words – has in modernity been a preferred and meaningful way of transforming the individual into a subject (Linke 1996). Through the rhetorics – of confessing, writing and sharing your bodily experiences – you become a subject. Furthermore, the detailing of diets, side effects, emotions etc. places the member in a self-imposed system of control and regulation, which is a derivate of the modern medical 'gaze' (Foucault 1973, 1976). Reporting a weight loss might, for instance, elicit respect from other members.

11.7 The End Justifies the Means

Many posts, from both male and female members, display concern about counterfeit drugs:

> Something tells me you have been cheated once again. You can feel Clen almost the moment you have taken it. (*1, M)
>
> bloody hell you can feel clen an hour after taking the shit…you get an amphetamine like effect, muscles are all flexing and you're just lying there feeling hot. (*2, M)
>
> (06.-13.02.10; # 4)

The personal bodily experiences are often used to determine whether the drug taken is counterfeit or not. The body is ascribed a certain authority and (side) effects are the pivotal point. The effect has to be explicit and preferably almost instantaneous:

> I have just started on an ephedrine cure. My pills are only 5 mg, so I have taken 3 this morning, but feel no difference…[…] (*1, F)
>
> Which brand is it? 5 mg is bloody little. (*2, M) […] are they fake…?? (*1, F)
>
> […] Don't know. I have only seen them in 25 mg, 30 mg and 50 mg. What do you pay apiece? (*2, M) […] Hmm….can't really feel anything…maybe a little…maybe just fooling myself… hmm…I will grab 4 (20 mg) before boxing today and see if I can feel anything. (*1, F)
>
> Try taking 4 now and then you MUST bloody feel it. (*2, M)
>
> […] just don't get it, if you say there are no 5 mg pills…what am I eating then…?? (*1, F)
>
> Upload a picture of your pills? (*3, M)
>
> […] Ephedrin Atom are small white tablets with 'atom' stamped onto them, an English product. They come in 30 mg and you will NOT be in doubt that they're working if you have taken 4 of them at a time. As far as I know Ephedrin is only made as low as 8 mg […] (*4, M)
>
> (9.-12.12.09; # 14)

The decision whether a drug is counterfeit or not is based on several signs: the bodily effect of taking the pill, the physical appearance of the pill, and the strength of it. These parameters are, however, often not brought up until a bodily effect is not experienced – that is *after* the person has used the drug. This applies to both male and female members. Taking the risk seems to be part of the game.

Reflections upon counterfeit drugs support a general tendency in this community to make a kind of 'cost-benefit' analysis and reveal a willingness to take risks as long as the end justifies the means. In a thread a member asks about weight loss drugs on behalf of his girlfriend who is dissatisfied with her body and the respondents suggest ephedrine and Therma Power:

> [...] Take 1–2 30–60 minutes before cardio training, begin with 1 they are pretty strong for a beginner, and you might feel excited, in other words you might be feeling a little light headed and shaky in your hands but it will pass, it is not as bad as it sounds, it goes for ephedrine as well by the way... (M, 30.01.03; # 42)

Unpleasant – and serious – side effects are endured, accepted and normalized:

> Okay, the first day went all wrong. Mildly speaking!! After injecting, with the needle still half way in (couldn't press it any further!?) I had to run for the toilet, and then it came out of both my ends, everything went black and in 30 seconds my whole body was covered in sweat, it ran down my body, at the same time I was freezing and couldn't get warm again. I took the day off, went back to bed and awoke with the worst headache ever, worse than the worst hangovers! [...] Then I decided to inject in my tummy since the needle will probably get in easier, and if the same happens I WILL SELL THE SHIT! But the second day all went well, I have just done it five minutes ago, I pinched my tummy and there were no problems at all – THANK GOD! (F, 19.09.11; # 58)

Risks are explicitly weighed up and each 'case' is regarded in relation to person and drug. The risk assessment is, however, not neutral:

> In my opinion it depends on the person. [...] What is wrong in principle in this situation is that your first thought shouldn't be to take metabolic hormones [...] Had you written 'I have a girlfriend who wants to cut a little extra before going to her 3. BB-competition' no one in here would have responded in a harsh way. [...] In principle I think it is fucked up to use juice but that doesn't mean I haven't thought about it myself... But you shouldn't go looking in your medicine cupboard when you want to change your body unless you have spent at least x (with this I mean 5) years on doing it in a natural way. For too many body builders having the right look, being strong etc. comes to mean so much that their risk assessment, from 'a rational point of view', goes by the board. (M, 05.-11.02.03; # 42)

The estimation of risk depends upon the user's gender and seriousness. Using drugs to advance in your sport or to gain a competitive advantage seems in this quoted post (and others) to be acceptable because it is a sign of being serious, while using drugs as a short cut is deemed irresponsible and unacceptable.

Women in particular are often advised not to use drugs:

> In my opinion women should keep away from hard drugs, everything harder than ordinary ephedrine is no go. Sometimes the woman knows what she is doing, of course [...] (M, 15.01.05; # 44)

It seems implied that women in general do not know what they are doing, at least concerning drugs. Furthermore, while several of the members articulate the problem of using drugs in order to obtain a desired goal, the critique is often addressed at women:

> [I] think that it is a sad turn if people have to do illicit drugs just to gain a little...another example is the melanotan-thread, where 16 year old chicks are running melanotan just to get a little tan... what next? Can see why you go on juice if you want guns 50+, but the other way, that's bullshit.. (M, 23.02.10; # 45)

Obviously some drugs – and reasons for using them – are acceptable and legitimate, while others are discredited. A young woman is advised not to use ephedrine – and mocked – but responds:

> But if it is wrong to do eph, how come it is ok to do AS? Eph is probably not the most dangerous... (F, 22.10.02, # 43)

As the quotation indicates, focus on AAS seems to have influenced the general conception of performance enhancing drugs. In comparison with AAS, ephedrine is perceived here as less risky to use. Several female members challenge this gender difference in drug use and emphasize that it is their own choice and their bodies, not to be controlled by others.

11.8 Respecting the Body

Both male and female members report constantly switching from one method or product to a new one. Bad experiences, no results, or advice not to use a certain drug push them forward in their pursuit of a solution:

> What about ephedrine? Would it turn up burning during training? [...] Would I be totally hopeless if I try ephedrine during training? Or maybe just stacker, which I have heard should be a wonder drug? What about Cheater Relief?? (F, 05.01.08; # 55)

They experiment, turn the doses up (mostly) or down, combine drugs etc. or use other drugs to boost energy while on a diet or to counteract side effects. Often their rationale seems to be based on 'common knowledge', narratives, myths, hear-say or on dubious information from their community. This willingness to take risks is paradoxical when considered in relation to the simultaneous underlying discourse of respect for the body and human nature: The body has its own ways, which should be respected, and which, at least rhetorically, are non-negotiable. The body might "be saturated and in need to recover" (M, 16.12.09; # 54) or simply "not in a hurry" (F, 30.08.10; # 31) – or maybe it is just its nature:

> notice your whole body?...does it behave strangely, do you have pains in your chest, pains anywhere else, do you feel ill etc.... you can read about all the side effects on the internet, but first rule is – OBSERVE if anything feels different, except from a little extra energy + a little heart-throb, it can't be avoided [...] (M, 25.12.10; # 21)

> And if she isn't built to be 45–50 kilos, with her minimum food intake...well then she will gain weight until she reaches the weight most favourable for her. It is a battle the body will take far to win. (M, 12.01.11; # 33)

> And yeah the body is constructed, especially for women to take care of the woman so that she can protect a child in pregnancy...Glad I am not a woman. (M, 12.01.11; # 33)

> It's your genes that determine it and you just have to live with it... (M, 13.01.11; # 33)

In other words, there are things about the body that you cannot change: Anatomy is destiny. The body is described as an independent being which by nature is constructed in a reasonable way – nature knows best. This 'nature' should not be "contaminated with medicine" (M, 13.05.11; # 10), or messed with, unless – of course – you know what you are doing:

> Doesn't sound neither healthy nor reasonable, and definitely not with the hcg diet. You should reconsider the natural way. (M, 11.01.11; # 33)

> The medicine we are discussing here is meant to be supplementary to the natural way of doing things – that has to work first (M, 12.01.11; # 33)

> Understand how the body works, per definition diets don't work! (*1, M)

> [...] I agree with D, it doesn't work to fill your self with drugs if the rest doesn't work. Learning how to use your body in the right way is a way of living, and there is no easy way out, not even ephedrine... (*2, M)
> (22.10.02; # 43)

11.9 Rhetoric of Precaution

These articulations of concern for the body, the advice to listen to the body, go easy etc. reveal a rhetoric of precaution which seems inconsistent with using drugs and celebrating endurance, strength of character etc., and with the actual choice of enduring side effects and continuing with a diet. The emphasis on respect for the human body is most explicitly articulated in responses concerning young people's and women's use of performance enhancing drugs. When a young woman (18) asks which drug diet she should take in order to reach her goal of being trimmed, the response is that she is too young to begin on "juice" and that she can easily reach her goal without drugs (# 51). The advice not to use drugs, when young – and especially female is recurrent:

> there is no reason at all to advise a girl of 18 to do juice when boys at 18 are told to wait another 2 years. [...] (M, 03.10.11; # 25)

> what's happening...according to your profile you are 19, and then you are sitting here asking advice about t3 for your girl...who is probably even younger than you...you deserve a good hiding and being sent to bed (M, 03.01.08; # 59)

Several of the male members who ask on behalf of their girlfriends express a wish to protect them, for example: "the princess shouldn't be harmed" (M, 05.12.10; # 11). Concern for and protection of women and children/youngsters are traditional values in paternalistic cultures and seem in this context to refer to an underlying, though not explicitly articulated, idea of something almost sacred. While, for instance, it seems acceptable for grown up men to use drugs which counteract side effects of other drugs, the irreversible side effects of women's use of for example AAS are good reasons for strongly recommending not to use these drugs:

> [...I have] got something...'Levotiron tablet' [...] I don't know how to take it or what it is? (*1, F)
>
> [...] You shouldn't take your 'Levotiron'. It is T4 and is a very strong weight loss drug, which will give you big side effect. Pull yourself together. (*2, M)
>
> (10.11.08; # 59)

> Hello everybody! I am a girl of 22, I have been training for 4 years and would like to get a little more cut. I have got some Rexobol stanzolol tablets 50 mg [...] Would like to know about your experiences with it? Thanks! (*1, F)
>
> Stoooooooooooooooooop!!!!! Do you want to fill yourself with medicine that might potentially kill you? The hormones you've got your hands on are extremely androgenic and will make you even more masculine than your father. (*2, M)
>
> [...] Winstrol and girls is not a particularly good combination, rather anavar, then. (*3, M) (06.10.11; # 8)

The advice for women not to use drugs is based on inherent bodily differences between male and female biology, and focuses on side effects, especially irreversible ones:

> It is completely braindead to put a girl on t3. [...] (*1, M)
>
> why insist on t3 when it is a bad drug, especially for girls? There are much better and easier solutions, which won't destroy her abdomen. [...] t3 is not a wonder drug which rips off 10–15 kilos of a chick without sides, my own wife has had her ugly experiences with her abdomen after using t3, but she loves superclen [...] she only shakes a little, but that's it. (*2, M)
>
> [....] t3 fucks up your period if you're taking too large doses, in women that is [...] (*2, M)
>
> [...] Bottom line: There are far too many variables in doing metabolic hormones to recommend them to 'normal' body builders and recreational athletes. [...] BUCK up and get a hold on your food before you destroy yourself and even worse you foolish guys who project your own abuse onto your poor girlfriends who have no idea at all about how dangerous it is and will take anything, to make their boyfriends and the rest of the world think that they are not 'too fat'. (*3, M) (03.01.-21.02.08; # 59)

The risk of becoming infertile, in particular, is emphasized by several members in answers about weight loss drugs to women, and men requesting advice on behalf of girlfriends or wives:

> Echo that! On a bad day she may even become sterile by it [T3]... [...] [...] I forgot to mention that both hypothyroidism (too low T3/T4 production) and hyperthyroidism (too high T3/T4 production) is associated with infertility. (M, 30.01.-02.02.03; # 42)

The risk is articulated directly and is expressed as a fundamental concern. A rhetoric of precaution permeates. The melodramatic, emotional loaded responses indicate that this part of human nature is and should remain untouched. These limitations to an acceptable use of performance enhancing drugs indicate a concern about enhancements which cause irreversible changes to the body and/or which impinge upon reproduction. In her book, *Sport, Technology and the Body*, Tara

Magdalinski presents a similar line of thought and links it to fears of the monstrous feminine:

> Whilst these technological modifications are clearly accepted as a necessary part of 'repairing' or 'restoring' the body to an accepted femininity, the use of technologies to take women beyond feminity is rejected. As such, the consumption or use of performance enhancing substances or methods for the specific purpose of altering the female shape is regarded as an infringement against, and even an obliteration of, the very essence of femininity. A female body transgresses the margin that distinguishes male from female by disfiguring the physical attributes that a woman should possess, and is thus regarded as unnatural, abnormal and monstrous for ignoring the nature/artifice and male/female binaries as well as for potentially impairing its culturally ascribed role of reproduction. (Magdalinski 2009: 95).

11.10 Drug Use: A Symbolical Issue?

From a historical point of view, it is, however, not remarkable that the rhetoric of precaution concerns other people's bodies (women and children) and not the human nature of the primary group of members in this community (men), and it is furthermore noteworthy that these 'other people' traditionally have been subjected to control, regulation etc. by exactly the primary group (men) (see for example Braidotti 1994; Haraway 1991; Foucault 1973, 1975, 1976). The exclusion of women from drug use is recurrent in different cultures and often secures a social system (Jay 2011: 28). Regulating and controlling drug use seems to be not just about the drugs, as Kohn stresses:

> [...] the modern discourse about drugs is about far more than drugs [...] Drugs permit the terror of the social subconscious to be voiced. It is an eloquent panic. (Kohn 2001: 1)

Kohn is writing about the British drug underground in the 1920s, but even though styles, issues and the social context have changed, the drug theme is recurrent:

> [...] the rhetoric that directly concerns drug use itself has survived more or less unmodified. The same imagery and the same level of hysteria can be found in British newspapers of the 1920s and the 1980s – two decades in which notable drug panics occurred; there seems no reason to suppose that they will not recur in the future, when the connotations of drugtaking once again offer a way to speak about more profound social anxieties. (Kohn 2001: 9)

Based on his studies Kohn points out that women's drug use was a symbolic issue in which more fundamental crises might be reworked in microcosm (Kohn 2001: 5). Due to the First World War, women's positions were transformed and boundaries between male and female domains were disrupted. But even though the social context has changed today, the issue of gender, played out in the drug theme, is still present (Kohn 2001: 10–11).

While the issue of women and drugs and especially of excluding women from drug use may be about 'something else', it seems almost ironic that the reverse strategy has historically been used in relation to medicine. Women have traditionally been medicalized in order to fulfil their duties and obligations. In spite of the passive form, this should not be seen as simply a discourse of victimization. The complexity of choice, medical advice, social norms etc. is far more entangled. An example of this 'drugging' of women is, as John Hoberman has convincingly demonstrated, the onset of the use of hormones in the 1940s (Hoberman 2005). A recent Danish example is the launching of Premalex by the pharmaceutical firm Novo Nordic. The product, which is a selective serotonin reuptake inhibitor, is meant to relieve women suffering from premenstrual syndrome, but can in a sociocultural context be interpreted as a means to medicalize a natural state and keep women sociable and able to manage job, family etc. In other words, a medical drug to make women cope with and fit into daily life anno 2012.

11.11 Concluding Remarks

Within the limitations concerning representability, reliability and validity, the presented analysis of the Bodyhouse community's attitudes towards and reflections upon the use of performance enhancing drugs provide us with a hint on what is actually going on among (potential) users. Summing up, women as well as men use drugs to enhance their performance. Although the use may not always be linked to an explicit athletic enhancement, there is often a connection with recreational sports. The short version of the Bodyhouse community's general attitude towards use of performance enhancing drugs is that you should not use drugs if it is possible to achieve the goal in another way, and you should only use them if you know something about them. As to women and youngsters: no way. The attitude reveals a non-articulated hierarchy between those who (at least claim to) know how to use drugs in a responsible way and those who do not know. As this chapter has attempted to show, this hierarchy is clearly gendered, and furthermore, the members' identity and place within the symbolic hierarchy of the community is reinforced by the confessional mode of many posts. The general attitude in the community towards the use of performance enhancing drugs indicates that, even in a community where performance enhancing drug use is legitimate, there are boundaries that should not be transgressed. The precautionary discourse concerns in particular women and youngsters. An idea of respect for nature and the body grounds these limits, which – next to the risk of fatal heart failures and thrombosis in the brain – concern irreversible side effects, in particular those relating to biological sex and gender: infertility and androgenic, virilising effects in women. These boundaries suggest two elements in relation to enhancement technologies: Culture – in particular the destabilization of gender (articulated through terminology such as 'bitch tits'; infertility, masculinized female bodies); and nature – biology, here expressed as concern for reproduction. Though my close reading has focused on the enhancement technologies themselves and on the forum members' attitudes, it is interesting that the neither male nor female members in this community question the practice of optimization – they may discuss and disagree on the means and methods, but the general attitude is that it is an individual choice. The precautionary discourse concerns side effects and not the use per se of enhancement technology. It is, for instance, not *explicitly* discussed whether the use of drugs is for example 'unnatural' or 'inhuman'. The general approach is pragmatic and not principle-based. This pragmatic approach is, however, in line with the imperative of performance and the culture of optimization.

References

Anti Doping Danmark. 2004. *Doping. Forfængelighed eller præstationsangst – en debatbog om doping og livsstil*. Brøndby: Anti Doping Danmark.

Anti Doping Danmark. 2009. *Håndbog for fitnesscentre*. Brøndby: Anti Doping Danmark.

Bach, A.R. 2005. *Mænd og muskler*. København: Tiderne Skifter.

Braidotti, R. 1994. *Nomadic subjects: Embodiment and sexual difference in contemporary feminist theory*. New York: Columbia University Press.

Breivik, G., D.V. Hanstad, and S. Loland. 2009. Attitudes towards the use of performance-enhancing substances and body modification techniques. A comparison between elite athletes and the general population. *Sport in Society* 12(6): 737–754.

Brooks, P. 1993. *Body work. Objects of desire in modern narrative*. Cambridge/London: Harvard University Press.

Brown, D.E. 1991. *Human universals*. New York: McGraw Hill.

Carlsen, L.T. 2002. Lene Jennis mor: Jeg har altid mistænkt pillerne Den farlige slankepille. *Tænk+test* 28: 10–19.

Christiansen, A.V. 2009. Doping in fitness and strength training environments – Politics, motives and masculinity. In *Elite sport, doping and public health*, ed. V. Møller, M. McNamee, and P. Dimeo, 99–118. Odense: University Press of Southern Denmark.

Conrad, P. 2007. *The medicalization of society. On the transformation of human conditions into treatable disorders*. Baltimore: The Johns Hopkins University Press.

Coomber, R. 2007. Introduction to the special issue. *Journal of Drug Issues* 37(4): 749–754.

Evans-Brown, M., and J. McVeigh. 2009. Anabolic steroid use in the general population of the United Kingdom. In *Elite sport, doping and public health*, ed. V. Møller, M. McNamee, and P. Dimeo, 75–97. Odense: University Press of Southern Denmark.

Foucault, M. 1973. *Naissance de la clinique*. Paris: Presses Universitaires de France.

Foucault, M. 1975. *Surveiller et punir: Naissance de la prison*. Paris: Éditions Gallimard.

Foucault, M. 1976. *Historie de la Sexualité I: La volonté de savoir*. Paris: Éditions Gallimard.

Franke, W., and B. Berendok. 1997. Hormonal doping and androgenisation of athletes; a secret program of the German Democratic Republic government. *Clinical Chemistry* 43(7): 1262–1279.

Furedi, F. 2011. The dissolutions of the anti-drugs claims making from the grammar of morality. In *Conference: Anti-doping – Rational Policy or Moral Panic*, 18–19 August, 2011. Aarhus: Aarhus University, International Network of Humanistic Doping Research.

Giddens, A. 1991. *Modernity and selfidentity. Self and society in the late modern age*. Cambridge: Polity.

Gillham, B. 2005. *Research interviewing – The range of techniques*. Berkshire: Open University Press.

Haraway, D. 1991. *Simians, cyborgs and women: The reinvention of nature*. London: Routledge.

Hartgens, F., and H. Kuipers. 2004. Effects of androgenic-anabolic steroids in athletes. *Sports Medicine* 34(8): 513–554.

Hoberman, J. 2005. *Testosterone dreams. Rejuvenation, aphrodisia, doping*. Berkeley/Los Angeles/London: University of California Press.

Jay, M. 2011. *High society. Mind altering drugs in history and culture*. London: Thames & Hudson.

Kohn, M. 2001. *Dope girls. The birth of the British drug underground*. London: Granta Books.
Kulturministeriet. 1999. *Doping i Danmark. En hvidbog*. København: Kulturministeriet.
Linke, A. 1996. *Sprachkultur und Bürgertum. Zur Mentalitätsgeschichte des 19-Jahrhunderts*. Stuttgart/Weimar: Verlag J.B. Metzler.
Magdalinski, T. 2009. *Sport, technology and the body. The nature of performance*. London/New York: Routledge.
Mogensen, K. 2004. *Unge mænds brug af doping*. København: Center for Ungdomsforskning.
Møldrup, C., and R.R. Hansen. 2006. Public acceptance of drug use for non-disease conditions. *Current Medical Research and Opinion* 22(4): 775–780.
Nielsen, M. 2002. Mistanke mod slankepille. *Tænk+test* 26: 4–6.
Nielsen, M.W., E.H. Hansen, and N.K. Rasmussen. 2002. *Meidicinforbrug og social position i den danske befolkning. Lægemiddelforskning 2002*. København: Danmarks Farmaceutiske Højskole.
Pinker, S. 2002. *The blank slate*. London: Penguin.
Porta, M. 2008. *A dictionary of epidemiology*. Oxford: Oxford University Press.
Savulescu, J., B. Foddy, and M. Clayton. 2004. Why we should allow performance enhancing drugs in sport. *British Journal of Sports Medicine* 38: 666–670.
Shilling, C. 2005. *The body in culture, technology & society*. London/Thousand Oaks/New Delhi: SAGE Publications.
Shilling, C. 2006. *The body and social theory*. London/Thousand Oaks/New Delhi: SAGE Publications.
Shilling, C. 2008. *Changing bodies. Habit, crisis and creativity*. Los Angeles/London/Thousand Oaks/New Delhi: SAGE Publications.
Singhammer, J., and B. Ibsen. 2010. *Motionsdoping i Danmark. En kvantitativ undersøgelse om brug af og holdning til muskelopbyggende stoffer*. Odense: Syddansk Universitet.
Thiblin, I., and A. Petersson. 2004. Pharmacoepidemiology of anabolic androgenic steroids: A review. *Fundamental and Clinical Pharmacology* 19: 27–44.
Verroken, M. 2003. Drug use and abuse in sport. In *Drugs in sport*, ed. D.R. Mottram. London/New York: Routledge.
Wilkinson, D., and M. Thelwall. 2011. Researching personal information on the public web: Methods and ethics. *Social Science Computer Review* 29(4): 387–401.

Chapter 12
Toward a Situated and Dynamic Understanding of Doping Behaviors

Denis Hauw

12.1 Introduction

Reflections on ethics and performance enhancement in sport may range from sports development in general to athletes' dreams of setting new records and the methods to improve training to achieve this end. A major challenge for sports professionals has long been the problem of doping technology. Endless attempts are made to control doping behavior out of a desire to protect both the fundamental principle of fairness in sports competition and athletes' health. Yet evidence is growing that this specific normative approach is insufficient to deal with the complex processes involved in doping. Another approach to dealing with the relationships between athletes, performance enhancement and doping technology is to think of athletes' activities. The psychosocial sciences have been helpful in this regard, offering various models of these relationships that provide deeper insight into doping behavior (for a review, see Hauw et al. submitted). Most of the psychosocial studies have portrayed doping athletes as rational and accountable for the acts they commit in their search to enhance performance. The emphasis has been placed on the compromise between the intention to maximize the advantages of performance-enhancing drugs and the need to minimize the risks of being sanctioned. According to these studies, such behavior reflect the endeavor to reach athletic and financial goals after a period of reflection and planning mediated by attitudes, personal subjective norms and perceived behavior control (see for example Donovan et al. 2002; O'Donnell et al. 2006; Strelan and Boeckmann 2003; Tsorbatzoudis et al. 2009; Petróczi and Aidman 2008). Hence, athletes emerge as quite similar to others

D. Hauw (✉)
Institute of Sport Sciences, University of Lausanne,
Quartier UNIL-Mouline Bâtiment Géopolis, 5441 CH - 1015 Lausanne, Switzerland
e-mail: denis.hauw@unil.ch

J. Tolleneer et al. (eds.), *Athletic Enhancement, Human Nature and Ethics*, International Library of Ethics, Law, and the New Medicine 52, DOI 10.1007/978-94-007-5101-9_12,

who are fundamentally goal-oriented and who have personalities that show more or less moral development. One conclusion that can be drawn from such studies is that there are few reasonable options for limiting, controlling, or preventing the links between doping technology and performance enhancement, except for the naïve and optimistic belief that "pure moral athletes" will prevail.

However, considerable evidence about athletic performances in the context of everyday life suggests that some of these behavior conceptions can be challenged. For example, the testimonies collected from the free and anonymous French call-in service *Ecoute dopage* indicate that the reasons for doping or non-doping and the mechanisms involved in the decision-making are complex, contingent, and embedded in the specific culture of a given sport (e.g., Bilard et al. 2011). The analysis of these data suggests that doping is the consequence of far more than personality characteristics or rational decision-making about sports goals. Indeed, inextricable psychological, biological and social mechanisms underlie the act of doping to enhance performance. One epistemological approach to analyzing such complexity is the dynamic and situated theoretical approach, which emphasizes the importance of analyzing acts and how they change over time rather than the determinants or the supposed aims of the acts (Clark 1997; Doris 2002; Hendriks-Jansen 1996; Joas 1992; Juarrero 1999; Nowak and Vallacher 1998). The use of this framework shifts the perspective on doping and performance enhancement from "human nature" as enduring personal and general dispositions to the "nature of human activity" that focuses reflection on the dynamics of acts and thereby, provides the opportunity to understand the relationships between performance enhancement, ethics and doping from a pragmatic viewpoint.

In situated approaches, the models generally used to analyze the nature of human activity take into account actors' everyday activities and their contexts, emphasizing their lived experiences of participation in a specific social world (O' Connor and Glenberg 2003; Rogoff 2003). This perspective focuses on human activity as a constant and dynamic process of co-determination between actions and situations, which gives rise to the enacted meaning of an individual's world (e.g., Bruner 1990; Kirshner and Whitson 1997; Sannino and Sutter 2011; Theureau 2006). This view of dynamic and situated activity gives primacy to the changes in everyday existence: their improvisation, coordination, and interactional choreography that form specific and evolving organizations of human activity (Clancey 1997; Maturana and Varela; 1987; Varela 1992). In this sense, reflection on the relations that link athletes, performance enhancement and doping technology should be based on a specific and broader analysis of athletes' lived experiences of participation in their social world, all of which constitute the elements for a description of the nature of human activity.

Several authors have underlined that human activity displays circumstantial, embedded, autonomous, dynamical, extended and embodied characteristics (Clark 1997; McGee 2005a, b; Robbins and Aydede 2009). Each of these characteristics will be explained briefly here, and at length in the subsequent sections. The *circumstantial* thesis holds that activity is an accomplishment in specific circumstances that are not pre-determined, but rather uncertain and undetermined (e.g., Dewey 1938). According to the *embedded* thesis, activity and context are coupled, meaning that activity builds context, which in turn allows the accomplishment of the activity (e.g., Varela 1992). The thesis of *autonomy* suggests that activity is not under the control of a causal agent but emerges from the interaction of agents in a specific context (e.g., Clark 1997). The *extended* thesis describes the boundaries of activity as beyond the boundaries of individual organisms, whereas the *embodied* thesis expresses the idea that activity depends on body involvement as well as mind or cognitions (Robbins and Aydede 2009; Varela et al. 1991). In the following sections, our analysis will be focused on substance use in sports in general and doping in particular. Our aim is to provide some grounds for reflection on the organization of this activity as it relates to athletic performance enhancement, the relationship between this activity and human nature, and the ethical consequences in the context of sports.

12.2 The Circumstantial Nature of Doping Activity

This analysis takes a context-specific view of the nature of human activity (e.g., Bronfenbrenner 1979; Wozniak and Fisher 1993). Behavior is not understood on the basis of a few simple universal principles but is instead considered in terms of its ecological properties. A circumstantial analysis thus emphasizes the particulars of acting in specific environments. It specifically considers the interactions between activity and its current context for accomplishment or the immediate, face-to-face setting in which the actor exists (Suchman 1983). It also makes use of the concept of microsystems from Bronfenbrenner's theory. From this perspective, doping can be considered to be one component of an athlete's activity-in-a-specific-situation.

Two short clinical cases taken from the calls received by *Ecoute dopage* illustrate this idea. The first tells the story of an elite athlete experiencing a difficult period of weariness, tiredness, and feelings of emptiness. He heard on the radio about a magic pill that brings happiness to those in distress. Without consulting anyone, he decided one weekend to try this pill, completely disregarding the usual questions that elite athletes are taught to ask before ingesting a substance: what is the composition of this substance?; what is its impact on performance?; is there a risk of being harmed?; and is there a risk of being sanctioned for a doping violation? After 1 week with no effect, he suddenly became anxious at the thought of possibly being involved in doping and called *Ecoute dopage* for counselling. The second case concerns the story of an elite athlete who explained that for the past 7 years he had been using cannabis extensively during his off-seasons, but only during this period. During training and competition periods, he completely stopped. However, in the year during which he called *Ecoute dopage*, his cannabis use had continued into the beginning of the season and this athlete was now anxious about having a positive control.

One of the main conclusions that can be drawn from these cases, which are typical of those received by *Ecoute dopage*, is that specific circumstances make substance use and doping possible, even when the athletes claim that they respect the rules against doping in sports. At a specific moment, the accepted norms are eclipsed by powerful needs. Just as specific aspects of the physical environment can affect cognitive development (Wachs 1987) or specific circumstances can affect the use of stereotypes (Sinclair and Kunda 1999), specific circumstances can modify the protective safety belt of an anti-doping athlete's general attitudes. According to this assumption (e.g., Schwartz 2007), we should think of attitudes as evaluative judgments that are formed when needed, rather than as enduring personal dispositions. This context-sensitivity allows us to see that even though doping use may appear to be the outcome of a person's personality traits or attitudes, it is in fact a particularity of specific situations. One step in developing a strong anti-doping ethic is thus to define the characteristics of such situations. The easy solution would be to identify the high-risk situations: for example, loneliness is an acknowledged risk for doping (e.g., Skärberg 2009). However, it is impossible to define these situations without taking into consideration how they interact with the athlete's activity. Loneliness is risky because the athlete's activity is linked to a situation of distress and his or her ability to find, imagine, or select solutions other than doping has not been mobilized. The complexity resides in the interaction of a component of activity that occurs in specific circumstances and makes illicit substance use possible. At this level of analysis, the first step to controlling doping is to follow the course of interactions between the activity and the situation. The solution is to break the connection between doping and performance enhancement by resolving the problematic situations in the athlete's life with the offer of better options for his or her activity. If these options for activity offer the possibility of overcoming circumstantial needs, athletes will be less at risk of turning to doping or prohibited substances.

12.3 The Embedded Nature of Doping Activity

The embedded nature of activity can be observed at different levels in the organization of activity but here we will develop a socio-cultural view. As many studies of the history of doping have shown, this practice is not recent and has evolved through many periods associated with specific substances (e.g., Hoberman 1992; Waddington 2000). Taking account of this historical evidence, socio-cultural analyses have provided a wider view of the embedded character of doping in athletes' activity. Human activity implies constraints and possibilities stemming from a long history of practices, continuously revised and fine-tuned by practitioners in response to current events and concerns. Hence, doping to enhance performance is an embedded activity, part of a culture in which this practice has been continuously reinforced and transformed by its practitioners. Recent research has focused on this aspect by using an interactional approach to examine substance use and doping in cyclists (e.g., Brissonneau et al. 2008). Cultural patterns of activity were identified and linked to the meaning attributed to substance use and doping. For example, these authors observed different levels of consumption in relation to performance level, training load, competitive goals, and the manner of performing. They described different periods in which the activity was transformed from "learning the trade", during which the cyclists discovered legal substances (e.g., magnesium, iron, vitamin B12), to "doing the trade" and "becoming a champion", during which they began using the professional pharmacopoeia (i.e., amphetamines, cortisone, anabolic steroids) and eventually moved on to using substances specifically to win (e.g., erythropoietin, growth hormone). They also identified a final step of reconversion associated with the use of substances to forget the distress associated with the end of athletes' career (e.g., amphetamines, Belgian pot).

These studies show some of the properties of human activity that contribute to the ethical problems relating to performance enhancement and doping. Substance use needs to be considered within the culture in which it takes place. The example of cyclists showed that this culture is in part defined by activities that organize relationships between the type of substance to use, the type of involvement (e.g. performance aims, training), and the rank attained in the athlete's career. In addition, participating in these activities becomes a habit and eventually the use of substances typical for each step in the career becomes "normal". Different substances are needed for different forms of effort, involvement, and training load, and athletes progressively modify their activity in relation to the level of performance they are seeking to attain. In the process, they acquire habits concerning substance use that make these athletic activities possible. The link between substance use and the other components of activity is such that one is both the cause and the consequence of the other. Similar interactions characterizing the embedded nature of human activity were also observed in bodybuilders (e.g., Bilard et al. 2011).

One strategy for anti-doping policies would be to re-organize how substance use is linked to other activity components, but this would be very delicate because substance use is constitutive of the cyclist's identity at different moments in his career (Brissonneau et al. 2008). It should also be underlined that doping is an element of the community of practice for these athletes. As Wenger suggested (1998), this activity helps to maintain a tight community through the characteristics of mutual engagement, joint enterprise and shared repertoire. Being an athlete means being engaged in a community of practice at each step of one's career, being recognized by others as skilled, and adopting specific ways of thinking and acting. Progressively, what it means to be an "athlete" changes and personal histories about substance use are created, shared and mutually recognized by the community members. A challenge for anti-doping policies will be to stop the use of doping substances as a marker of membership and a "positive deviance" (meaning that the use is a normal and essential need for these athletes) (Hughes and Coakley 1991). Given the intertwining of identity, activity and community, the difficulty with this type of practical doping culture is the strong conviction that doping is a "necessary" activity that cannot be avoided. Also, because it does not appear to be the consequence of rational deliberation but rather the progressive development of a habit, there is very little space for imagining how to protect against doping. Doping may be one of the strongest elements that unite this community in its identity and activity and, even though the athletes cannot make this public, it becomes inconceivable or impossible to stop. Moreover, athletes who do not take part in this community activity are rapidly excluded and pressured to stop competing. This context is a reference point for specific practices and athletes who do not comply are "outsiders" (Lave 1988; Lave and Wenger 1991). The well-known French cyclist Christophe Bassons explained that in 1999, after he mentioned his suspicions of doping in the pack to the media, he was forced to stop competing: "For several days, I felt rejected by the pack. Several cyclists made it very clear that they didn't like seeing me in this Tour. (…) I talked with (…) Marc Madiot, who asked me to never speak with journalists about doping again (…)." (Le Monde, 18/07/1999, author's translation).

Hence, at a cultural level, the perspective of the embedded nature of human activity suggests that ethical thinking about the relationship between performance enhancement and doping should take into account what the members of a community build through carrying out their practices. Cultures of performance, enhancement, and doping use are tightly linked to an activity that creates a contextual and dynamic system in which the designing of anti-doping strategies could be grounded.

12.4 The Autonomous Nature of Doping Activity

The autonomous nature of activity refers to the assumption that activity emerges from the interaction between its constitutive agents or components (Clark 1997; Varela 1980). This means that doping should be considered as a potential component in the organization of athletes' activity. This perspective of an emergent and autonomous structure associated with doping is based on the shift in modern socio-psychological theory whereby activity is conceived of as a set of components that interact in specific situations to form an emergent structure (Juarrero 1999; Vallacher and Novak 1994). One can thus hypothesize that various emergent structures of activity will correspond to descriptions of specific forms of doping use.

We analyzed these emergent structures (which we call "configurations of activity") to determine the combinations of activity components in doping athletes during periods of substance use and doping (Hauw and Bilard 2011, 2012). The components assumed to be important for understanding the interactions at stake were: (a) the mode of involvement, with various levels of implication (e.g., an open focus, which corresponded to a moderate involvement in sports performance and training because the athlete was staying open to other concerns, or an addictive focus, which corresponded to excessive involvement and concomitant detrimental effects in other areas of life); (b) the type and form of the substances; (c) the evolution in performance (e.g., increased or unstable); (d) the characteristics of training changes (e.g., changed or stable); and (e) the types of lived experience (e.g., experiences of distress, confidence or well-being).

We compared the combination of these components during a period of doping with other non-doping configurations from other periods of the same career. The results showed specific activity configurations that allowed us to define some signatures of activity emerging and corresponding to doping technology use. Put briefly, these configurations of activity were characterized by an addictive focus, various performance-related components, changes in training, and experiences of distress (see Hauw and Le Meur in press, for a detailed analysis). Because each component was observed separately at different periods over the athletes' careers, no single component was sufficient to characterize the activity linked to performance-enhancing substance use. For example, although "addictive focus" was closely linked to doping activity, it was also linked to non-doping activity in 50 % of the cases observed. The same observations could be made for other components that seemed to be implicated in doping activity, such as a "decrease" or an "irregular" evolution in performance, or the experience of a period of personal "distress". However, the activity configurations that linked these components in a specific way could be considered as representative of doping because some of them were observed only in doping cases. Hence, we have identified configurations of an autonomous system linking components in specific circumstances that pointed to specific signatures of doping activity (Hauw and Bilard 2012).

The conclusion from these results is that thinking about the ethics of performance enhancement and doping is enriched by the analysis of specific interactions of particular components of human activity. Each doping case should be considered as the result of a system in which components interact to such an extent that no single component can be considered as the cause of a final "tendency" to dope or not. Moreover, because doping is not an insignificant component of the activity of performance enhancement, the "configurations of doping activity" should also be considered as attractors in the self-organizing process of athletes' activity, whatever the intention to use that mobilizes the athlete. An effective anti-doping policy should thus identify the signatures of doping activity and construct educational or prevention programs to address them. By adding these components (e.g., specific prevention messages, timely threats of sanction, etc.) to the domain of athletes' activity, new activity could emerge in place of the doping activity attractors, just as "kicking a horse to move faster can produce a sudden transition from a trot to a gallop in the horse's gait" (Nowak and Vallacher 1998).

12.5 The Dynamic Nature of Doping Activity

The dynamic nature of activity refers to the time-linked view of human activities and nature. This formulation emphasizes an essential aspect of dynamical system meta-theory as it is applied to the analysis of everyday activities (Nowak and Vallacher 1998). The assumption is that any understanding of a complex system without explicit considerations of time is incomplete. To identify the essence of a complex system, the change in the system over time must be tracked (Abraham 1990). This emphasis on dynamics suggests that doping occurs as a process of change in the stream of athletes' activity. To observe and analyze this phenomenon, Hauw and Bilard (2011, 2012) compared the life courses of activity in doping and non-doping athletes. The activity components linked to the contexts of training, performing and living, and the changes in substance use were identified and mapped on a year by year basis. The results revealed different time scales for understanding the dynamics that lead to doping.

On a short time scale (less than 5 years before use), the organization of athletes' activity was characterized by relatively stable perturbations (i.e., distress, misunderstandings, fruitless searches, etc.) for two consecutive years (year-1 and -2 before using prohibited substances), associated with a very intense mode of involvement (i.e., "addictive" or "closed focus"). Changes in training modes were also observed (e.g., increased load or change in method). In addition, performance changes were distributed over various modalities (i.e., "increased", "decreased", "stable", or "irregular"). These results suggested that doping emerged from a specific period that progressively could be defined as problematic. Athletes were searching for something they could not find over a certain period of time. This period of searching acted as a catalyst that pushed them toward doping attractor activity.

The analysis over a longer timescale revealed that doping began after a long period of practice (e.g. Hauw 2011, on average 17 years). All doping athletes had been regular consumers of legal substances such as vitamins and supplementary substances for no fewer than 2 years before doping activities. However, these patterns were also noted in the non-doping athletes. During their life course of activity, doping athletes transformed their mode of involvement, spending an average of 3–4 years at a moderate level of sporting concerns (i.e., "undetermined", "oriented", "open focus"), 6 years with closed concerns, and more than a year with an addictive focus. An important observation regarding the shift to an exclusive involvement in the sport was that it appeared faster in doping athletes than in non-doping athletes. These results suggest that doping should be considered as related to something that at one moment may be innocuous but which later may take on a very different meaning. For example, increased involvement in training to improve one's performance is normal and fruitful at a certain level for young athletes, but it may appear later as the origin of psychological saturation. Because the consequence of an early involvement is a longer time spent focused exclusively on training and the search for high performance, athletes could be seen as having been more exposed to the negative effects of a long period of high-level commitment. In short, an innocuous activity at a specific time may produce a cumulative effect that transforms a problematic situation into a dramatic one, with radical qualitative changes in activity.

The conclusion from these results is that the ethical issues that link performance enhancement to doping should closely related to the evolution of human activity. The time-linked organization of doping can be summarized by the image of a doping athlete falling into a hole or a "doping basin of attraction". This fall describes the suddenness of the transition in the qualitative organization of doping activity, but it cannot be understood without considering athletes' vulnerability to this event. Our results tend to show that this vulnerability grows with the decline in athletes' psychological flexibility as they encounter problematic situations in their sports careers. Although these problematic situations are normal in an elite athlete's sporting life course, they are not experienced in the same way by all athletes at all points in their careers because of differences in experience and personal histories. Many life stories of doped athletes illustrate this point (e.g., Agassi 2009; Chouki 2009). Efficient anti-doping policies should thus specifically target the parameters of control that orient the development of the sporting life course in order to prevent the development of dangerous trajectories (e.g., repetitions of problematic situations, injuries, etc.).

12.6 The Extended Nature of Human Activity

The extended nature of activity refers to the observation that the boundaries of activity lie outside the individual organism, encompassing features of the physical and social environment (Robbins and Aydede 2009). This assumption differs from the traditional psychosocial approach to doping behavior, which assumes that doping resides mainly in the athletes' "heads". Yet when activity is examined in real-life problem-solving situations, athletes appear to think and act in conjunction with others and with the help of culturally provided tools (e.g., Hutchins 1995; Lave and Wenger 1991). The term "extended cognition" not only indicates the relationships between activity and social, cultural and technological environments but also suggests that these environmental elements may improve the activity. Individual activity is related to material or social others that enhance the resources of the activity, which is considered as a complex system. This means, for example, that computers, significant others, coaches, and parents can play a major role in the activity of athletes by helping them to think, reason, remember and so on, or even do so in their place. Situated cognition theorists also speak about this conception as "distributed cognition", which characterizes one of the main shifts proposed by these approaches order to better understand the nature of human activity.

By considering doping as an "extended activity," analysis is not focused on the individual, as attention is also given to the resources available in the activity system. Let us look at the cases of athletes who decided to use a certain substance because other athletes had told them it was good. The testimonies received by *Ecoute dopage* often describe this situation both for licit and illicit substance use. Here, the significant (often older) athletes can be considered as models. But the situation can also be analyzed as more than a copying activity, with the other athletes serving as an unconscious resource in the decision system. From an extended activity approach, the decision to take the substance is made in and through a specific social context. The decision process involves two steps: the problem arises in a setting (i.e., the possibility of using the substance emerges) and the problem is solved (i.e., the decision to use the substance or not). In these examples, we do not know the settings for the emergence of the problem, but we can assume that the solution emerged from the social interactions in this context. The athletes did not have to deliberate much because the choice had already been made by others. They only saw the possibility to use substance in a field of widespread use by other athletes.

This situation of extended rather than personal reasoning has many consequences for the analysis of doping activity and anti-doping policies. Many studies in the social sciences focus on the problem-solving step in the decision-making process. But the notion of the extended nature of activity suggests that there are also concerns linked to opening the field of possible use (i.e., the setting of the problem). If certain resources can push an athlete to use a doping substance, anti-doping strategies must be based on an understanding of how this occurs (i.e., how the field of possible use opens up) before athletes can be helped to build "resistance" to negative resources. The "setting of the problem" in the decision-making process is also an extended activity (e.g., Schön 1983). For example, the analysis of calls to *Ecoute dopage* showed that laws are resources that may prevent the emergence of the possibility of doping, despite opposition. One of the most prevalent factors that protected against doping was "doping controls from the environment (drug testing)" (Bilard et al. 2011). This finding supports the view that drug testing is a strong deterrent, resonating at the unconscious level and supporting the operation of the athlete's conscience to block the emergence of doping use. This is all the more noteworthy in that our sample was mainly composed of regional-level athletes, for whom the likelihood of drug testing was extremely low, if not inexistent (AFLD 2009). In contrast to this protective action, the shortcomings of laws and control systems could be exploited in favor of doping activity. For example, we also heard reports from doping athletes who trained for the upcoming season during the wintertime in isolated training centers in Africa or Eastern Europe, where controls were not efficient and they were able to use any substance they wanted. They had no ethical problems with this. Clearly, their decision was strongly influenced by the system that offered the possibility to engage in this activity.

Another important consequence of the "extended" nature of activity is illustrated by the cases of isolated athletes trying to decide whether or not to take a doping substance. Being isolated meant they had no possibility to turn to direct social resources. But indecision pushed some of them to external resources, such as the internet. Many websites explain quite well the hows and whys of purchasing and using a variety of substances, and thus the limited and inefficient available resources for solving the athletes' dilemma were rapidly extended. Once again, the decision was not made solely in the head of an isolated athlete but within a complex high-tech system with distributed competencies. Donati (2007) has described how the internet market has increased the possibility of using substances prohibited in sports and other domains of social activity. Recently, the French affair of Jeannie Longo and her husband-coach Patrice Ciprelli made this practice public. Hence, we do not agree with the view that the availability and affordability of substances are secondary factors in the decision-making process (e.g., Donovan et al. 2002). In an extended activity conception, they are strong contributors to the decision to use doping technology.

Finally, athletes' use of external resources in making a decision about doping raises the ethical issue of how to help them deal with uncontrolled resources and develop a proper resistance. Several questions arise, such as: What are the situations that promote an unlimited use of external resources?; and why did these athletes show no free will and rely on the context without precaution? Some researchers have underlined "risk behavior" and "sensation seeking" as causal determinants of this activity (e.g., Zuckerman 1986). Our observations suggest that doping athletes are also vulnerable individuals in a problematic situation. In this situation, they are at the mercy of unethical coaches, greedy for financial gain and ready to promise heaven to those who follow their advice. Once again, it is not really the influence of these coaches that is responsible for the shift to doping, but the impossibility for the athletes of solving their problems on their own. The use of resources from a social system where someone else makes the decisions for them is a very efficient and economical coping strategy. These circumstances and the distribution of competencies are fundamental to understanding doping practices and to define anti-doping policies.

12.7 The Embodied Nature of Doping Activity

The embodied nature of activity refers to the idea that the body shapes the activity in which we are engaged. This approach suggests several ways to understand the nature of doping activity. Doping is first of all a practical action. Doping athletes swallow pills, drink powdered mixtures, and inject with syringes. Behind these actions, they have their own rules that are unknown to us. For example, our interviews with doping athletes revealed that when athletes decided to use a substance, they often doubled the recommended dosage. Other observations indicated that when athletes decided to use nutritional supplements, most of them forgot to follow the prescribed dosage over the long term. Few people realize that many athletes are afraid of injections or that most quickly tire of managing medication. All these examples reveal a multitude of practices that are well known in athletics but not seriously and thoroughly studied and discussed. Doping activity is practical and bodily interaction with a substance. The entire activity that accompanies this must be kept in mind, but the foremost aspect is the practical component and not a cognitive process involving rational deliberation.

Embodied activity also suggests that substances may be progressively absorbed as part of the athletes' identity. In line with Brissonneau and colleagues' observations based on their study of cyclists' careers, we observed that some athletes cannot perform during competition and training if they do not have their "magic potion" or power drink in their bag. The substance plays the role of scaffold or framework for performance, just as other rituals do. As a means to deal with the indeterminate characteristics of performance, substances are taken into the body to create a structure adapted to the specific world of performance and thus give confidence to athletes. Many advertisements use the image of elite and victorious athletes to express this embodied aspect of substance use and create a real incentive for use in the culture of practice: there is no chance for the body to be at its best without using a substance.

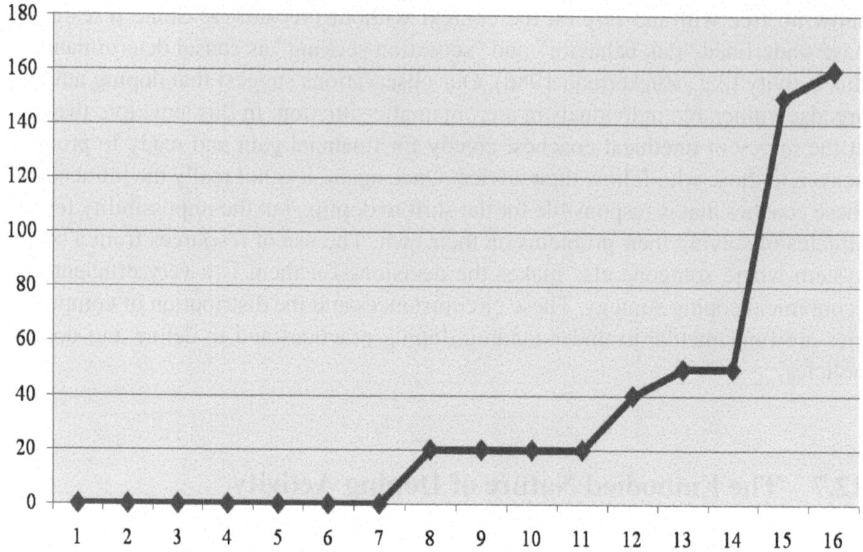

Fig. 12.1 Typical "step by step" pattern in the use of licit substances over 16 years of practice (x axis: years of practice; y axis: coefficient of substance consumption)

Another form of embodiment is observed in the transition from the use of a legal substance to a prohibited one (Hauw and Bilard 2010). Twenty elite athletes' (12 sanctioned for doping violations and 8 not) profiles of substances intake during sporting life course were analyzed using a coding system that contained the description of the number and the form (i.e., occasionally, curative, continuous) of substance (coefficient of consummate substances). For the calculation of athletes' coefficient, occasional consumption of one substance was coded 1, while curative was coded 10 and continuous 100. The addition of the value for each substance use provided the final coefficient for each athlete and year after year. The evolution of each athlete's coefficients defined the profile of consumption during career before the 17 years of practice and was represented by a figure. A qualitative comparison was devoted to the form of the figure (e.g., linear, discontinuous, with stages, increasing, decreasing). Specific patterns of licit substance use was observed that led to the use of prohibited substances (i.e., step by step and with a pattern of hesitation, as shown in Figs. 12.1 and 12.2) (Hauw and Bilard 2011; Martinez et al. 2012).

From these two figures, we can infer that these typical changes in use reflect a specific process of substance embodiment that leads to doping. Once again, it is not the use of a licit substance that leads to doping, but specific forms of use associated with other components of athletes' activity. The notion of embodiment suggests that specific forms of consumption are able to modify the nature of the relation to the substance, as, for example, drug use can create the drug addict's progressive dependence. One of our hypotheses is that increasing and inexorable use, characterized by

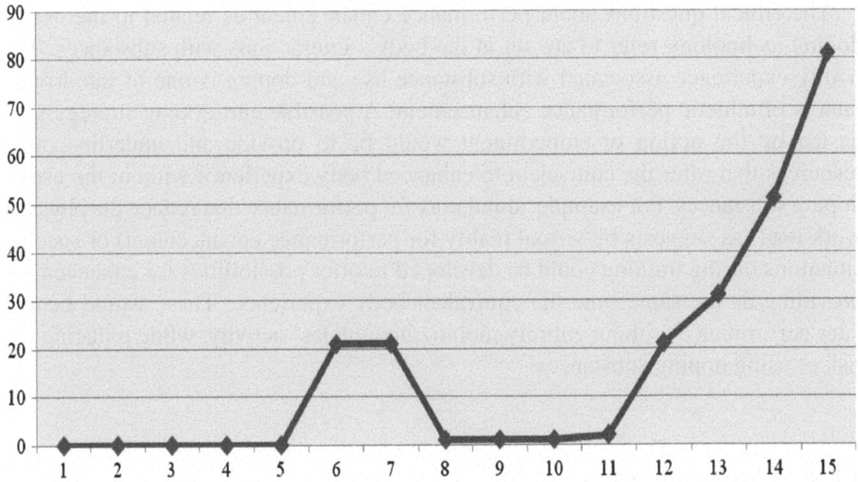

Fig. 12.2 Typical "hesitation" pattern in the use of licit substances over 16 years of practice (x axis: years of practice; y axis: coefficient of substance consumption)

the number of substances, or the form of use (expressed by the coefficient of consumption in the figures), and/or both, define this process of substance embodiment. Prohibited substance use is preceded by a state of poly-consumption that appears progressively during the doping athlete's career and not for non doping athletes. We can thus hypothesize that the need for doping substances is progressively written in the "body" and not in the "head".

Another major consequence of taking embodied activity into account is the need for phenomenological investigations of lived experience as a necessary complement to standard scientific inquiry (Varela and Shear 1999). Phenomenological investigation aims to capture how athletes deal with the substance in relation to their bodily "views from within". This was one of our aims in the analysis of calls received by *Ecoute dopage*, where the psychologists offers athletes the possibility to explain their feelings or emotions linked with their sports activity and how the need to use substances fits in. We showed, for example, that cyclists used substances not primarily as much with the aim of performing well but in order to deal with the small daily hassles of elite athletic training. The substance was used in response to negative bodily experiences associated with effort, exhaustion, and recovery. We did the same in self-confrontation interviews concerning the pattern of the life courses of doping athletes, that consisted of provoking the re-emergence of elements from past experience when the participant was bodily face to face with traces of his/her own activity (e.g., Hauw and Bilard 2012; Hauw and Le Meur in press). This method offers the possibility of putting athletes in a situation where they can re-enact past situations and their relationships with these situations, and then explain, comment on, and testify about their entire experience, including the physical, affective, practical and cognitive elements that composed the actions.

The ethical questions about performance enhancement as related to the use of doping technology refer to are set in the body's interactions with substances. The bodily experience associated with substance use and doping is one of the driving factors of athletic performance enhancement. A possible anti-doping strategy suggested by the notion of embodiment would be to provide and underline other resources that offer the equivalent to enhanced body experience without the use of doping substances. For example, simulators for performance that reduce the physical work load (as suggests by virtual reality for performance enhancement) or specific situations during training could be developed to offer possibilities for enhancement providing as the same time the equivalent body experience. These would benefit later performance without entirely mobilizing athletes' activity, while reducing the risk of using doping substances.

12.8 Conclusion

"Dynamic and situated activity" analysis offers a pragmatic way to explore ethics-related issues in the relationship between athletic performance enhancement and doping. By using an alternative analysis, rather than the usual deliberative approach, the key question that emerges appears to concern how athletes' activities are organized in relation to performances and doping. From this perspective, doping is not a factor of athlete's progressive dehumanization. It can instead be seen as one of the circumstantial and cultural actualizations of the possibilities for enhancing their resources, emerging in a dynamic, extended and distributed athletic activity system. In contrast, but not in opposition, to an ethical approach of "clean" sport based on general moral principles, such as respect for equal opportunity in competition, a dynamic and situated anti-doping ethics in sport can thus be built in close relationship with athletes' activity. The ideas and findings discussed in this article indicate that athletes' "sensitivity" to doping technology is situated in both previously accomplished activity and the activity they are currently engaged in. Anti-doping strategies should directly target the identified risk factors by offering possibilities to stop the emergence of doping practice. Following this line of thought, four major directions have been identified. First, specific attention should be given to critical periods in the athlete's career so that other healthier pathways can be introduced, with psychological support. As we have demonstrated, specific attention should be focused on periods of instability, indicated by inquiries with no helpful response, turning inward, or chronic tiredness. Second, strategies for reaching elite performance and performance enhancement should be analyzed with regard to the proximal and distal effects on the stability or instability of the athletes' development. Early specialization can be questioned in relation to the athlete's development and his or her maturity with regard to dealing with the drive to achieve high performance. Many examples showed that early achievement of high performance was associated with an attraction to doping. This suggests that the relationship between age and performance level should to be taken into account in estimating the time needed to reach elite performance, on a sport by sport basis. This approach could be fruitful for defining the ethical issues regarding performance enhancement during an athlete's career. Third, the use of legal substances like nutritional supplements should be limited and controlled because of the problems associated with the "embodiment" that they create. Specific attention should be directed toward self-medication, which can lead to the use of substances that are not under limited access (e.g., available over the internet and in sports supermarkets). Fourth, athletes need to know clearly where to go for help when needs arise in the course of their careers. By considering doping as a negative drift in the evolution of an athlete's activity, sports professionals will be better equipped to spot opportunities to limit this drift by stepping in with alternative solutions for enhancing performance.

The analysis provided here also shows that human activities generate personal experiences that progressively transform the actor's meaningful relationships to events. This provides a good foundation for a better understanding of the relationship between athletic performance enhancement and doping. Doping can be considered as the result of a complex activity over the course of which social, cognitive and bodily relationships with enhancement technology have evolved. These relationships are not only influenced by cognitive processes, as suggested by the major theoretical models in the social sciences, but they also express a linkage between experience and situations during activity that transforms the athlete's world of acting and meaning. This type of analysis suggests that doping activity can be limited by offering athletes the possibility to learn from experience and to develop skills for distancing themselves from certain dangerous or critical circumstances in the course of their sporting life. By teaching them to engage in reflective practice (individually, in discussions with other athletes or with sports psychologists, etc.), they will be better equipped to deal with negative events or feelings in healthier ways.

References

Abraham, F.D. 1990. *A visual introduction to dynamical systems for psychology*. Santa Cruz: Aerial Press.

AFLD. 2009. *Rapport d'activité de l'agence française de lutte contre le dopage*. Paris.

Agassi, A. 2009. *Open*. Paris: Plon.

Bilard, J., G. Ninot, and D. Hauw. 2011. Motives for illicit use of doping substances among athletes calling a national anti-doping phone-help service: An exploratory study. *Substance Use & Misuse* 46(4): 359–367.

Brissonneau, C., O. Aubel, and F. Ohl. 2008. *L'épreuve du dopage. Sociologie du cyclisme professionnel*. Paris: PUF.

Bronfenbrenner, U. 1979. *The ecology of human development*. Cambridge, MA: Harvard University Press.

Bruner, J.S. 1990. *Acts of meanings*. Cambridge, MA: Harvard University Press.

Chouki, F. 2009. *Ma course en enfer*. Paris: Hachette.

Clancey, W.J. 1997. The conceptual nature of knowledge, situations and activity. In *Expertise in context: Human and machine*, ed. P.J. Feltovich, K.M. Ford, and R.R. Hoffman, 274–291. Cambridge: MIT.

Clark, A. 1997. *Being there. Putting the brain, body and the world together again*. Cambridge, MA: MIT Press.
Dewey, J. 1938. *Logic: The theory of inquiry*. New York: Henry Holt & Company.
Donati, A. 2007. *World traffic in doping substances*. Technical report. http://www.wada-ama.org/Documents/World_Anti-Doping_Program/Governments/WADA_Donati_Report_On_Trafficking_2007.pdf. Accessed 28 Mar 2012.
Donovan, R.J., G. Egger, V. Kapernick, and J. Mendoza. 2002. A conceptual framework for achieving performance enhancing drug compliance in sport. *Sports Medecine* 32: 269–284.
Doris, J.M. 2002. *Lack of character*. Cambridge: Cambridge University Press.
Hauw, D. 2011. Commande et autonomie dans l'activité des sportifs de haut niveau: des savoirs pour une autre éthique du dopage. In *De l'amélioration des performances sportives: éthique et philosophie du dopage*, ed. J.N. Missa and P. Nouvel, 217–234. Paris: les cahiers du Centre G Canguilhem.
Hauw, D., and J. Bilard. Nov 2010. *Dynamical analysis of substance use before doping*. Presentation at the international social sciences doping conference, Copenhagen.
Hauw, D., and J. Bilard. 2011. *Cours de vie sportive et dopage* [Sporting Life course and doping]. Technical report, French Ministry of Sport, University of Montpellier 1, Montpellier.
Hauw, D., and J. Bilard. 2012. Situated activity analysis of elite track and field athletes' use of prohibited performance-enhancing substances. *Journal of Substance Use* 17(2): 183–197. Early online at http://informahealthcare.com/toc/jsu.
Hauw, D., and Y. Le Meur. In press. Dynamiques de l'expérience sur des temps longs : l'exemple de l'apprentissage du métier de sportifs de haut niveau. In *L'expérience au travail et en formation*, eds. Albarello, L., Bourgeois, E., Barbier J.M., and Durand, M. Paris: PUF.
Hauw, D., J. Bilard, and M.J. McNamee. Submitted. A critical analysis of three psychological research programs of doping behaviour. *New Ideas in Psychology*.
Hendriks-Jansen, H. 1996. *Catching ourselves in the act. Situated activity, interactive emergence, evolution and human thought*. Cambridge, MA: MIT Press.
Hoberman, J. 1992. *Mortal engines: The science of performance and the deshumanisation of sport*. New York: The Blackburn Press.
Hughes, R., and J. Coakley. 1991. Positive deviance among athletes: The implications of over conformity to the sport ethic. *Sociology of Sport Journal* 8(4): 307–325.
Hutchins, E. 1995. *Cognition in the wild*. Cambridge, MA: MIT Press.
Joas, H. 1992. *La créativité de l'agir*. Paris: Cerf.
Juarrero, A. 1999. *Dynamics in action. Intentional behavior as a complex system*. Cambridge, MA: MIT Press.
Kirshner, D., and J.A. Whitson. 1997. *Situated cognition social, semiotic and psychological perspectives*. Mahwah: Lawrence Erlbaum Associates.
Lave, J. 1988. *Cognition in practice*. Cambridge: Cambridge University Press.
Lave, J., and E. Wenger. 1991. *Situated learning. Legitimate peripheral participation*. Cambridge: Cambridge University Press.
Martinez, D., J. Bilard, and D. Hauw. 2012. Prévenir, informer et mesurer le dopage. In *Le dopage dans le sport. Etat des lieux et nouvelles perspectives*, ed. D. Oswald, 37–49. Neuchâtel: Edition CIES.
Maturana, H.R., and F. Varela. 1987. *The tree of knowledge*. Boston: New Science Library.
McGee, K. 2005a. Enactive cognitive science. Part 1: Background and research themes. *Constructivist Foundations* 1(1): 19–34.
McGee, K. 2005b. Enactive cognitive science. Part 2: Methods, insights, and potential. *Constructivist Foundations* 1(2): 73–82.
Nowak, A., and R.R. Vallacher. 1998. *Dynamical social psychology*. New York: The Guilford Press.
O'Connor, K., and A.M. Glenberg. 2003. Situated cognition. In *Encyclopedia of cognitive science*, ed. L. Nadel, 19–25. New York: Macmillan.
O'Donnell, T.M., J. Mazanov, and T. Huybers. 2006. Towards a choice model of athletes' decision to use performance enhancing substances or methods: Factors and covariates. In: ACSPRI

social science methodology conference 2010, presented at ACSPRI social science methodology conference 2010, University of Sydney, Australia, December 2010.

Petróczi, A., and E. Aidman. 2008. Psychological drivers in doping: The life-cycle model of performance enhancement. *Substance Abuse Treatment, Prevention, and Policy* 9: 2–34.

Robbins, P., and M. Aydede. 2009. *The Cambridge handbook on situated cognition*. Cambridge: Cambridge University Press.

Rogoff, B. 2003. *The cultural nature of human development*. New York: Oxford University Press.

Sannino, A., and B. Sutter. 2011. Cultural-historical activity theory and interventionist methodology: Classical legacy and contemporary developments. *Theory & Psychology* 21: 557–570.

Schön, D. 1983. *The reflective practitioner. How professionals think in action*. London: Temple Smith.

Schwartz, N. 2007. Attitude construction: Evaluation in context. *Social Cognition* 25(5): 638–656.

Sinclair, L., and Z. Kunda. 1999. Reactions to a black professional: Motivated inhibition and activation of conflicting stereotypes. *Journal of Personality and Social Psychology* 77(5): 885–904.

Skärberg, K. 2009. Anabolic-androgenic steroid users in treatment: social background, drug use patterns, and criminality. *Örebro Studies in Medicine*, Series 28. http://www.steroids.dk/da/Bibliotek/~/media/steroids_dk/infotek/Forskning/Skrberg_AASusersintreatment_09.ashx. Accessed 28 Mar 2012.

Strelan, P., and R.J. Boeckmann. 2003. A new model for understanding performance-enhancing drug use by elite athletes. *Journal of Applied Sports Psychology* 15: 176–183.

Theureau, J. 2006. *Le cours d'action: Méthode développée* [Course of action: Developed method]. Toulouse: Octarès.

Tsorbatzoudis, H., A. Rodafinos, H. Spiliopoulou, V. Barkoukis, and L. Lazuras. 2009. *Determinants of doping intentions in sport*, WADA final report, Aristotle University of Thessaloniki, Salonika.

Vallacher, R.R., and A. Novak. 1994. *Dynamical systems in social psychology*. San Diego: Academic.

Varela, F.J. 1980. *Principles of biological autonomy*. New York: Elsevier North Holland.

Varela, F. 1992. *Un Know-how per l'ettica*, The Italian Lectures 3, Editrice La Terza. Roma.

Varela, F.J., and J. Shear. 1999. *The view from the within: First-person approaches to the study of consciousness*. Thorverton: Imprint Academic.

Varela, F.J., E. Thompson, and E. Rosch. 1991. *The embodied mind: Cognitive science and human experience*. Cambridge, MA: MIT Press.

Wachs, T.D. 1987. Specificity of environmental actions as manifest of environmental correlates to infant's mastery motivation. *Developmental Psychology* 23: 782–790.

Waddington, L. 2000. *Sport, health and drugs: Critical sociological perspectives*. London: E & FN Spon.

Wenger, E. 1998. *Communities of practice. Learning, meaning, and identity*. Cambridge: Cambridge University Press.

Wozniak, R.H., and K.W. Fisher. 1993. *Development in context. Acting and thinking in specific environments*. Hillsdale: Lawrence Erlbaum Associates.

Zuckerman, M. 1986. Sensation seeking and the endogenous deficit theory of drug abuse. *National Institute on Drug Abuse Research Monography* 74: 59–70.

Chapter 13
Restoring or Enhancing Athletic Bodies: Oscar Pistorius and the Threat to Pure Performance

Tara Magdalinski

13.1 Introduction

In early 2012, organisers of the London Olympics confirmed that the Games would be the 'most tested' ever, with at least 1,000 tests being performed to identify athletes who had surrendered to the temptations of illicit enhancement. Such pronouncements, some 6 months prior to the Opening Ceremony, were an early reminder that the 'war' on illegal enhancement is on-going and that the vigilance to identify transgressors is unwavering. Despite ever more sophisticated testing systems and propaganda campaigns aimed at 'cleaning up sport', there remains concern that elite performance sport is under constant threat of contamination from various synthetic technologies. At the heart of such fears is a romanticised belief that sport is a natural activity that expresses human mastery over the elements (Bale 1994), evoking images of physical activity located in green environments, from grassy fields to pristine waterways, where bodies run, jump, swim and move through nature, an image that is disrupted by technological intrusion.

Despite the concern that technologies, particularly at the level of the body, confound popular conceptions of athletic endeavour, modern sport is predicated on technological intervention. From advances in equipment, clothing and facilities through to the regulation of diet and training interventions, the reality is far from the popular imagery of sport as 'free' and 'liberating'. But whilst the incursion of technology into areas that facilitate sport may no longer conjure up the same level of fear, the interface between the body and technology continues to present a perilous moment where the integrity of the former and the validity of the latter might be compromised. A body that succumbs to the temptation of illicit enhancement or is

T. Magdalinski (✉)
UCD Centre for Sports Studies, University College Dublin, Dublin 4, Ireland
e-mail: tara.magdalinski@ucd.ie

J. Tolleneer et al. (eds.), *Athletic Enhancement, Human Nature and Ethics*, International Library of Ethics, Law, and the New Medicine 52, DOI 10.1007/978-94-007-5101-9_13,

tainted by the presence of threatening technology typically has no place within sport where pure performance demands uncontaminated participants strive to establish the limits of the body's physical capacity. This chapter examines the interface between nature and technology to determine whether technologically enhanced bodies disrupt mainstream conceptions of pure performance. The triadic relationship between sport, the body and performance is interrogated with reference to Oscar Pistorius, a liminal figure whose presence in mainstream sport represents a threat to the integrity of sports performance.

13.2 The Nature of Sport

Any examination of the notion of enhanced bodies within the context of performance sport needs to begin with a discussion of the nature of sport. Sport rests on a very specific set of ideologies that suggest social advantages are gained by engaging in structured physical activity. Young children, adolescents and adults alike are encouraged to participate, not merely to gain well-established physical benefits, but to develop social norms and elicit behavioural changes that are thought to be inherent to the practice of sport. Specifically, the 'spirit' of sport invokes notions of fairness, morality and exemplary lifestyles; governmental agencies and sporting organisations cultivate the image of sport as a mechanism to communicate life lessons in leadership, time management and integrity; elite athletes are offered as role models for the young to emulate; and sponsors support the clean, moral image of sport in the hope that some of the gloss rubs off on their goods and services.

Although the 'true' spirit of sport is commonly referenced, few agree on what that 'spirit' is. For some, sport represents the playful expression of the human condition, a feature endemic in all societies. For others, sport is 'amusement solely' and its essence is 'relaxation', a moment 'when we disport ourselves from labour and our usual daily work' (Allison 2001: 1), or a 'challenge', a quality that is closely associated with 'courage' (Reid 1998). Sport may also be understood as 'a physical competition between opponents', the outcome of which is determined by 'ability, strategy, and chance' (Eitzen 2006: 1), or the 'recreation of the human spirit through the sheer joy of play' (Wigglesworth 1996: 152). Furthermore, sport is thought to reflect those traits favoured in modern, western society, including 'character building, health promotion, the pursuit of competitive excellence, and enjoyment' (Jenkins 2002: 99). What these descriptions have in common is the fundamental assumption that embedded in sport are positive or even redemptive qualities that are apparently absent from other recreations.

The reality of sport, for many, is quite different as high-profile athletes are paraded in front of the world's media for sexual transgressions, gambling, racism and a range of other immoral if not criminal behaviours, and the desirable quality of sportsmanship is often absent from a sports field where exploiting the weakness of an opponent is a fundamental tenet of the game. Nevertheless, sport remains, at least in the popular mind, an ideal means to impart a sense of commitment, discipline, dedication and sacrifice, and to strengthen character and fortitude in the face of adversity (Verroken 2003; Jenkins 2002; Butcher and Schneider 2001; Reid 1998).

The notion of sports as a 'carefree' and 'joyous' expression of humanity, has its roots in the nineteenth century, where sport served as an antidote to the threats of industrialisation and urbanisation. Not only did these jeopardise the health and hygiene of individuals but they were feared as detrimental to society more broadly. Sport, with its images of green pastures and open spaces, offered a link to the natural realm, away from the confines of the filthy industrial city, and thus was inscribed with many of the qualities that were attributed to nature (Bale 1994). Nature, it was agreed, offered not only a site of rejuvenation but possessed an inherent morality that could inspire and instruct human society. Through its close association with this 'untouched' realm, sport was consequently imbued with a similar purpose.

Locating sport within a discourse of nature emerged from the Romanticism of the late eighteenth and early nineteenth centuries that celebrated the natural realm. Nature was positively regarded as wild and untamed, untouched by human activity, and was juxtaposed against the rationality of industrial culture (Adam 1998). Wilderness and natural areas were thought to be a refuge from industrial life, as the countryside represented healthy, organic values that offered the potential for liberation from the rationalised industrial centres. Not only was nature a retreat from urban life, it also offered the opportunity to restore and rejuvenate the weary body and soul. Yet, the body could also be revitalised through playing sport, which, through its replication of rural spaces, represented the antithesis of the confined, dirty and unwholesome world of work.

Nineteenth century sporting ideologues, such as modern Olympic founder Baron Pierre de Coubertin trusted in the potential for sport to remedy social ills (Holt 1989). The development of sports, parks and other recreational spaces and activities not only distracted the masses from their daily plight, but could also be instilled with a higher moral purpose that would educate and reform the working classes of the day (Bednarek 2005; Crawford 1984; Whorton 1982). The amateur philosophy that underpinned middle class sport created expectations of how sport should be played; not just the rules and style, but the philosophy of play was prescribed (Allison 2001). Sport, it was reasoned, should be played with a particular spirit and thereby aim towards some higher moral imperative.

Although this popular philosophy is well established, there remain significant concerns that the fundamental essence of sport is being undermined by a number of threats, such as commercialisation, professionalism and technology. The amateur underpinnings of sport that imbued sport with a moral imperative, whilst no longer realistic, are privileged as a more noble engagement, yet are endangered by those who seek personal glory and financial enrichment rather than the pursuit of sport to advance honour. Whilst these factors are thought to blemish the integrity of sport, technology is thought to have a more marked impact on the validity of sports performance and is held responsible for slowly confounding the primacy of the body. Anxieties about the increasing reliance on technology are provoked by a more general technophobia that values 'natural' products over human-made or artificial ones (Barilan and Weintraub 2001), and this is confirmed by the expectation that sport is a natural activity.

13.3 Technological Enhancement of Performance

Whilst many profess to being uncomfortable with its influence, technology is a fixture in contemporary sport. Innovation in product and material design as well as complex engineering have improved equipment, created artificial scapes and surfaces, and generated clothing that helps rather than hinders performance. Athlete health and safety have also been provided for with the construction of sophisticated helmets, mouthguards and padding. Furthermore, sport requires technological intervention to regulate time and space, so that even the locations where sport is played are the consequence of the deliberate modification or reproduction of the landscape (Magdalinski 2008). Nevertheless, the introduction of new techniques or equipment is closely scrutinised to ensure that the integrity of sport is not undermined. Training regimes, improved equipment and nutritional substances, for example, are examined to ensure they do not 'enhance' athletes, and instead merely allow for an accurate measure of an individual's actual capacity by reducing external influences that may obscure or hinder the performance.

Determining the pure performance capacity of an athlete is a fundamental principle that underpins elite sport (Magdalinski 2008). Changes to rules and regulations as well as modifications to playing surfaces, clothing and equipment often focus on removing uncertainty from sport so that an athlete's measured performance accurately reflects the competitor's actual bodily capacity. Extraneous influences are mitigated as far as possible or even eradicated to ensure that only the body itself contributes to the final outcome, recognising that athletic competition is not merely an arena in which to test bodies against one another, but to establish the absolute limit of human capacity. In view of this principle, it is clear that athletes cannot simply compete against one another without preparation. Over the last century, training systems have changed markedly, such that a range of legal chemical concoctions, radical training techniques and innumerable other advances have been ingested, followed or applied to generate greater results. The never-ending pursuit of records and personal bests would suggest that on some level, modern sport is predicated on improving athletic performance.

The concept of 'performance enhancement' is marginalised in contemporary sport, conjuring up images of steroid-fuelled, 'mannish' women with deep voices, facial hair and bulging muscles, or freakish bodybuilders, injecting drugs into their thighs. Suspected drug abusers are exposed in the sports media even before their guilt is determined, whilst confirmed dopers are paraded as public warnings to those who transgress acceptable bodily limits (Magdalinski and Brooks 2002). Using illicit technologies to provide an 'unfair advantage', represents the antithesis of all that is considered meaningful about organised sport, and conceiving elite performance as nothing more than the consequence of extreme technological intervention is similarly unthinkable. As such, training methods, supplements and other applications are particularly criticised if they are defined as an 'unnatural' or 'artificial' way of enhancing performance, as they are thought to negate sport's basic 'natural' or 'human' tenets. In fact, given the vigour with which suspected transgressors are pursued, it would seem that there is almost no greater evil in sport than to supplement the body artificially, despite the fact that the competitive structure of sport pressures competitors to consider any and every means possible to enhance their performance.

There are many technologies that are used legitimately to improve performance, and exercise scientists, coaches and other relevant personnel are entrusted with the task of finding mechanisms to generate the 'winning edge' in competition. Despite these very obvious attempts to 'enhance' the performance of the athletes in their charge, there are many technologies that are deemed to be inappropriately intrusive and thus a threat to the purity of the athlete. Interventions that alter the body and its capacity can detract from determining an athlete's physical ability, for in the presence of technologies it is feared that one can never be sure whether an individual's performance is a genuine reflection of their own physicality or unduly enhanced by the internal or external aid (Magdalinski 2008). Whilst technologies can be utilised both internally and externally, internal technologies are typically more problematic, as their integration into the body makes them invisible to scrutiny and, unless detected, they may be unaccounted for in the final outcome. External technologies, by contrast, are more easily reconciled with dominant notions of sport and performance. A bike, for example, is a necessary technology for the sport of cycling, and whilst an advanced piece of machinery certainly offers additional benefits to the rider, it is understood that s/he must still produce the requisite physical performance to win. The bike might be lighter or remove friction, so that what is ultimately measured is the athlete's capacity to move her/himself around a track with few, if any, external impediments that may detract from that performance. In essence, an external technology can be accounted for in the final performance. It is visible and thus knowable.

Artificial enhancement that extends the body beyond its natural limitations is of significant concern, so in essence, the notion of performance enhancement per se is not at issue, but rather the *manner* in which performance is improved (Gardner 1989). Athletic performance is typically only considered of merit if it is the culmination of a body's natural capacity coupled with visible hard work, discipline and sacrifice (Reid 1998). Technological enhancements are, by contrast, often represented as 'shortcuts' or 'easy fixes' that negate the need for commitment. This suggests that utilising 'passive' methods to improve performance does not garner as much respect as hard, sweaty activity, though, conversely, few acts are as admired in sport as the seemingly effortless performance of the 'naturally' talented athlete. In this context, it is important to note the juxtaposition of 'doped effortlessness' against 'natural effortlessness'. Performance results that come from changes in the body are thus only legitimate if the physiological changes to the body are 'earned' through work, rather than 'received' through passive means. Technologies such as hypoxic chambers where athletes need only sit or sleep are deemed to contravene the spirit of sport, as an athlete does not have to work to earn a benefit (Levine 2006).

In many respects, performance enhancing drugs are regarded as perhaps the clearest evidence that the spirit of sport is at risk, and their use is passionately labelled a 'crime' that 'undermin[es] the very essence of sport' (O'Leary 2001: 29). Doping is popularly understood to disturb the 'level playing field', a basic principle of sport, and offers 'unfair' advantages to those who ingest or inject them. Illicit performance enhancement is thought to impact the authenticity of the sports performance by creating an 'inevitable' outcome where the doped competitor is assured of victory (Reid 1998). This, in turn, seemingly lessens the value of the contest as an accurate measure of the capacities of individual competitors. Those who take a 'chemical shortcut' have their characters and morality questioned, are thought to lack discipline and courage, and are regarded as incapable of respecting 'natural capacities and limitations' (Reid 1998). It is apparent that not only the health and well-being of athletes are jeopardised by the presence of illicit performance technologies, but the very moral fibre of sport itself is at risk.

13.4 Prosthetic Technologies

Unlike performance enhancing substances, the apparel worn by an athlete has typically provoked fewer concerns that the boundary between nature and artifice is being obscured. Whilst banned substances are thought to corrupt the athletic body internally, the application of technologies to the body's surface does not seem to threaten its integrity in quite the same way (Magdalinski 2000). It would appear that the externality of these aids allows the athletic body to remain 'natural', though, of course, this is not to say that any and every external device is readily accepted without debate, as evidenced by the recent anxiety over, and subsequent rule changes to, new generation swimsuits. These 'skins' sheath the body to improve performance, whereas technologies with specific functional objectives can also be affixed to competitors. Whilst it may be argued that prostheses provide an 'enhancement' to the disabled athlete, they have not, until recently, been subject to the same level of controversy as those applied to the able-bodied athlete.

Given the vocal and sustained disquiet about 'unnatural' athletic enhancement, it may seem extraordinary that prosthetic devices are, to a large degree, embraced by the sports community as necessary and desirable. Whilst their artificiality is clearly apparent – sports prosthetics are not designed to replicate the look of the missing limb – they are nevertheless regarded as a kind of natural artifice: an acceptable substitute for the real thing rather than an inappropriate extension of the body's capacity. These technologies offer competitive advantage to the user; an amputee would certainly be incapable of running a 200 m race without a prosthesis, yet they are not regarded as an illegal enhancement. Their status as prosthetic devices, supplements that add to, but do not fully integrate on a cellular level with, the body, allows these technologies to reside comfortably alongside the body without any threat to its integrity or to the legitimacy of the resulting performance.

Part of the reason that these technologies are accepted lies in the fact that 'enhancing what is already nearly perfect and repairing what is seriously damaged are qualitatively different undertakings' (Hood 2005). The former offers advantage over competitors, whilst the latter provides the very means to compete. Shilling (2005: 178) argues that the body is not simply enhanced, because prosthetic limbs and similar technologies are thought to "*restore* rather than extend people's capacities", and are thus less confronting than those that seek to go beyond what the 'natural' body is capable. Technologies that replicate the form and function of missing limbs, for example, do not confound our understanding of the body, whilst the replacement of body parts by higher order technologies provide a worrisome step towards the creation of science fictive cyborgs. Athletes with artificial limbs are not typically thought to be 'unnaturally' enhancing their bodies, and thus are not disparaged as athletes who take drugs might be, as long as they remain within the confines of their own athletic arena. Even the use of prostheses inspired by the movement of wild cats is still mostly regarded as returning the body to a state of 'normal functioning'. As these replacements become more advanced, augmented with bionic technologies and visually less aligned with human limbs, they may become regarded as more than simply bodily restoration, reaching a point where they cease enabling participation and begin producing performances beyond the expected physical capacity of the athlete.

Whilst the area of disability sport is well established, there has been only limited focus on examining the body/technology juncture. In many Paralympic sports, participation is entirely reliant on artificial aids, such as prosthetic limbs or wheelchairs, and yet the liminal nature of the athletic cyborg only threatens the sports world order when it directly confronts athletic bodies not in need of such assistance. In this domain, the case of Oscar Pistorius presents an interesting dilemma for sports administrators. His participation in mainstream sporting events reinforces sport as an inclusive practice that 'breaks down barriers', and yet his 'imperfect' body threatens the imperative of performance sport to find the elusive endpoint of human capacity. It is thus no surprise that Pistorius' quest for entry into the Olympic arena was fraught, and decisions to variously ban or permit his participation have been controversial. Pistorius visually represents the interface between the two, and thus offers a useful insight into how the two might coexist.

Prosthetics can compensate for the loss of a limb by providing increased functionality and allowing the wearer to enjoy greater movement and independence. Within the context of sport, technological aids are often critical for those who are without certain functionalities to engage in recognised athletic activity. Indeed, a double amputee would be unable to participate in a race without the use of technology that allows for human-propelled forward motion. Sports are, of course, modified to suit the abilities of participants, and alternative versions of mainstream events emerge at the point where body and machine fuse to create a bionic competitor. Although early prostheses were designed to restore function whilst trying to resemble the missing limb (Ott 2002), the mechanics of athletic movement typically do not require an anatomically correct replica. Initial prosthetics were heavy attachments that ultimately detracted from athletic competition. Liberated from the need to replicate form as well as function, in recent decades engineers have tried to create 'bionic' arms, legs and other prosthetics that better serve the athlete's biomechanic needs (Shilling 2005), without being constrained by their appearance. New generation technologies are producing devices that bear little visual relationship to the body part they seek to replace, whilst the cyber athlete is built through the incorporation of cybernetic and bionic components. As these innovations are intended to do much more than simply replace a limb, biomechanists and engineers are designing prosthetic devices to exceed 'normal' performance and allow the wearer to 'close the gap' on their able-bodied counterparts. Unsheathing the operating mechanism of prosthetic limbs has, in some cases, caused disquiet, and although Pistorius' quest to run in the Olympic Games has prompted questions about whether or not his running blades are 'fair', athletic prosthetics are typically not regarded as an inappropriate 'enhancement' of the body and are understood more as a means to 'restore' the body to 'normal' (Shilling 2005).

The notion of returning bodies to a 'normal' state is a common theme within the bioscientific model of disability, yet some argue that the presumption that a 'disabled' body deviates from the norm presupposes a "normalcy system" that "devalu[es] bodies that do not conform to cultural standards" (Garland-Thomson 2002: 5). Like other minority groups, bodies with physical impairments are defined against a standard or "norm that is assumed to possess natural physical superiority" (Garland-Thomson 1997: 19), implying that disabled bodies are necessarily inferior specimens. Significantly, disabled bodies can only exist in a binary relationship with able bodies, as without the 'norm', there is no disability. A disabled body can only be understood not "for what it is, but for what it fails to be" (Shildrick 2005: 756), and for that reason offers a mirror against which the normal body can be measured. Carrie Sandahl and Philip Auslander (2005: 8) suggest that disability cannot be reduced to merely the level of the body for it is more accurately a "disjuncture between the body and the environment." A blind person presented with a text in Braille is not impaired; only when s/he is given printed words does a 'disability' emerge. Elite sport similarly exposes disability as its practice is based on comprehensive physical and mental functionality, embedded in rules that specify appropriate physical engagement and discard those who cannot comply. A swimmer who is unable to rest both feet flat on the starting blocks owing to a shortened right leg is repeatedly disqualified from her able-bodied competition (Andrews 1999). Whilst her minor disability is largely immaterial to her performance, it is nevertheless exposed through a regulation that assumes physical symmetry.

Disabled bodies are positioned uncomfortably within competitive sport, where skilled bodies are pushed beyond established performance limits, and where the 'less competent' body, whether disabled, overweight or unfit, is juxtaposed against the proficiency of the trained elite. Whilst athletes with disabilities are, of course, similarly engaged in pursuing their performance limits, their endeavours and achievements are measured against those of able-bodied competitors, with the former typically appearing second rate for "obviously …. Paralympians cannot hope to equal the performances of Olympians" (Cowley 1999: 14). Yet, these comparisons are largely theoretical, as able and disabled bodies almost never compete directly against one another, with disabled bodies more likely marginalised in specially established events and occasionally intruding on 'mainstream' sporting events as 'demonstration' sports. Instead of representing strong, aggressive and powerful bodies striving to achieve success whilst pushing themselves to the limit, the para-athlete is more commonly framed by a mixture of pity and courage, portrayed as 'brave' souls, competing for the 'love of sport' (Goggin and Newell 2000, 2005), a living reminder of the 'true values' of sport. Indeed, the origins of disability sport, initially conceived as part of a broader physical and social rehabilitation program for those suffering spinal cord injuries, reinforce such paternalistic treatment (de Pauw and Gavron 2005).

Despite efforts to increase awareness of patronising media coverage (Smith and Thomas 2005), athletes with disabilities are presented as the physical embodiment of the 'true spirit' of athletic competition, where participation is more valued than performance. Such discourses are particularly evidenced by the shock that accompanies revelations when Paralympians are found to be enhancing their performance, or 'boosting', as it is commonly known. Positive drug tests are returned at Paralympic events, and various methods to elicit a performance response from the body are used. Wheelchair athletes, for example, may deliberately sit on pins or tacks or block their catheters to fill their bladders till bursting. These do not cause any physical discomfort, but rather provoke the body to release adrenalin, the 'flight or fight' hormone. There are estimates that there is up to a 15 % improvement in performance (Reilly 2000; Holtz 1996). It is no surprise that Paralympians will also utilise methods available to them to secure a competitive advantage over their opponents, but what is revealing is the preference to refer to illicit tactics as 'boosting' rather than 'performance enhancing'. This suggests that athletes with disabilities only need a little extra help to 'boost' their bodies, whilst athletes with no disabilities can rely on their bodies and thus seek to enhance their performance.

The 'brave new world' of prosthetic limbs has prompted further anxiety about technology's role in creating fair/unfair 'advantage' in sport (Longman 2007; Hood 2005), and raises additional concern about the relationship between technology and the body in the construction of athletic cyborgs. Whilst the notion of the cyborg is not new, prosthetic limbs, wheelchairs and other devices traditionally signified the body was *deficient*, rather than *unfairly enhanced*. Rather than raising questions about bodily purity and the legitimacy of their inclusion in mainstream sporting events, their imperfection provided for the relegation of these bodies to a separate competitive sphere. Although many forms of discrimination have been identified in sport, the area of disability remains reasonably untouched, as long as paralympic athletes remain separate from able-bodied athletes. Rather than confirming the body as organic and natural, these athletes happily derive their functionality from a body/machine interface that exposes the body as vulnerable, open and impure. Despite their presence on, rather than in, the body, prosthetic limbs blur the edges of the corpus to which they are attached (Wilson 1995) and, with the construction of state of the art dissipative prosthetic knees or energy-storing prosthetic feet, usher in a "new age when [orthotic and prosthetic] appendages will no longer be separate, life-less mechanisms, but will instead be intimate extensions of the human body, structurally, neurologically and dynamically" (Herr et al. 2003: 133). This reveals the instability of bodily borders that will eventually collapse the body and machine into the neither-one-nor-the-other cyborg. Prosthetics will then no longer simply be a supplement to the body, whose "operating system [is] different from the body's organic processes" (Wilson 1995: 243), but rather a fully integrated and integral part of those organic processes, allowing the wearer to "more readily accept their new artificial appendages as part of their own body, rather than foreign objects that must simply be tolerated" (Herr et al. 2003: 133).

13.5 Oscar Pistorius

Whilst there are a few instances of athletes with disabilities competing in the Olympics (de Pauw and Gavron 2005), as prosthetic technologies become increasingly advanced, there may be an increasing number seeking entry to mainstream sporting events, such as the Olympic Games. Having already broken a host of world records, most of them his own, South African sprinter Oscar Pistorius, hailed as the 'Fastest Man on No Legs', set his sights on besting the 200 and 400 m qualifying times for the 2007 IAAF World Championships and the 2008 Beijing Olympic Games. He had already performed impressively in non-para-sporting events, including a second place in the 400 m at the 2007 South African Senior Track and Field Championships. Running on adjustable carbon fibre, heel-less, and unfortunately named, 'Cheetah' legs, Pistorius' quest to compete in able-bodied events by the 2008 Olympics seemed achievable, until a decision rushed through by the International Association of Athletics Federations (IAAF), track and field's governing body, appeared to thwart his efforts. In March 2007, the IAAF Council decided to implement an immediate rule change that prohibited the use in competition of "any technical device that incorporates springs, wheels or any other element that provides the user with an advantage over another athlete not using such a device" (IAAF 2008a). This stipulation, although they denied it, appeared to target the use of prosthetic devices, effectively disqualifying athletes such as Pistorius.

The timing of the rule change is significant. At the same meeting, the Council had examined over 100 other proposed amendments and had developed a set of recommendations to be circulated to member federations for consideration at the IAAF Congress just prior to the 2007 World Championships. That this was the only change implemented with immediate effect and without discussion, suggests that the IAAF believed it needed to act promptly to 'protect' the World Championships. Elio Locatelli, the IAAF's Director of Development at the time, argued that the regulation was a matter of "purity", which was specifically threatened by "something that provides advantages", including his prediction that, without a specific rule to prevent it, athletes might be tempted to try "another device where people can fly with something on their back" (Longman 2007). Although the technology to enable people to fly has certainly been in place for quite some time, its potential incorporation into track events did not seem to be much of a concern to the IAAF only 2 years previously.

In 2005, when questioned specifically about the eventuality that Pistorius may qualify for mainstream competition, the then General Secretary of the IAAF, Istvan Gyulail, indicated explicitly that "it would seem inhuman and against the sport to say 'go away and compete in events for the disabled'" (Hood 2005). Indeed, that very year, the IAAF had invited Pistorius to complete in the Grand Prix event in Helsinki (Philip 2005), though he had to decline the invitation. Given the 2007 rule change, it would seem that the practice of excluding athletes was really only ever 'inhuman' or 'against the sport' when it was highly unlikely that those with disabilities would ever actually try to gain entry to mainstream events. It is not cynical to suggest that the offer of inclusion was made at a time when Pistorius would probably not be able to qualify for high-profile track and field meets. As the likelihood increased, the IAAF moved to prevent his qualification after a commissioned study concluded that 'Cheetah' blades were 'technical aids' that provided "clear mechanical advantages" to the wearer (IAAF 2008b) as well as a metabolic advantage. The IAAF thus decided that the prosthetics contravened Rule 144.2, and Pistorius was deemed ineligible to compete in events run under IAAF rules, including, of course, the Olympic Games (IAAF 2008b). What the IAAF essentially tried to do was establish a physiological and biomechanical definition of running that relied on measures that were certain to confirm their pre-existing position that Pistorius' eligibility should be denied. This decision was overturned in the Court of Arbitration (CAS) in May 2008, which noted that not only the research was flawed, but that the IAAF appeared to have already reached its decision prior to voting. The CAS's ruling reinstated Pistorius' eligibility for IAAF events but only approved those blades that had been tested. Any improved running prosthetics would need to be retested before gaining approval for their use in mainstream sport (CAS 2008).

Whilst Pistorius did not make the qualifying times for the 400 m at the 2008 Beijing Olympics, he was hopeful of securing a place on the relay team, yet even then, the IAAF communicated their concern about his participation, citing reasons of 'safety' for discouraging the South African Olympic Committee from selecting Pistorius and then noting their lack of resources to monitor the blades that were being used in each race (BBC Sport 2008). They later were forced to release a statement confirming that Pistorius was more than welcome to seek qualification to any IAAF-run event and that they do not desire to influence the South African Olympic Committee. In July 2011, Pistorius was finally successful in achieving the A-qualifying time for the 400 m, which prompted his selection in the South African team for the 2011 IAAF World Championships. Pistorius made the semi-finals of the 400 m, and competed in the heats for the relay team, earning him a silver medal in this event. At the London Summer Olympic Games, Pistorius became the first amputee sprinter to compete, reaching the semi-finals of the men's 400 m race and the final of the men's 4×400 m relay.

Whilst Pistorius had appeared in the mainstream media on the odd occasion, it is significant that he generated a media storm only after he seemed set to enter able bodied competitions at the international level (Longman 2007), however, his running legs had already been controversial within the Paralympic arena for a number of years. His ability to "manipulate [his] anatomical structure" by adjusting the length of his legs to create a greater stride had prompted an anatomical formula, based on a competitor's armspan, to determine the 'true height' of double amputees (Hood 2005). This process was designed to mitigate 'unfair' advantage and to ensure that the artificial limb only restores, rather than enhances (Hood 2005). Whilst intended to regulate 'fairness' within Paralympic sport, as Marlow Hood (2005) observes, rather than trying to 'level' the playing field, this measure may in fact prevent double amputees from ever racing as fast as able-bodied athletes. He suggests that the "the underlying and unspoken prejudice may be that if a disabled sprinter is able to match the times of the world's best able bodied runners, then, almost by definition, he must somehow have an 'unnatural' advantage" (Hood 2005). It is perhaps ironic that after losing to him in the 200 m final at the London Paralympic Games Pistorius' criticism of fellow Paralympian Alan Oliveira centred on the length of the Brazilian runner's blades. Pistorius argued that they were not running a "fair race" and suggested that Oliveira's blades raised him four inches above his "natural height", and was indignant at how "unbelievably high" they were (Gibson 2012; Chowdhury 2012). Significantly, the International Paralympic Committee (IPC) formula that calculates "natural height" for double amputees would have allowed Pistorius to use longer blades himself. However, in non-Paralympic events means Pistorius is only allowed to use the blades approved by the CAS in 2008. Given it would be counterproductive to train on different blades for each event, it appears that Pistorius' mission to run in the Olympics came, perhaps, at the expense of success in the Paralympics.

The relationship between technology and the athletic body remains fraught, and the gradual integration of 'cyborg' athletes like Oscar Pistorius into mainstream sporting events creates friction between the mission of elite sport to establish the limits of human athletic performance and the ideal of sport as an inclusive activity. If elite sport has the objective of establishing the outer limits of performance, then bodies that are compromised by illicit technologies confound those boundaries and render their achievements meaningless. Similarly, an 'imperfect' or 'restored' body offers few insights into what the human organism is capable of achieving, and instead only confirms what such an athlete may be able to accomplish once it is returned to 'normal' functioning. Yet, this is only part of the anxiety.

As external technologies are visible, there are typically fewer concerns that they disrupt the integrity of the athletic body, whereas interventions imperceptible to the human eye provoke fears about the unnatural influence these have on performance. Whilst prosthetics have typically been regarded as necessary aids that permit wearers to engage in athletic activity, the case of Oscar Pistorius reveals that there is increasing suspicion about the ability of these technical aids to enhance performance beyond 'natural' expectations both in mainstream and Paralympic sport. Determining what his 'natural' capacity would be is, of course, a futile endeavour, so the IAAF created measures that would calibrate 'normal' running in order to determine how far Pistorius, when wearing his blades, deviates from these invented standards. They argued that the blades offer an unnatural advantage over competitors, and provided statistics to demonstrate that the South African required less energy to run. Whilst their decision to make Pistorius ineligible to run in IAAF events was overturned on appeal, their furious attempts to exclude him suggest that the interface between the organic body and inorganic prosthetic has caused increasing disquiet about the 'humanity' of performance. Inconceivably, the efforts to label Pistorius' running blades as 'unfair' or 'advantaged' and to scientifically render his performance not 'proper' running, overlook the fact that regardless how efficient his prosthetics may be, Pistorius is still missing his two lower legs, and his technological aids can perhaps not be accurately classed as an 'unfair advantage' until athletes are willing to forgo their legs in favour of these running blades.

13.6 Conclusion

Enhancing bodies in the pursuit of elite performance is fraught with suspicions about the appropriateness of selected technologies, techniques and supplements. Whilst 'pure performance' remains privileged, its integrity is threatened by a range of bodily enhancements, interventions or manipulations that are thought to destabilise not just the notion of the body, but of sports performance as the outcome of hard work, discipline and sacrifice. Concerns about the ingestion of illicit substances have been well documented, however, there is increasing disquiet over the use of technologies that seek to extend athletic capacity. External devices, namely those that are attached to, rather than integrated with, the body, have traditionally provoked less anxiety than drugs that intimately modify the body. The mechanical binding of prosthetic to body has not been thought to disrupt corporeal integrity as its visibility allows for its presence to be accounted for in the final performance. Nevertheless, this simple division between internal and external technologies is undone by the liminal figure of Oscar Pistorius, who exposes the juncture between body and technology to be as problematic in terms of assessing the validity of performance as ingested substances. Rather than serving merely to restore functioning, his running blades appear to offer him the opportunity to outperform not only other able-bodied athletes, but also, potentially, an able-bodied version of himself. Being enabled to do more than he would 'normally' be able certainly disrupts the ideal of pure performance, but it his transgression into the mainstream athletic arena that is most trepidatious. It is here that the 'flawed but restored' body threatens to outshine the disciplined bodies of elite performers, confirming that the cyborg athlete relies on technology to enhance rather than simply enable performance.

References

Adam, B. 1998. *Timescapes of modernity: The environment and invisible hazards*. London: Routledge.

Allison, L. 2001. *Amateurism in sport: An analysis and a defence*. London: Frank Cass.

Andrews, M. 1999. Blood, sweat and well-earned tears. *The Newcastle Herald*, 17 July.

Bale, J. 1994. *Landscapes of modern sport*. Leicester: Leicester University Press.

Barilan, Y.M., and M. Weintraub. 2001. The naturalness of the artificial and our concepts of health, disease and medicine. *Medicine, Health Care and Philosophy* 4(3): 311–325.

BBC Sport. 2008. Pistorius fails to make Olympics. *BBC Sport Website*, 18 July. http://news.bbc.co.uk/sport2/hi/olympics/athletics/7510774.stm. Accessed 10 Jan 2012.

Bednarek, J.R.D. 2005. The flying machine in the garden: Parks and airports, 1918–1938. *Technology and Culture* 46(2): 350–373.

Butcher, R.B., and A.J. Schneider. 2001. Fair play as respect for the game. In *Ethics in sport*, ed. W.J. Morgan, K.V. Meier, and A.J. Schneider, 119–140. Champaign: Human Kinetics.

Chowdhury, S. 2012. Oscar Pistorius beaten by Alan Oliveira in 200m. *BBC Sport*, 2 September. http://www.bbc.co.uk/sport/0/disability-sport/19460868. Accessed 10 Oct 2012.

Court of Arbitration for Sport. 2008. *Arbitration CAS 2008/A/1480 Pistorius v/IAAF*, award of 16 May 2008. http://jurisprudence.tas-cas.org/sites/CaseLaw/Shared%20Documents/1480.pdf. Accessed 10 Oct 2011.

Cowley, M. 1999. O'Neill, Thorpe may be next to cover up in go-fast 'scoot suit'. *Sydney Morning Herald*, 3 August.

Crawford, R. 1984. A cultural account of 'health'. In *Issues in the political economy of health care*, ed. J.B. McKinlay, 60–104. London: Tavistock.

De Pauw, K.P., and S. Gavron. 2005. *Disability sport*. Champaign: Human Kinetics.

Eitzen, D.S. 2006. *Fair and foul: Beyond the myths and paradoxes of sport*. Lanham: Rowman & Littlefield.

Gardner, R. 1989. On performance-enhancing substances and the unfair advantage argument. *Journal of the Philosophy of Sport* 16: 59–73.

Garland-Thomson, R. 1997. *Extraordinary bodies: Figuring physical disability in American culture and literature*. New York: Columbia University Press.

Garland-Thomson, R. 2002. Integrating disability, transforming feminist theory. *National Women's Studies Association Journal* 14(3): 1–32.

Goggin, G., and C. Newell. 2000. Crippling Paralympics? Media, disability and Olympism. *Media International Australia* 97: 71–83.

Goggin, G., and C. Newell. 2005. *Disability in Australia: Exploring a social apartheid*. Sydney: University of New South Wales Press.

Gibson, O. 2012. Oscar Pistorius erupts after Alan Oliveira wins gold. *The Guardian*, 3 September. http://www.guardian.co.uk/sport/2012/sep/03/paralympics-2012-oscar-pistorius-alan-oliveira. Accessed 10 Oct 2012.

Herr, H., G.P. Whiteley, and D. Childress. 2003. Cyborg technology: Biomimetic orthotic and prosthetic technology. In *Biologically inspired intelligent robots*, ed. Y. Bar-Cohen and C. Breazeal, 103–144. Bellingham: SPIE Press.

Holt, R. 1989. *Sport and the British*. Oxford: Oxford University Press.

Holtz, A. 1996. Paralympic officials keep eye out for 'boosting'. *CNN Website*. http://www.cnn.com/US/9608/16/paralympics.booster/index.html?eref=sitesearch. Accessed 14 Aug 2006.

Hood, M. 2005. Born to run. *IEEE Spectrum Online*, November. http://www.spectrum.ieee.org/nov05/2189. Accessed 20 Dec 2006.

International Association of Athletic Federations. 2008a. Competition rules 2008. *IAAF*, Monaco. http://www.iaaf.org/mm/Document/imported/42192.pdf. Accessed 30 April 2008.

International Association of Athletic Federations. 2008b. Press release: Oscar Pistorius – independent scientific study concludes that cheetah prosthetics offer clear mechanical advantages. *IAAF*, 14 January. http://www.iaaf.org/news/kind=101/newsid=42896.html. Accessed 30 Jan 2008.

Jenkins, P. 2002. Doping in sport. *The Lancet* 360(9327): 99–100.

Levine, B.D. 2006. Should 'artificial' high altitude environments be considered doping? *Scandinavian Journal of Medicine & Science in Sports* 16: 297–301.

Longman, J. 2007. An amputee sprinter: Is he disabled or too-abled? *The New York Times*, 15 May.

Magdalinski, T. 2000. Performance technologies: Drugs and Fastskin at the Sydney 2000 Olympics. *Media International Australia* 97: 59–70.

Magdalinski, T. 2008. *Sport, technology and the body: The nature of performance*. London: Routledge.

Magdalinski, T., and K. Brooks. 2002. Bride of Frankenstein: Technology and the consumption of the female athlete. *Research in Philosophy and Technology* 21: 195–212.

O'Leary, J. 2001. *Drugs and doping in sport: Socio-legal perspectives*. London: Routledge Cavendish.

Ott, K. 2002. The sum of its parts: An introduction to modern histories of prosthetics. In *Artificial parts, practical lives: Modern histories of prosthetics*, ed. K. Ott, D. Serlin, and S. Mihm, 1–42. New York: New York University Press.

Philip, R. 2005. Pistorius masters quick step. *The Telegraph*, 27 April 2005. http://www.telegraph.co.uk/sport/olympics/2358775/Pistorius-masters-quick-step.html. Accessed 10 Jan 2012.

Reid, H. 1998. *Sport, education and the meaning of victory*. Paper presented at Twentieth world congress of philosophy, Boston, August 1998. www/bu.edu/wcp/Papers/Sport/SportReid.htm. Accessed 31 Oct 2006.

Reilly, R. 2000. Paralympic paradox. *Sports illustrated*. http://sportsillustrated.cnn.com/inside_the_game/magazine/life_of_reilly/news/2000/12/05/life_of_reilly/. Accessed 5 June 2006.

Sandahl, C., and P. Auslander. 2005. Introduction. In *Bodies in commotion: Disability and performance*, ed. C. Sandahl and P. Auslander, 1–12. Ann Arbor: University of Michigan Press.

Shildrick, M. 2005. The disabled body, genealogy and undecidability. *Cultural Studies* 19(6): 755–770.

Shilling, C. 2005. *The body in culture, technology and society*. London: Sage.

Smith, A., and N. Thomas. 2005. The 'inclusion' of elite athletes with disabilities in the 2002 Manchester Commonwealth Games: An exploratory analysis of British newspaper coverage. *Sport, Education and Society* 10(1): 49–67.

Verroken, M. 2003. Drug use and abuse in sport. In *Drugs in sport*, 3rd ed, ed. D.R. Mottram, 29–62. London: Routledge.

Whorton, J.C. 1982. *Crusaders for fitness: The history of American health reformers*. Princeton: Princeton University Press.

Wigglesworth, N. 1996. *The evolution of English sport*. London: Routledge.

Wilson, R.R. 1995. Cyber(body)parts: Prosthetic consciousness. In *Cyberspace, cyberbodies, cyberpunk: Cultures of technological embodiment*, ed. M. Featherstone and R. Burrows, 239–260. London: Sage.

Part V
Practices and Policies

Chapter 14
Sports Physicians, Human Nature, and the Limits of Medical Enhancement

John Hoberman

14.1 Athletic Performance and 'Human Nature'

The purpose of this essay is to explore the role of the physician within what we can imagine as two concentric spheres. The first and "smaller" sphere is a social institution, while the "larger" sphere that encloses this social sphere is what we think of as the "human nature" that underlies, empowers, and limits all human activities. The social institution in this case is the global subculture of high-performance sport as it is exemplified by the *citius, altius, fortius* ideal of the Olympic Games and the practices the pursuit of this ideal has produced. These concentric spheres are related through the physical and mental demands that high-performance sport imposes on the human organism, since the purposes and consequences of making these demands, which originate in the ambitions of the social institution, can be evaluated in relation to the capacities and limitations that belong to any reasonably comprehensive conception of human nature. The sports physician, who treats (or manipulates) the elite athletes who operate within the high-performance athletic subculture, is in a position to make judgments about the medical consequences of the stressful demands of training and competition. These judgments will be influenced by a conception of human nature that medical personnel have both absorbed and helped to construct, along with other members of society. It is in this sense that an operative idea of human nature "encloses" *any* social institution, whether it is a sports league, a factory, or an army, which imposes physical and mental demands upon the people who perform inside it. Their performances are both made possible and constrained by the parameters of human nature as we understand them, a challenging topic we will address below. Our major focus, however, is on how the medical practitioner interprets

J. Hoberman (✉)
Department of Germanic Studies, University of Texas,
BUR 336, Mailcode C3300, Austin, TX 78712-0304, USA
e-mail: hoberman@austin.utexas.edu

J. Tolleneer et al. (eds.), *Athletic Enhancement, Human Nature and Ethics*, International Library of Ethics, Law, and the New Medicine 52, DOI 10.1007/978-94-007-5101-9_14,

the resources and the limitations of "human nature" as he or she treats and counsels the patient who is also a performer. Sports teams, factories, and armies employ athletes, workers, and soldiers who are served by sports medicine, industrial medicine, and military medicine, respectively. Does the Hippocratic Oath imply an adherence to a specific model of "human nature" in relation to the stress and performances that social institutions demand? Should this oath restrain the doctor's willingness or ambition to boost human performances by medical means? If it is true that sports medicine "is also interested in the enhancement of a normal body so that it can perform extraordinary, often intrinsically dangerous, feats of physical achievement and endurance," then where are we to set the ethical limits to this sort of medical behavior? (Heggie 2011).

There is general agreement that a definition of human nature must consist of multiple perspectives on human beings as well as interpreting the experiences and achievements of which they are capable. Systematic assessment of human nature must, therefore, employ a multi-disciplinary approach that includes biological, sociological, psychological, and philosophical perspectives. Theological interpretation, while optional for many analysts of human nature, has the advantage of foregrounding the issue of creation, whether by the natural world or by a Supreme Being. The creation of an organism implies a blueprint that prescribes its capacities and limitations and to what degree its "nature" has an invariable or protean essence. A blueprint should also indicate laws or principles that govern the development of the organism as well as the degree to which interventions into the development of the organism can or cannot alter the forms it might eventually take.

This very brief exercise in philosophical anthropology should be useful to those who attempt to define the proper and the improper, the ethical and the unethical roles of high-performance sports medicine in the care and development of elite athletes. Disciplined speculation about the role of high-performance sports medicine surveys a broad spectrum of possible roles that range from libertarian permissiveness to a refusal to alter the human organism in any way that deviates from traditional care, meaning the treatment and prevention of disease. The current fragility of this notion of 'traditional' medicine becomes clear if we define traditional care as the restoration of a physiologically deficient organism to its former, 'normal' state. Invoking the concept of normality here opens the door to 'replacement' or 'substitution' therapies that both sports physicians and 'anti-aging' doctors have made controversial in recent decades. In fact, medical intervention always alters the human organism in some way. This alteration becomes an ethical risk if the alteration violates a norm that is based on a defined 'normal' functioning. The ethical risk is thus a construction that originates in rules that aim at preserving an anthropological model of the athlete but that can also be violated by medical personnel who aim at maximizing athletic performance by altering the body in certain ways. The anthropological perspective can present the human organism as either mutable or immutable. The conservative position against altering the body of the athlete is a basic principle of the Vatican's theology of sport.

14.2 A Vatican Anthropology of Sport

From the standpoint of a theological doctrine that regards the human being as the product of a Creator, doping represents the violation of a divinely inspired biological creation that is endowed, according to Vatican doctrine, with an inviolable form that is subject to divinely sanctioned limits on what may be done with it. Pope Pius XII (1939–1958), called "the sportsmen's Pope," was the first pontiff to offer a critique of scientific sport. In his *Address to Roman Athletes* of May 20, 1945, Pius declared that: "The human body is the crowning achievement of the divine Creation," and that the body could only be allowed to exhaust itself "within healthy limits." This implicit critique of scientific sport became more explicit during the 1950s. On October 9, 1955, Pius aimed his criticism directly at doping practices, and specifically the use of amphetamines, when he deplored "the error of claiming the right to dispose unconditionally of his body and thereby to submit it to obvious risks and to exhausting efforts, or else, in order to obtain results that are beyond his own natural forces to absorb gravely noxious substances such as the case when consuming highly stimulating drugs, which besides being likely to cause harm to the body of a possibly irreplaceable nature, are considered as fraudulent by specialists! (Pius XII 1955)." This condemnation of physiological stimulation of the human body prefigured the deeper critique of biological manipulation that Pius delivered toward the end of his life to an audience of doctors. The Church, he said, "disapproves of every experiment of genetics which would make light of the spiritual nature of man and would treat him on a level with any member whatsoever of the animal species (British Medical Journal 1958)." In other words, Pius equated biological manipulation of the human body for athletic purposes with a desacralizing of the body whose ultimate form would be an illegitimate man-made alteration of the human genome.

Papal protests against the instrumental use of the human body for athletic purposes have continued to the present day. In his *Evangelium vitae* of March 25, 1995, John Paul II decried a "culture of death" in which the human body "is simply a complex of organs, functions and energies to be used according to the sole criteria of pleasure and efficiency." The day before the Giro d'Italia in May 2000 John Paul gave an audience to the professional cyclists who would be competing in the race. "On your knees before the pope!" thundered the *Gazzetto dello Sport*, a co-sponsor of the event. The pontiff went straight to the point: "Clarity, sincerity, and a respect for one's own limits," he demanded of the dopers who sat quietly before him. The next day before the start of the race a papal deputy, Monsignor Antonio Cecconi, did not mince his words: "Many of you," he told the riders, "have taken illegal substances to squeeze everything possible out of the human machine." For this transgression there was only one remedy: "Throw off the chains of slavery, oppose injustice, refuse compromises. And, dear riders, do not forget to confess (Süddeutsche Zeiting 2000)." Ten years later, on November 15, 2010, Pope Benedict XVI condemned doping in athletic competition, calling such "illicit means" incompatible with "human and Christian values." In November

2009 the pontiff had stated that the Church could not support a sport culture "that disfigures its nature with practices that can even damage the body, as in the case of doping (Rocca 2010)."

An anthropological model of the athletic life that is based on moral absolutism of this kind cannot address the real lives of high-performance athletes. The modern Vatican has valued sport as a school of self-control and self-abnegation, and it has made use of professional cycling as an allegory of purposeful suffering. But suffering acquires meaning only within specific contexts that may differ profoundly. Since the fifteenth century popes have applied the term *Athleta Christi'* ('Champion of Christ') to military heroes who defended the Church. The conversion of the *Athleta Christi* into a more sportive metaphorical figure has come easily in the modern era, as in: "The Olympian trains himself every day to reach perfection in his game, the athlete of Christ struggles daily to reach God through unceasing prayer (Tudora 2008)." But the shared theme of strenuous exertion is an inadequate basis for equating the *Athleta Christi* with his athletic counterpart. In *Sport at the Service of the Creator* (July 29, 1945), Pius XII argued that: "Sport, rightly understood, is an occupation of the whole man, and while perfecting the body as an instrument of the mind, it also makes the mind itself a more refined instrument for the search [for] and communication of truth…" High-performance sport, by way of contrast, is "an occupation of the whole man" only in the sense that every faculty is enlisted on behalf of producing the body's optimal performance.

For this reason the papal goal of "perfecting the body as an instrument of the mind" is the precise inverse of training the mind as an instrument of the body, as advocated by the sports psychologists who work with elite athletes. What is more, the athlete pursues, not truth, but a technical and functional perfection that could not be more irrelevant to "the service and praise of his Creator," as Pius phrased it. (Claims by "muscular Christians" since the Victorian era that athleticism is a path to holiness should thus be taken *cum grano salis*.) Every pope who has commented on sport has subordinated performance to the pursuit of a holiness that is wholly indifferent to the Olympic ideal of *citius, altius, fortius*. The perfect opposition between religious devotion and Olympic success has been gradually overlooked or strategically disregarded by the pontiffs who started to embrace sport as moral hygiene and the drama of sport as an analogue to religious zeal. At the same time, the genius of Olympic spectacle has been to endow high-performance sport with a pseudo religious aura.

Juxtaposing the *Athleta Christi* with the high-performance athlete is useful because it presents the sharpest possible contrast between competing conceptual models of human nature. On the one hand, the idea of divine creation deemphasizes what man can make of himself; on the other, high-performance athleticism adopts an instrumental approach to the human organism the Vatican rightly views as a form of high-tech paganism. Given that the Christian sports physician has yet to appear in the world of sports medicine, our analysis will focus on sports medicine's compatibility with the conception of human nature that has come to prevail, not only in sports medicine, but in a globalized world in which competition, efficiency and performance outrank any other value system that might propose a competing ideal of human nature. This is the modern world in which human nature is subject to the supposedly improving effects of 'enhancement'.

14.3 The Concept of Enhancement

The performance principle [*Leistungsprinzip*] that animates our technology-based civilization has created an anthropological model of human functioning that presupposes a human organism that must adapt itself, physiologically and psychologically, to the requirements of efficiency and productivity within the economic sphere. Inherent in this model is the idea of a performance deficit that appears as soon as the human organism confronts a performance norm it has not (yet) shown it can meet. This performance principle projects its authority and charisma well beyond the economic dimension of human life, thereby creating a generalized expectation of enhanced capacities and performance(s) across a wide range of activities that may be competitive or non-competitive in nature. This model suggests, in fact, that the enhancement ideal has an essentially unlimited mandate to inspire human ambitions to enhance capacities and performances of various kinds, and this mandate becomes clear when we contemplate the sheer range of fantasies about, and attempts to achieve, enhanced human functioning that can be documented in the public record. So the question to be answered is: How many of the dimensions and capacities of what we call 'human nature' have been recruited into the ongoing project that sees the human organism as a functional system that can be enhanced or improved or boosted or recalibrated in ways that serve some kind of productive or otherwise enhancing purpose? It should be noted that human ambitions to enhance functioning have also been directed at non-human organisms. Racing greyhounds and carrier pigeons, for example, have been doped with anabolic steroids to enhance their physical performances (running for speed and flying for distance, respectively). Such practices confirm that the mandate to enhance embraces, at least theoretically, the entirety of the organic world and its biological potential. We will, however, confine this collection to aspirations and experiments to enhance human traits and achievements, while excluding genetic manipulations of the human genome that at this point in time remain hypothetical.

Enhancement as it is now imagined embraces the functioning of the entire human organism, meaning both physical and 'cognitive' enhancement. The fact that cognitive enhancement remains at this point more of a fantasy than a technology is less of an impediment than an incitement to the belief in its effectiveness, thanks to the underlying assumption that, eventually, the sky is the limit. Indeed, much of the power of the enhancement ideal originates in the confident assumption that it is unstoppable in the longer term. The anthropological significance of the prospect of unlimited enhancement is that it imagines both a human organism that can be continuously enhanced for the foreseeable future as well as an expanding repertory of methods for enhancing human performance. This fantasy is rooted in modern confidence in the scientific method and in the idea of untapped human potential that conflicts with the idea of human limits (see below).

Cognitive enhancement is far more problematic than physical enhancement, because the brain is infinitely more complex than the skeletal muscles that power athletes. Cognitive enhancement ('neuroenhancement') drugs differ from therapeutic mood control drugs in that some people believe that they produce quantifiable results. A drug that enables a mental worker to "maintain intense focus" experiences an effect that is, at least in theory, measurable. A drug-catalyzed improvement of short-term memory or "verbal and visual episodic memory" can be observed as a quantifiable change (see Talbot (2009). An implantable electrical device is another possible enhancement technique; the vagus nerve stimulator tested in 2005 failed to relieve clinical depression, but its potential for other applications is in principle unknown (Carey 2005; Hubbard 2004). The anthropological questions concern how far one is willing to extrapolate from very modest experimental results, as in the case of the 'memory drug,' and whether one accepts the principle of unlimited possible applications of, for example, a device that sends electrical signals to the brain or to a muscle. The neuro-utopian perspective imagines an indefinitely improvable and higher-functioning brain that will produce quantifiable performance gains.

This is an attractive scenario for a civilization that makes productivity an ideal human behavior, but it is not necessarily an accurate conceptualizing of human potential. There is, in fact, a choice between a utopian or optimistic model of human functioning and a different model that is predicated on a principle of natural limits and a physiological homeostasis that imposes limits on the indefinite expansion of human faculties. This choice between two physiological, or psychophysiological, models of human functioning is medically significant in that the sports physician requires a theory of human functioning to formulate therapeutic interventions in response to the physiological state of the athlete. Most sports physicians assume that there are natural (physiological or anatomical) limits beyond which performance becomes medically pathological. Some physicians will be more inclined than others to test whatever limits may apply to athletic performance, and it should be noted that the motivation to do so may well come from the athlete-client rather than the physician. At that point the physician confronts an ethical dilemma to which he will respond in accordance with his own standards and, perhaps, in response to pressure from the ambitious athlete or from the equally ambitious "entourage" that surrounds him or her. These auxiliaries may include, along with physicians, managers, trainers, nutritionists, chiropractors, and others.

The expanding repertory of enhancements is a dynamic aspect of the enhancement phenomenon itself, since the continuous appearance of unpredictable (and often surprising) enhancement methods makes its own contribution to making enhancement seem like a limitless and inexorable process. Current enhancement techniques are often routine practices and include both familiar and lesser-known enhancements. Drugs are the most familiar enhancement technology. After more than two decades of intense and virtually global scrutiny of the doping practices of elite athletes, there is a familiar repertory of doping drugs. The best-known pharmacological enhancements for athletes include stimulants, anabolic steroids, and blood-boosting drugs (e.g., EPO). At the same time, athletic doping with drugs is only one of many forms of 'workplace doping' that include the widespread and largely unpublicized steroid doping practiced by policemen.

Cosmetic procedures are now widespread and include injections as well as surgery on a variety of body parts that range from noses to female genitalia, which can be surgically altered to 'improve' their appearance or sensitivity (Myung-Ok Lee 2011). Some procedures, depending on the context, may or may not be enhancements. Laser eye surgery, for example, can restore 'normal' vision or 'enhance' the eyesight of an athlete. The same practical approach to surgery produces face-lifts for actors and actresses (Dargis 2005).

But the power and prestige of enhancement as an ideology, as a cultural force, is based on more than the proliferation of enhancing techniques. The current prestige of the enhancement ideal also derives from the fact that it cannot be reduced to a stable process that simply expands by adding new enhancement methods to its repertory. Enhancement is rather an expanding set of fantasies that entices us to redefine and re-imagine what 'enhancement' itself is. This process of redefinition takes several forms.

It is possible, for example, to imagine that a familiar phenomenon possesses hitherto unrecognized performance-enhancing potential. There is, for example, the recommendation that athletes listen to classical music "so that exercise begins at a low pulse-rate to aid the blood flow to the muscles (Alan Hubbard 2004)." A German sports scientist has claimed that athletes can train by imagining movements in 'lucid dreams' (Koch 2004). It has been speculated that a pregnancy, aborted or completed, might boost the performances of female athletes, and that some women may have planned and aborted pregnancies for this purpose (Schüller 1988). In a less speculative vein, researchers have reported that players of video games that involving the hunting and killing of imaginary enemies greatly improve their visual attention skills (Blakesley 2003). The proliferation of such fantasies and reports transforms a wide range of human activities into potential, or imaginary, sources of enhancement, thereby intensifying the sense that enhancement is a ubiquitous dimension of modern life.

There are many 'therapeutic' drugs, including all the classic sports doping drugs, that have been used for the purpose of enhancement, thereby blurring the therapy/enhancement distinction. The apparent disappearance of the difference between therapy and enhancement creates in turn a sense that enhancement has absorbed, and thereby put an end to, purely therapeutic treatment as a medical ideal. In the realm of sports medicine, the shift from therapy to enhancement as the framing concept of 'treatment' makes the athlete more of an empowered 'client' than a 'patient' who accepts the advice of his or her doctor. This loss of authority on the part of the doctor will diminish his ability to insist on therapeutic treatment rather than accept the performance-enhancement demanded by the athlete-client. At this point the sports physician is operating inside the ethos of enhancement and has relinquished the therapeutic mandate as a check on the athletic ambition of his client. There are undoubtedly sports physicians who will refuse to pay this price for the privilege of attending to the needs of elite athletes. Other sports physicians, however, will collaborate with the athlete; some of these doctors identify so strongly with the athlete's goals as to abandon medical norms in favor of more charismatic enhancement-related ambitions.

Enhancement can also acquire a semantic dimension; for example, calling a food product 'functional' transforms its imagined status from 'nourishment' to 'enhancement,' as if nourishment were not already functional. Food takes on an extra cachet because it, too, is now enlisted in the campaign to boost the human organism to a higher level of functioning (Katan 2004).

The sphere of imagined and imaginable enhancements also expands when we recognize that other cultures use certain drugs for performance-enhancing purposes. Disapproval or 'moral panic' about such drug use arises (a) when the alien drug threatens to affect performances and competitions within one's own cultural sphere or (b) when such alien drug use is condemned on the (culture-bound) assumption that prohibition of such drug use is justified in accordance with a universal norm. One 'foreign' enhancement technique of this kind is the chewing of coca leaves to mitigate hunger and boost stamina in the workplace (Forero 2004). Khat is chewed by Yemeni workers for stamina and by students to enhance concentration on their school work (Fahim 2006).

Pharmacologically 'innocent' energy-boosting products that fall into the category of what were once called 'tonics' are now seen as performance-enhancing. The shift from innocent tonic to enhancement gives the substance a new social significance, since it now belongs to a civilizational project that aims at boosting the human organism in various ways. In the sports world such substances can be categorized as tonics or as stimulants that are banned as 'doping.' The World Anti-Doping Agency (WADA), for example, has gone back and forth on whether to regulate the intake of caffeine by the elite athletes who fall under its authority (MacMichael 2010).

One need only consult the standard definitions of human nature to see that it is a hopeless abstraction until it finds a social context and thus the human purposes that make human nature comprehensible to people who recognize their own capacities and goals in this construct. In a world of enhancement, the preeminent form of human nature is the animating principle of a human organism that can respond to the various demands for performance that are made upon it. High-performance sports medicine is an excellent prototype of this model.

14.4 Medical Thinking About Human Athletic Limits

The beginning of scientific speculation on human athletic limits coincides with the development of the biological sciences in Europe during the last decades of the nineteenth century. The extreme outcomes of sportive exertion – stress, trauma, and death –had been studied by many physiologists before doctors began to investigate the medical consequences of extreme athletic effort. The intensification of athletic effort at this time created medical concerns. The idea of human limits was closely bound up with the concept of fatigue. Increasing concern with the consequences of extreme physiological stress coincided with a transformation of the concept of fatigue from being a natural and accepted condition to being a problem that required a practical and scientific solution.

Successive eras in the development of exercise physiology have imagined the consequences of extreme exertion, and the related potential for limits on performance, in accordance with the state of contemporary scientific knowledge or ignorance. A century ago some medical observers saw physical exhaustion as a pathological, and potentially fatal, experience. A generation later the pioneering sports physiology of A.V. Hill described a healthy and self-regulating human organism that was not vulnerable to catastrophic over-exertion (see below).

During the 1880s a static view of human physical limits was the norm. "The blacksmith's arm," an American physician wrote in 1883, "cannot grow beyond a certain limit. The cricketer's quickness cannot be increased beyond this inexorable point.… The limit is fixed at different points in each man in regard to his various powers, but there is a limit beyond which you cannot go in any direction in each faculty and organ." This limit, he maintained, was nothing less than "a law of Nature (Clouson 1883)."

During the last decades of the nineteenth century there occurred a struggle between two opposed theories of human potential: the older doctrine of natural limits and a new doctrine of expanding biological limits. The conflict between these two perspectives on human limits reflected in turn the destabilization of the idea that 'normal' physiological norms existed at all (see Canguilhem 1989; Hoberman 1998; Jarry 1902).

The doctrine of natural limits assumed the human body would pay a physiological penalty for over-exertion. In 1892, for example, the British medical journal *The Lancet* reported on the fearful consequences of one of the "long and sleepless" bicycle rides that were coming into fashion. Tormented by episodes of vomiting, one exhausted rider "lost the power of his senses, and for some miles tugged on as if he were blind, tearing away, in fact, in a kind of trance, his higher nervous centers paralyzed and his body retaining its life and mere animal power." Such efforts, this author wrote, were "dangerous up to the verge of insanity," and they did not encourage contemporary observers to dream of extending the limits of an already overtaxed organism (Scientific American Supplement 1892). In the same year a contributor to the *British Medical Journal* warned against "the frequent repetition of severe muscular effort" and, especially, "the practice of cycling against time," described as "a particular pernicious practice." But a question remained as to whether the hypertrophy of the left ventricle observed in the hearts of "trained athletes" was "to be regarded as *physiological* or *pathological* [emphasis added] (Collier 1892)."

The pioneering French sports physician Philippe Tissié repeatedly described the consequences of athletic exertion as a medically pathological phenomenon. In 1894 he claimed that prolonged muscular exertion produced a state of "auto-intoxication" that amounted to being poisoned (Tissié 1894a). In 1896 Tissié claimed to have established that "intensive training provokes the same psychopathological phenomena that occur in hysteria, a hypnotic trance, and neurasthenia, such as boredom, disgust, impulsive behavior, split personality, hallucinations, illusions, phobias, paramnesia, echolalia, obsessions, etc (Tissié 1894b)." In 1919 Tissié bluntly stated that: "The athlete is a sick person [*un malade*]." His prophetic claim was that "athletic fatigue produces a kind of experimental disease in the healthy man," "a brief experimental illness on the level of cellular physiology" that causes a rise in body temperature and even a degree of immune deficiency – an intuition modern sports physiologists have confirmed (Tissié 1919).

The distinction between the physiological and the pathological also appears in a 1908 German medical dissertation on the Olympic athlete of this period, in which the author calls competitive cycling "a very dangerous experiment." The nature of the stress inflicted on the kidneys by athletic exertion, he writes, makes it difficult "to draw a clear distinction between physiological and pathological states" of the organ. He and other physicians, he reports, were now seeing in their practices "professional sportsmen who frequently contract arteriosclerosis and wind up invalids or dead at relatively early ages" – overambitious athletes who became debilitated patients. Here, too, the physiological penalty to be paid for excursions beyond human limits is portrayed in the most extreme terms (Hoberman 1992, 29).

The melodramatic pathology-centered approach to athletic exertion was displaced during the 1920s by scientific progress in the field of exercise physiology. The key figure in the creation of this more advanced science of physical exertion was the Nobel Prize-winning physiologist A.V. Hill, who as a young man had competed in cross-country and track in England. As an applied physiologist Hill emphasized not enhancing performance, but what he regarded as the mammalian body's capacity for self-regulation and, therefore, self-preservation. First, he proposed a dynamic relationship between the respective strengths of muscle and bone. "There is a resistance," he wrote in 1927, "increasing with the speed, inherent in the muscle-substance itself. This acts as an automatic brake, preventing an animal from moving too quickly and so developing such high speeds in his limbs that they would be apt to break under their own inertial stresses (Hill 1927a)."

Second, Hill and his colleagues had argued in 1923 that oxygen uptake "reaches a maximum beyond which no effort can drive it[.] The oxygen intake may attain its maximum and remain constant merely because it cannot go any higher owing to the limitations of the circulatory and respiratory system[.]"(Bassett 2002). The self-regulating human body was its own guarantee against harmful self-exploitation. "A man may be able to force himself to a greater degree of exhaustion than his rider can force a horse," Hill wrote in 1925, but Hill was not concerned about physiological penalties resulting from this kind of extreme effort (Hill 1925). For this reason, and contrary to the alarmism of certain predecessors, Hill maintained that athletes "can be experimented upon without danger," because he did not expect athletes or scientists to behave in extreme ways (Hill 1927b).

The theme of the physiological penalty for extreme exertion was revived during the public discussion of amphetamine drugs in sport during the 1950s in England and the United States (see Hoberman 2006). As early as 1953, as British sports officials struggled to find a definition of 'doping,' Sir Adolphe Abrahams, Britain's most prominent sports physician, was warning in *The Times* of London about "drugs which by inhibiting or paralyzing the protective mechanisms that normally guard against over-exhaustion could contribute to a greater physical output (Adolphe Abrahams 1953)." Such drugs constituted the exception to Abrahams' surprisingly open-minded attitude toward performance-enhancing drugs (Sunday Times 1957).

Four years later, he reiterated his concern about any "drugs of narcotic character which by diminishing sensitiveness paralyse the faculties of judgement and remove the inhibitions that ensure self-preservation," thereby removing healthy limits on human performance. It was beneath the dignity of any British athlete, he said, to seek advantage by inflicting such a physiological penalty upon himself. And it was beneath the dignity of any British physician to risk the health of an athlete in the name of sport. Within this athletic subculture, self-imposed limits on athletic performance were a matter of honour.

Late twentieth-century sports science produced new ways to theorize about the limits to human athletic performance and the potentially pathological consequences of exceeding certain physiological variables. In a lecture presented to the Congress of German Sports Physicians in October 1986, the prominent German sports physician and scientist Wildor Hollmann bluntly stated: "The *fact that we have arrived at the biological limits* can be clearly demonstrated." He addressed the historical origins of this problem in 1987: "Due to the inherent logic of high-performance sport the founders of the modern Olympic movement had unintentionally laid the foundation for a gigantic biological experiment on the human organism. By the 1960s at the latest, these human specimens had already reached their biological limits in relation to athletic performance (Hollmann 1987)." In 1984 Hollmann and two other well-known sports physicians, Heinz Liesen and Alois Mader, referred to "the biological border zone" where human athletic potential reached its limit (Der Spiegel 1984). In 1977 Alois Mader had claimed that high-performance sport was already subjecting "the biological structures of the human being" to maximum levels of stress and adaptational challenges (quoted in Sehling et al. 1989). In 1988 Heinz Liesen stated that: "The body of a high-performance athlete is no longer comparable to the body of a normal person (Süddeutsche Zeitung 1988a)." Many years later, Liesen described high-performance athletes as "sensitive people who are under extreme stress (Von Hacke and Ludwig 2011)." In the eyes of these physicians, the athletes who entered "the biological border zone" were now biologically fragile "human specimens" who risked paying a physiological penalty for having reached the limits of human performance. This situation presented the elite sports physicians with a medical and moral dilemma. Was their role to preserve the health of these athletes at the possible cost of their athletic careers, or was it to repair these people in such a way that their performances would continue to improve? In a word, was the athlete a *patient* or a *client*?

14.5 How Sports Physicians Think About Human Athletic Limits

Partnerships involving high-performance athletes and doctors have evolved during the course of the doping epidemic that began in earnest during the 1960s. During the 1970s Heinz Liesen encountered "athletes who were experimenting on themselves and knew more than we did." It is well known that most sports physicians knew little about doping drugs during the 1970s and 1980s. The celebrity sports physicians who became public (and politically connected) figures in West Germany during the 1980s and 1990s were conspicuous exceptions to this rule. Published accounts of their relationships with athletes emphasized their reputed expertise and the often charismatic effect this had on the many elite athletes who wanted their attention. Such relationships always involve a balance of power between doctor and client, and we may assume that in most of these relationships it is the doctor who has the upper hand. This is certainly the case when athletes have in recent years sought out celebrity sports physicians such as Hans-Wilhelm Müller-Wohlfarth of Munich or Anthony Galea of Montreal (on Müller-Wohlfarth see Fish 2011, on Galea see Van Natta et al. 2009). It is important to keep this balance (or imbalance) of power in mind when we examine how doctors think about treating their elite athletic clients. Is the doctor's goal to restore the athlete to his or her 'normal' level of functioning, or is it to extend performance limits beyond what has been normal for that athlete? And to what extent can the doctor influence the choices made by a client who may be emotionally dependent on the doctor?

The role of high-performance sports medicine is thus ambiguous, in that it attempts to promote the viability of the athlete's body for the purpose of enabling it to adapt to various kinds of potentially hazardous stress. At the same time, "elite athletes push themselves so close to their physical limits, that illness or injury may be viewed as an unavoidable occupational hazard (Kirby et al. 2011, 220)." Despite the fact that these extreme circumstances create a doctor-client relationship that is incompatible with the medical obligations of the Hippocratic Oath, many doctors have been willing to deviate from this traditional standard: "medical members of the athletes' support personnel were often found to be influential in facilitating doping. This facilitation can take a number of forms, including monitoring of athletes' health while doping, provision of knowledge to enable doping or administration of doping substances and methods (Kirby et al. 2011, 221)." However, such behaviours do not in themselves demonstrate that these physicians are interested in extending the limits of human performance.

A few modern sports physicians have publicly identified with the goal of pushing back human limits. "The drive toward performance is inherent in man," Wildor

Hollmann said in 1985, thereby legitimating the practice of reparative sports medicine "in the biological border zone" as well as his own refusal to become a conscientious objector to high-performance sport (Süddeutsche Zeitung 1985). In a 2006 interview given shortly before he confessed to having served as a doping doctor for German cyclists, Lothar Heinrich confessed to his own infatuation with the sportive drama in which he played a special role: "when you watch sports you are always hoping for a miracle. Aren't you? As a spectator, and I am one of them, you hope to participate in something of historic significance (Kirstner and Burkert 2010)." It is important to recognize that such sentiments are a form of romantic longing for the transcending of human limits. The physician and historian of medicine Georges Canguilhem (1904–1995) indulged in athletic romanticism of this kind in his masterwork *The Normal and the Pathological* (1966). A year before the 1924 Olympic Games came to Paris, as he notes, the literary critic Albert Thibaudet had quipped: "It is the record figures, not physiology, that answer the question: how many meters can a man jump?" Canguilhem appreciated this sort of physiological bravado because it embodied his own longing "for life to go beyond the codified biological constants" and into a biological *terra incognita* (Canguilhem 1989). But I suppose that romanticism of this kind is not what motivates most doping doctors to get involved with high-performance athletes. And romanticism is certainly not what motivates the athletes who collaborate with them.

Most elite athletes who dope do so to prolong their careers, not to set records or transcend human limits. Doping is a pragmatic choice rather than a romantic one. Even the athlete's ambition to set records can derive from his or her perception that such performances are what the public demands. The Spanish 1,500-m runner Sergio Sanchez said in 2010 that the elite athlete can "feel the need to continue entertaining the fans and putting up faster times" and then succumb to "that addictive feeling of being the centre of attention (Martinez 2010)." In a similar vein, a 2011 analysis of what motivates elite athletes to dope refers to "the pressure to please fans who [want] to see progressively better performances and have records broken on a regular basis (Kirby et al. 2011, 215)." Here the romantic ambition to set records originates, not in the athlete, but in the spectator's wish for the emotional fulfilment that comes with witnessing what the doping doctor Heinrich called the "historic" performance that can alter one's sense of human nature and its potential for self-transcendence.

Looking back at the last three decades of the best-known elite sports physicians, their celebrity, and their mystique, one sees less romanticism than professional narcissism.[1] Hans-Wilhelm Müller-Wohlfahrt once spoke of his empathy with his patients, his need to listen to their personal stories, the superiority of sensitive hands to technological diagnosis, and the 'courage' required of the doctor to 'really engage' with his patient (Von Halter and Schlimmöller 1991). "We are both front-line people", Armin Klümper said of himself and his fellow sports physician Heinz

[1] On the cult of the elite sports physician in Germany, see Hoberman (1992: 252–265).

Liesen, “who really take care of athletes instead of sitting in the ivory tower of science (Süddeutsche Zeitung 1988b).” A year earlier the hundreds of injections Klümper had administered to the German septathlete Birgit Dressel contributed to her agonizing death.[2] While medical arrogance is a common aspect of this syndrome, there is a difference between the self-blinding and potentially dangerous arrogance of a Klümper and an unshakable self-confidence based on the adulation of patients who are not put at risk of medical harm. Müller-Wohlfahrt and Anthony Galea, both specialists in repairing athletic bodies, appear to belong in this second category. What both groups share is a sense that they are exceptional doctors who are not bound by the normal rules of medical practice or by the standards of the peer-reviewed scientific literature. The limits these doctors defy are, therefore, not the physiological limits of human performance, but the limits the medical profession has been unable to impose on the charismatic doctors’ sense of entitlement to perform unorthodox procedures on, and in some cases seduce, the many clients who trust them.[3]

References

Abrahams, Adolphe. 1953. ‘Doping’ of athletes (letter), *The Times* [London], 10 July.

Associated Press. 2011. Canadian doctor Galea remorseful ahead of sentencing. *Associated Press*, 15 December.

Athletic Exercises as a Cause of Diseases of the Heart and Arteries. *The British Medical Journal* (December 3, 1892): 1, 235–236.

Bassett, David R. 2002. Scientific contributions of A. V. Hill: Exercise physiology pioneer. *Journal of Applied Physiology* 93(2002): 1574, 1575.

Blakesley, S. 2003. Video-game killing builds visual skills, researchers report. *New York Times*, 29 May.

British Medical Journal. 1958. Pope Pius XII: One of His last talks to doctors. No. 970, 18 October.

British Medical Journal. 2004. Health claims for functional food. *British Medical Journal* 328: 180–181.

Carey, B. 2005. Implant device for depression nears approval. *New York Times*, 21 May.

Canguilhem, G. 1989. *The normal and the pathological [1966]*. New York: Zone Books.

Clouson, T.S. 1883. Female education from a medical point of view. *Popular science monthly*, December, 215.

[2] On the death of Birgit Dressel, see Hoberman (1992: 1–2).

[3] The Spanish runner Sergio Sanchez has described one seduction scenario as follows. “Now I’m an elite athlete. And I get a lot of uncomfortable offers from doctors. When a doctor approaches you and before he even begins treating you he asks you a long list of questions like: “Have you ever used anabolic steroids? Have you used EPO? Have you consumed human growth hormone before? Have you taken testosterone?” If you say no to everything, two things happen: (1) He doesn’t believe that you’ve gotten where you are without having ever having taken anything, and decides not to treat you. Or (2) that the same doctor gets the idea that you really are a diamond in the rough, and he says, “Man, there’s a ton of room for improvement. With what I can give you, you can become awesome.”” See Martinez 2010.

Collier, A. 1892, December 3. Athletic exercises as a cause of diseases of the heart and arteries. *British Medical Journal* 1: 235–236.

Dargis, M. 2005. One word: Plastics. *New York Times*, 13 January.

Der Spiegel. 1984. Typen wie aus dem Panoptikum, 23 July, 71.

Der Spiegel. 1988. Da reißen Mädels Bäume aus. *Der Spiegel*, 12 September, 216, 219, 221, 223.

Der Speigel. 1991. Ich mache meine eigene Medizin. *Der Spiegel*, 1 April, 196, 198.

Der Spiegel. 2004. Botschaft aus dem Traumreich. *Der Spiegel*, 180–181.

Der Spiegel. 2011. Rotwein im Kofferraum. *Der Spiegel*, 154–155.

Fahim, K. 2006. Crackdown makes the pleasures of 'a cultural thing' riskier. *New York Times*, 31 July.

Fish, M. 2011. The doctor they call 'Healing Hans'. *ESPN.com*, 15 December. http://espn.go.com/espn/print?id=7324261&;type=story. Accessed 23 Oct 2012.

Forero, J. 2004. Bolivia's knot: No to cocaine, but yes to coca. *New York Times*, 12 February.

Heggie, V. 2011. *A history of British sports medicine*, 5. Manchester/New York: Manchester University Press.

Hill, A.V. 1925. The physiological basis of athletic records. *The Scientific Monthly* 21: 420.

Hill, A.V. 1927a. Are athletes machines? *Scientific American*, August.

Hill, A.V. 1927b. *Muscular movement in man: The factors governing speed and recovery from fatigue*. New York: McGraw-Hill; cited in David R. Bassett. 2002. Scientific contributions of A. V. Hill: Exercise physiology pioneer. *Journal of Applied Physiology* 93: 1573

Hoberman, J. 1992. *Mortal engines: The science of performance and the dehumanization of sport*. New York: The Free Press.

Hoberman, J. 1998. Fin-de-Siècle Physiology as Sexual Farce:

Hoberman, J. 2006. Amphetamine and the four-minute mile. *Sport in History* 26: 289–304.

Hollmann, W. 1987. Risikofaktoren in der Entwicklung des Hochleistungssports. In *Sportmedizin–Kursbestimmung*, ed. Rieckert, H. Kiel: Deutscher Sportärztekongreß, 16–19 October 1986; 15–16. Berlin: Springer.

Hubbard, Alan. 2004. Inside lines: Why mozart is 'better than drugs' for sport. *The Independent on Sunday* [London], 11 April.

Huffinton Post. 2010. Pope Warns athletes on doping, 15 November. http://www.huffingtonpost.com/2010/11/15/pope-warns-athletes-on-do_n_783840.html. Accessed 30 Mar 2012.

Jarry, A. 1902. The supermale (1902). *Nineteenth-Century Prose* 25(Spring 1998): 154–166.

Katan, M.B. 2004, January 24. Health claims for functional food. *British Medical Journal*: 180–181.

Kirby, K, Aidan Moran, and Suzanne Guerin. 2011. A qualitative analysis of the experiences of elite athletes who have admitted to doping for performance enhancement. *International Journal of Sport Policy and Politics* 3(July 2011).

Kirstner, T., and A. Burkert. 2010. Meine Arbeit ist Teamarzt, nicht Dopingkommissar. *Süddeutsche Zeitung*, 17 May.

Koch, J. 2004. Botschaft aus dem Traumreich. *Der Spiegel*, 1 March, 180–181.

MacMichael, S. 2010. WADA president to urge re-banning of caffeine. *Road.CC*, 11 August. http://road.cc/content/news/21341-wada-president-urge-reconsideration-lifting-caffeine-ban. Accessed 23 Oct 2012.

Margaret, T. 2009. Brain gain. *The New Yorker*, 27 April, 32, 35.

Martinez, S.S. 2010. Spanish athletics doping scandal: Sergio Sanchez opens up about doping in Spain. *LetsRun*, 14 December. http://www.letsrun.com/2010/sergio-1214.php. Accessed 23 Oct 2012.

Müller-Wohlfarth see Mike Fish. 2011. The doctor they call 'healing hans'. ESPN.com (December 15, 2011). http://espn.go.com/espn/print?id=7324261&type=story.

Myung-Ok Lee, M. 2011. Designer vagina surgery: Snip, stitch, kerching!. *The Guardian*, 14 October. http://www.guardian.co.uk/lifeandstyle/2011/oct/14/designer-vagina-surgery. Accessed 23 Oct 2012.

New York Times. 2003. Video-games killing builds visual skills, researchers report. *New York imes*, 21 May, XXX.

New York Times. 2005. One word: Plastics. *New York Times*, 23 January.
New York Times. 2006a. Bolivia's knot: No to cocaine, but yes to coca. *New York Times*, 12 February.
New York Times. 2006b. Pleasure of 'a Cultural Thing' turns riskier. *New York Times*, 31 July.
New York Times. 2009. Doctor who treated top athletes is subject of a doping inquiry. *New York Times*, 15 December.
Pius XII. 1955. *Address to the centro sportivo italiano*. 9 October, found in Italian in *Discorsi e radiomessaggi di Pio XII* XVII(1955): 277.
Pope Pius XII: One of His Last Talks to Doctors. *British Medical Journal* (October 18, 1958): 970.
Rocca, F.X. 2010. Pope warns athletes on doping. *The Huffington Post*, 15 November. http://www.huffingtonpost.com/2010/11/15/pope-warns-athletes-on-do_n_783840.html. Accessed 23 Oct 2012.
Schüller, H. 1988. Da reißen Mädels Bäume aus. *Der Spiegel*, 12 September, 216–223.
Scientific American Supplement. 1892, September 24. Cycling and vitality, 13: 597–598.
Sehling, Michael, Reinhold Pollert, and Dieter Hackfort. 1989. *Doping im Sport: Medizinische, sozialwissenschaftliche und juristische Aspekte*, 99. Munich: BLV.
Should Athletes Take 'Pep' Drugs?. *The Sunday Times* (June 16, 1957): 27.
Spanish Athletics Doping Scandal: Sergio Sanchez Opens Up About Doping in Spain. (December 14, 2010). http://www.letsrun.com/2010/sergii-1214.php#ixzz11k2irmnm.
St. John the Baptist Greek Orthodox Church. 2011. The athletes of Christ, November. http://stjohndfw.info/the-athletes-of-christ.html. Accessed 18 Nov 2011.
Süddeutsche Zeitung. 1985. Der Sport ist wie ein ungepflügtes Land. *Süddeutsche Zeitung*, 30 January.
Süddeutsche Zeitung. 1988a. Sportmediziner uneins. *Süddeutsche Zeitung*, 2 February.
Süddeutsche Zeitung. 1988b. Frontmann, Guru, Zielscheibe: Klümper öffnet die Tür. *Süddeutsche Zeitung*, 19 July.
Süddeutsche Zeitung. 2000. Audienz für die Radsport-Sünder. *Süddeutsche Zeitung*, 13 May.
Süddeutsche Zeitung. 2006. Meine Arbeit ist Teamarzt, nicht Dopingkommissar. *Süddeutsche Zeitung*, 19/20 August, 547.
Sunday Times. 1957. Should athletes take 'Pep' drugs? 16 June, 27.
The Guardian. 2011. Designer vagina surgery: Snip, stitch, kerching! *The Guardian,* 14 October
The New York Times (NYT). 2005. Implant device for depression nears approval. *The New York Times*, 21 May.
Tissié, P. 1894a. Concernant un record velocipédique. *Archives de physiologie normale et pathologique*, 837.
Tissié, P. 1894b. L'entraînement physique. *La Revue Scientifique*, 518.
Tissié, P. 1919. *L'Éducation physique de la Race*, 156–157. Paris: Ernest Flammarion, Éditeur.
Tudora, Fr.V. 2008. The athletes of Christ. *Father Vasile's Monthly Reflections*. http://stjohndfw.info/the-athletes-of-christ.html. Accessed 23 Oct 2012.
Van Natta, D. Jr, M.S. Schmidt, and I. Austen. 2009. Sports medicine pioneer subject of doping inquiry. *New York Times*, 14 December.
Von Hacke, D., and U. Ludwig. 2011, October 31. Rotwein in Kofferraum. *Der Spiegel* 44: 154–155.
Von Halter, H., and H. Schlimmöller. 1991. Ich mache meine eigene Medizin. *Der Spiegel*, 1 April, 196–198.

Chapter 15
Anti-doping Policies: Choosing Between Imperfections

Bengt Kayser and Barbara Broers

15.1 Introduction

A series of widely publicized doping scandals and public outrage at the end of last century triggered an increasingly strong movement advocating doping-free sports, which led to the inception of the World Anti Doping Agency (WADA) in 1999. WADA aims at harmonizing anti-doping practices worldwide and is helped by the fact that a majority of UN member states signed the UNESCO anti-doping convention. Even though the principle of the anti-doping rule is simple, its implementation needs a complicated, highly technical and costly organisation. Still, anti-doping cannot achieve its declared objective, the eradication of doping. Even more, anti-doping policies are ethically problematic, and possibly lead to more harm to society than they prevent (e.g. Amos 2008; Kayser and Smith 2008). Anti-doping seems therefore not a solution, but an increasingly costly imperfection. We think that the discussions on doping and anti-doping should not ignore these imperfect practical outcomes of current anti-doping policies, in elite sports and in amateur and outside sports.

In this chapter we therefore suggest an alternative way of dealing with the problem of doping in sports. We do this by drawing parallels between anti-doping and the '*war on drugs*', a mainly USA-led worldwide effort to rid society from illicit psychotropic drugs. This comparison is not as farfetched as it initially appears. In fact, the '*war on drugs*' and anti-doping share various features, such as

---

This chapter is partly based on material published in a paper by the same authors: Kayser B, Broers B. 2012. "The Olympics and Harm Reduction?." *Harm Reduction Journal* 9 (1): 33. doi:10.1186/1477-7517-9-33. (Creative Commons Attribution License)

B. Kayser (✉)
Institute of Movement Sciences and Sports Medicine, Faculty of Medicine,
University of Geneva, 10, rue du Conseil Général, 1205, Geneva, Switzerland
e-mail: bengt.kayser@unige.ch

B. Broers
Unit for Dependency in Primary Care, Department of Community Health and Primary Care,
University Hospitals of Geneva, Geneva, Switzerland

J. Tolleneer et al. (eds.), *Athletic Enhancement, Human Nature and Ethics*, International Library of Ethics, Law, and the New Medicine 52, DOI 10.1007/978-94-007-5101-9_15,

similarities between policies based on repression and surveillance, black markets controlled by organized crime, attempts to shape internationally harmonized legal frameworks, ideology and political convenience anchored in media-fuelled moral outrages. Furthermore, as a detail, but an illustrative one, the presence of cannabis derivatives, drugs less dangerous for the user and society than alcohol or tobacco, on the list of forbidden substances in sports, even though there are no known proven performance enhancing effects but rather evidence for the contrary. We first discuss past and current policies for psychotropic drug use and then draw parallels with past and current anti-doping policies. We end the chapter with a rough outline of an alternative anti-doping policy. For our purpose we use the concept 'human nature' in an evolutionary perspective, in which 'good' and 'bad' human behaviour are seen as anchored in our evolutionary past but influenced by human culture (e.g. Buss 2001). In such a framework both altruistic and selfish behaviour can be seen as 'natural', depending on the socio-cultural context in which they occur.

15.2 A World Without Psychotropic Drugs?

If given the choice, many of us would probably prefer to live in a world without the significant societal problems of psychoactive substance use. Even though moderate controlled use of, for example, alcohol, cannabis, cocaine or heroin is possible, the addictive properties of these and other licit or illicit psychotropic substances are indeed associated, especially in certain socio-cultural economic circumstances, with an important potential for excess use and major social and health hazards (Degenhardt and Hall 2012). In the mid-last century an increasing prevalence of visible illicit drug use in the USA triggered a 'moral panic' (Garland 2008), which led to the UN Single Convention on Narcotic Drugs of 1961 and the declaration of the 'war on drugs' by president Nixon in 1971. Fuelled by a utopian vision of a world free of illicit psychotropic substance use – "*A drug free world—we can do it*", (cited in Jelsma 2003) – a huge international effort was launched to root out both psychotropic drug production and use. After 50 years of enormous and extremely costly efforts the endeavour was recently labelled a failure by the Global Commission on Drug Policy (GCDP), an international group of knowledgeable people including several former statesmen and a past UN president (GCDP 2011). The commission's report completed an article from Wood et al. (2009) entitled *The war on drugs: a devastating public-policy disaster*. The Global Commission confirms that the 'war on drugs' not only has not attained its objective of reducing illicit drug use, but also has induced many more problems than it has solved, quite akin to an earlier attempt in the last century to ban alcohol in the USA, the prohibition period. The negative consequences induced by the war on drugs, in producer, transit and consumer countries, include the development of a huge criminal black market, extensive policy displacement with spending of public money on law enforcement instead of other purposes, geographical displacement with, for example, cocaine plantations hidden in natural reserves, HIV and hepatitis epidemics, and the stigmatisation and marginalization of drug users (GCDP 2011; Wood et al. 2009).

The Global Commission states that there is a clear need for fundamental reforms in national and global drug control policies. Again, these alternative policies cannot yield a final 'solution' to the problem of psychotropic drug use, because these drugs will not go away and their use can be seen as part of 'human nature'.[1]

Alternatively, the use of psychotropic drugs can be regulated and the harm to society and the individual minimized. The Global Commission's report started with "*the recognition of the global drug problem as a set of interlinked health and social challenges to be managed, rather than a war to be won*" (GCDP 2011, p. 4).

The alternative strategy proposed is one of regulation of drug use, based on human rights and public health principles, with a combination of pragmatic policies taking into account local socio-cultural and economic specificities, and continuously adapted to on-going developments. High on the list of policies proposed are treatment and so-called harm minimisation measures. The principle of harm minimisation is to propose strategies that limit the health hazards associated with drug use. Instead of striving for a utopian ideal, i.e. a drug-free world, society regulates its use and limits associated harms for both drug users and non-users, with the utilitarian perspective of less cost to the individual and society as a whole. Countries like the Netherlands, Switzerland and lately Portugal, where such policies have been put in place, have shown that these strategies are accompanied by a reduction in the cost to society and the individual, by decreasing drug-related mortality and morbidity, as well as crime and its associated costs, without an increase in the prevalence of illicit drug use (GCDP 2011). Examples of, mostly evidence-based, harm minimisation measures include needle and syringe exchange programmes, safe use facilities, opiate substitution therapy, overdose prevention and chemical analysis of party drugs. Some of these pragmatic ways to deal with societal problems can be perceived as 'messy'. For example, in safe injection facilities drug users bring their own supplies obtained on the black market, an idea that may be repulsive for many. But such facilities have proved their utility beyond any reasonable doubt, by reducing hepatitis and HIV transmission rates, reducing the general health burden in injecting drug user cohorts and reducing the societal costs of open drug scenes, without increasing drug use or injecting behaviour (EMCDDA 2010).

One should remember that harm minimisation is a common way to regulate human behaviour in society. Examples include safe sex campaigns to prevent sexually transmitted infection, use of safety belts while driving, and also limiting hours or age for alcohol sale, reduction of legally accepted blood alcohol concentration for driving, provision of nicotine patches for smokers or inhalation devices for cannabis smokers.

[1] As Sullivan and Hagen (2002) posit, it appears likely that we carry the results of our ancestors' several million years long evolutionary relationship with psychotropic plant substances. Our viewing of human substance-using phenomena should therefore incorporate the evolutionary implications of a long time relationship between psychotropic substances and people. For example, selective benefits of substance use may have been that neurotransmitter-analog plant chemicals were exploited as substitutes for costly, nutritionally constrained endogenous neurotransmitters (Sullivan and Hagen 2002).

15.3 Barriers and Opportunities for Harm Minimisation

The difficulty of such a pragmatic approach to the societal problems of (il)licit psychotropic drug use is that it explicitly accepts this behaviour as part of human behaviour and allows the behaviour to continue, putting it politically in a much less easier defended position than the utopian promise of a drug free world. The USA has, despite a strongly repressive stance and the employment of extremely costly means, among the highest levels of psychotropic drug use in the world and the highest number of imprisoned subjects, most for illicit drug-related offences (Degenhardt et al. 2008). By contrast, in the Netherlands, where cannabis use is tolerated and regulated (and *de facto* but not *de jure* legalized), and heroin addicts have access to opioids substitution programs, the burden for society is clearly less and both heroin and cannabis consumption may have fallen, rather than increased (Room and Reuter 2012).

According to a recent thorough analysis from a legal-economical perspective of past and present USA anti-drug policies:

> much of the debate over illegal drug policy and potential reforms hinges on two contentious questions. First, by how much would the prevalence and intensity of a drug's use rise under a different regime? Second, would reductions in other social costs—particularly through lower rates of crime and criminal justice enforcement costs—outweigh the costs of increased consumption? (Donohue et al. 2011, p. 4)

This careful analysis convincingly argues, for the cases of cannabis and cocaine, that a combination of legal changes (with scenarios ranging from depenalization to legalization) and treatment options for users, would very likely result in clear societal benefits.

15.4 Barriers and Opportunities for Change in Drug Policies

Consequent to the observation of the failure of the 'war on drugs', the Global Commission formulated a series of recommendations (GCDP 2011). These include ending criminalization, marginalization and stigmatization of people who use drugs but who do no harm to others, and the experimentation by governments with models of legal regulation of drugs to undermine the power of organized crime and safeguard the health and security of their citizens. One of the reasons why these recommendations are difficult to implement is that the 'war on drugs' has gained a universal status through the UN Single Convention on Narcotic Drugs of 1961 and its follow-up treaties. The Convention's objective is to eliminate the illicit production and (non-medical) use of cannabis, cocaine, opioids and other substances. The problem is that such international treaties constrain national policy experimentation because they require nations to criminalise drug use (Room and Reuter 2012).

> The international system precludes policies that are more aligned to the risks of drug use and the adverse consequences of prohibition, such as the regulation of producers, consumers, and the conditions under which drugs are used. Policy experimentation requires changes to the international treaties, which are possible in principle but unlikely in practice. Other options include renunciation of the treaties and re-accession with reservations, or adoption of a new treaty. (Room and Reuter 2012, p. 84)

This constellation unfortunately slows the possible pace of change and suggests that the 'war on drugs' will not be abandoned soon.

In sum, zero-tolerance policies and a 'just say no' attitude are politically much easier to defend than a utilitarian approach that accepts the presence in society of psychotropic drugs and uses pragmatic policies to reduce the associated burden. The idea of a world free from the societal problems related to psychotropic drugs is appealing, still a zero-tolerance approach criminalizing psychotropic drug use in order to reach this objective has clearly proven ineffective, is associated with important unintended harms and is extremely costly to society. No solution to the problem of psychotropic drug use exists, but there are proven examples of pragmatically dealing with it in such a way that the overall burden for society remains limited.

15.5 A World Without Doping?

If given the choice, many of us would probably prefer a world without doping. Still, history tells us that doping has been part of sports from its very beginnings (de Mondenard 2000; Yesalis and Bahrke 2002). If during the first half of the last century, in the general public's view, doping was seen as part of sports, over the last decades doping has become increasingly seen as deviant behaviour. One probable reason for this increasingly negative public opinion of athletes who dope is the similarity between the image of the doped athlete and the negative image of the illicit drug user, possibly as a result of the media coverage of the 'war on drugs' (Amos 2008). A series of high profile doping scandals in the 1980s and 1990s, like the Festina case during the Tour de France, led to an important change in how sports and society deal with doping in sports (Fincoeur 2009). This string of widely publicized events led the International Olympic Committee (IOC) to establish the World Anti-Doping Agency (WADA) which aims at harmonizing anti-doping efforts worldwide from a zero-tolerance standpoint.

15.6 Anti-doping Policy Today

What is doping? Generally doping is understood by the public as the use of forbidden performance enhancing drugs or methods by athletes. Operationally, in elite sports, doping is today defined in the WADA Code (WADA 2009): "*the occurrence of one or more of the anti-doping rule violations set forth in Article 2.1 through Article 2.8 of the Code*". Violations of the anti-doping rule include not only the use or attempted use of prohibited substances, but also the presence of a prohibited substance or its metabolites or markers in an athlete's urine or blood sample; violation of the athlete's obligation to inform about his/her whereabouts; tampering or attempted tampering with doping control procedures; possession of prohibited substances or the means for performing prohibited methods; and trafficking or attempted trafficking in a prohibited substance or the means for performing a prohibited method (WADA 2009). In collaboration with international sports federations, national anti-doping agencies and official anti-doping laboratories, international and national political institutions, WADA orchestrates global anti-doping efforts. WADA's declared objective is the eradication of doping in sports. This objective is considered of such importance that it justifies increasingly costly, severe and punitive measures, imposing a growing burden on athletes and society – dubbed by some the 'war on doping' (Mendoza 2002) – based on a zero-tolerance approach, quite like the 'war on drugs'.

A majority of the public is ignorant of how exactly anti-doping is enforced and what it implies for athletes. Important aspects of anti-doping are surveillance practices which include a 'whereabouts' rule, or the obligation for a selected pool of elite athletes to inform the anti-doping authorities where they will be at 1 h per day, each day of the year, to allow unannounced out-of (and in)-competition testing. These athletes have to inform the authorities four times a year, 3 months in advance of their plans, and use electronic and paper-based means to inform of any changes (WADA 2012). This 'whereabouts' rule is seen as an essential cornerstone of anti-doping, as it prevents out-of-competition doping in preparation for competition. In order to force athletes to comply, a rule states that three missed tests within an 18-month period constitute a doping offence. Athletes regularly have problems keeping the authorities properly informed, and sanctions because of three missed tests are not uncommon. The actual testing involves providing urine samples (produced in full view by an anti-doping officer), consenting to blood sampling, and possibly also providing hair samples for doping history and tissue for gene profiling for forensic practices. Longitudinal testing, looking for suspicious fluctuations in certain blood parameters, is now also being introduced. This practice, known as the 'athlete biological passport', has recently led to the first indictments of athletes based on indirect indices of presumed doping rather than laboratory tests directly showing the presence of the forbidden substances or their metabolites in urine or blood. The athlete biological passport is hailed by the authorities as an improvement of anti-doping (Zorzoli 2011). Still, it has already been suggested that this indirect method of doping detection might be unreliable and produce false-positive results due to analytical variability and outlying individual patterns resulting from the effects of behaviour (e.g. training, altitude training) and genetics (Lippi and Plebani 2011; Sanchis-Gomar et al. 2011; Banfi 2011). Furthermore, anti-doping has a non-negligible cost (Kayser et al. 2007; D'Angelo and Tamburrini 2010). The IOC finances half the budget of the WADA, while the other half comes from national governments. National anti-doping agencies are mostly co-financed by national sports federations and governments. Overall the tendency is towards increasing costs with a rapidly developing anti-doping industry steadily asking for more. The application of new national anti-doping legislations also comes with an increase in cost.

Taken together, all of these costly surveillance practices seriously impinge upon the privacy of athletes and set them apart from the general population, for whom the protection of the private sphere and autonomy are generally respected in democratic societies, and are at odds with general relaxed attitudes of modern society towards human enhancement practices such as cosmetic surgery (e.g. Frederick et al. 2007) or caffeine use, and with an increasing prevalence of the use of cognitive enhancement drugs (Smith and Farah 2011).

15.7 Some Other Problems of Current Anti-doping Policy

In spite of all this surveillance to attain the objective – elimination of doping in sports – doping continues. Official statistics report a steady rate of positive tests in 1–2% of elite athletes (www.wada-ama.org), while social sciences research using special questionnaire techniques that allow correction for social desirability answering bias (Petróczi and Nepusz 2011), estimate the actual rate to be much higher, probably at least at 7% (Striegel et al. 2010), or even more (Pitsch and Emrich 2011), although even those techniques may not accurately reflect the true prevalence (Lentillon-Kaestner and Ohl 2011). Still, it seems likely that the actual prevalence is much higher than that found by means of anti-doping controls. A recent analysis of repeated blood test results (7,289 blood samples from 2,737 elite athletes collected over 10 years), identifying deviations from a non-doped population baseline, was also strongly suggestive of a higher rate of blood doping practices in athletes than is reflected by the official numbers from WADA (Sottas et al. 2011). The reason for these discrepancies is that testing is limited by testing density and laboratory technology, and that well-informed and medically assisted top athletes can remain undetected by using specific dosage schemes of certain substances that are missed by current testing procedures. Most biomedical tests, including those used for clinical reasons, have their limits. Anti-doping needs to keep false positives as low as possible, while striving for the highest sensitivity possible. For statistical reasons the number of false positives occurring rises with the number of tests performed, as well as with a drop in prevalence of actual doping (Pitsch 2009; Banfi 2011). For some the accusation and condemnation of innocent athletes may be justified as inevitable for the sake of sports, but we find it ethically hard to defend. Given the failure of the efforts to reach the objective – eradication of doping – anti-doping thus introduces a paradoxical twist. Among the winners on the podia there probably are dopers who remain undiscovered while, occasionally, an innocent athlete may be accused of doping. Contrary to general principles of regulatory systems where some rule-breakers will not be caught while some rule-abiders may be caught unjustly, the fact that precisely the champions may belong to the rule-breakers puts anti-doping apart from most other regulatory systems. The celebration of 'clean' athletes is the very purpose of anti-doping, the inherent limits of anti-doping surveillance policies thus paradoxically introduce elements of inequality instead of establishing what is considered important by the sports establishment, a level playing field. Certainly, anti-doping has led to important changes in doping practices; some sorts of doping like the injection of 'normal' doses of erythropoietin to boost blood oxygen transport, or of anabolic steroids to stimulate muscle growth and facilitate recovery, cannot be used anymore because detection has become easy. An indirect indication of an effect of such changes in doping practices comes from the observation of a reduction in the average speed in the Tour de France since the introduction of better surveillance practices (Perneger 2010). Still, in spite of these changes, doping continues as the field quickly adapts its doping behaviour according to what anti-doping achieves. So even though doping type and prevalence are reduced it thus seems quite unlikely that doping will ever disappear and any champion remains de facto a potentially undiscovered sinner (Fincoeur 2009).

Anti-doping uses public health arguments to promote its stance. Nevertheless, surveillance measures of anti-doping are only applied to elite athletes, a very small fraction of the world population and of all persons practicing some form of sports. Since amateur and sub-elite athletes are not or are rarely the subject of surveillance, it can be expected that doping continues at these levels, and also in the general population not engaging in competition but using non-prescribed substances to enhance performance for other reasons. Examples of the latter are anabolic steroid use in fitness-club members for reasons of appearance, or the use of non-prescribed stimulants like modafinil or methylphenidate by students to increase cognitive performance, phenomena that seem to be on the increase (Teter et al. 2006; Kanayama et al. 2010; Smith and Farah 2011). Anti-doping in sports might influence these practices in the sense that, due to the legal changes induced by the anti-doping movement, the behaviour is relegated into hiding and pushes users to purchase substances on the black market and use them without medical supervision (Aitken et al. 2002). In the USA the Steroid Act of 2004 categorises steroids in class III drugs, subject to criminal law. A first offense of simple possession without a prescription is punishable by up to 1 year in prison. The unlawful distribution or possession with intent to distribute anabolic steroids is punishable, as a first offense, by up to 10 years in prison. In spite of this policy, the prevalence of anabolic steroid use among non-competitive gym and fitness members is much higher than that among elite athletes, in some populations estimated to be morc than 15 %, while one user in ten may use unsafe injection techniques (Parkinson and Evans 2006; Melnik 2009). Thus, a perverse side effect of anti-doping efforts may actually be an increase of public health problems. In his review Melnik cites a German study (Striegel et al. 2006) in which 50% of androgen users received the substance through physicians' prescriptions and one-third through a pharmacist, often without prescription. Melnik argues that this is a serious problem (Melnik 2009). The contrary can be argued too, since the effect of more control and tougher legislation could possibly be that substances are more frequently obtained on the black market and used without medical supervision, leading to a potential increase in health hazards. There is a flourishing black market in doping substances controlled by organized crime (Kraska et al. 2010), quite similar to the commerce in illicit psychotropic drugs. Just as in the latter case, the doping substances on the market are very often of doubtful quality, sometimes do not contain what is on the label or do not contain any active substance at all (Kohler et al. 2010). Adulteration poses health hazards (Cole et al. 2010) exacerbated by inappropriate injecting techniques and risk of transmission of hepatitis and HIV (Larance et al. 2008).

A striking example of how anti-doping policies further directly influence society outside the scope of competitive sport is the recent extension of Danish anti-doping policy to commercial fitness clubs (gyms) in which clients engage into weight lifting and other types of exercise for health and appearance purposes but not sports competition (Christiansen 2011). Danish gyms have to put either a happy green smiley on the entrance, indicating compliance with anti-doping Denmark rules, including surprise urine testing of clients, or an unhappy red smiley, with the explicit message that the club does not adhere to anti-doping Denmark. This example illustrates the potential for generalization of anti-doping surveillance practices in society in general.

Because of these unintended effects we suggest that anti-doping should not be a matter of concern to elite-sports only. Still, anti-doping arguments are frequently formulated which explicitly or implicitly ignore the actual practical consequences of actual anti-doping policies in and outside sports (Loland and Hoppeler 2011; Wiesing 2011). We believe that this is ethically unacceptable. The consequences of the ‘war on doping’ should be fully taken into account when engaging into the debate on how to regulate the use of performance enhancing substances and methods in sports.

To summarize, doping continues in elite sport in spite of the increased surveillance, while doping-like behaviour appears to increase in the general population and amateur sports. The legislative changes imposed on nations by the sports establishment give doping-like behaviour an illicit status with potentially dangerous consequences.

15.8 Why Do Athletes Dope and Continue Doing So?

Doping has always been part of sports. Already at its early Olympic roots and also at the beginning of modern sports about a century ago, athletes have always looked for and applied any ways imaginable to increase performance, be it by taking substances or by using methods (Yesalis and Bahrke 2002; de Mondenard 2000). Taking as an example the Tour de France, it can be posited that in the last century, in virtually each year the winner and/or runner-ups were either known to have used doping or strongly suspected of having done so (de Mondenard 2011). Aspiring to organize a Tour without doping can therefore be seen as trying to invent a Tour that has never existed before, even though anti-doping voices have also been present from the very beginnings of the Tour. The difference between earlier and more recent Tours, admittedly important, is the kind of performance enhancing technology that has been made available by the biomedical revolution over the last decades. If in the early days of sports the arsenal of performance enhancing compounds was quite limited, today's advances of biomedical science have indeed opened up Pandora's box with unlimited possibilities but also increased health risks. Since a doping culture has always been part of cycling it is quite understandable that these new possibilities from bio-medical research were, and still are being exploited for performance enhancement practices by cyclists and their entourage. WADA's claim that a culture of doping-free sport will develop and help attain the eradication of doping in sports remains to be proven; for competitive road-cycling, recent publications suggest that although doping practices certainly have changed, a culture of doping in professional cycling still prevails (Lentillon-Kaestner et al. 2011). A provocative editorial in the journal *Nature* proposed that perhaps the Tour de France should be the first competition to accept pharmacological performance enhancement (Nature 2007).

Given the fact that in spite of the increasingly repressive means employed to combat doping, athletes are regularly caught – while others probably get away with it – the question arises: why do athletes continue taking the risk? The underlying question is whether current anti-doping policy in sports, striving for a world free of doping in elite sports, is commensurate with how human (doping-) behaviour is determined. Using a point of view from a sociocultural perspective, but interpreting behaviour as partly determined by our evolutionary past, we suggest that doping in athletes is perfectly logical human behaviour. Given what modern sport is today, i.e. an important entertainment enterprise in which large amounts of money go round, strongly biased towards the celebration of winners and applying the general concept of 'winner takes all', it is not surprising that athletes can be drawn to whatever promises an edge in competition. The official credo of the Olympic movement is '*Citius, Altius, Fortius*', or faster, higher, stronger. It is completed by the phrase '*The most important thing is not to win but to take part*'. But the latter is not reflected by reality. Today an important tenet of elite sport is the celebration of the winner. This puts a lot of pressure on athletes, who are perpetually seeking the competitive advantage. As Petroczi (2007, p. 10) states: "*using doping agents may be more of a rational, outcome optimizing behaviour than deviance*".

What does it take to be a champion? The ingredients are a combination of talent, hard work and some luck. Talent is a licit, albeit unmerited gift and results from the genetic lottery. Hard work is essentially training and using any other means that are allowed to improve performance. Performance enhancement is a natural and essential ingredient of competitive sport. Athletes naturally look for ways to get better, by changing their training paradigm, by eating differently, by taking vitamins, by taking licit medication, by taking supplements. A huge sports supplements industry exists and it is very common for athletes to consume a lot of substances that are not on the list of forbidden substances (McDowall 2007). In 2012 the British Journal of Sports Medicine published its 28th article on the A–Z of nutritional supplements (Currell et al. 2012) and the list continues to grow. Many of the supplements do nothing, a few have an effect, but quite many may pose doping problems because of adulteration, leading to accusation of doping because of positive blood or urine samples (Petróczi et al. 2011a) or health problems because of excess intake of some compound, erroneously seen as innocuous if taken in excess (Carlsohn et al. 2011).

In spite of WADA's definition of doping, the line between licit and illicit is often perceived as blurred. Transgression is not necessarily accompanied by a fundamentally different mindset as when keeping to the rules. In athletes' minds, doping may align with illicit behaviour or with functional licit use of chemical or natural preparations (Petróczi et al. 2011b).

In summary, modern sport puts athletes under enormous pressure to win and the use of licit substances and methods to improve performance is explicitly encouraged. The line between licit and illicit fluctuates and has dimensions that can be perceived as arbitrary.

15.9 The Justification for Anti-doping

Because of the problems of current anti-doping policy the question arises as to what the reasons for the anti-doping endeavour are. The main justification for anti-doping is formulated as follows in the 'Code', the central document published by WADA outlining anti-doping:

> Anti-doping programs seek to preserve what is intrinsically valuable about sport. This intrinsic value is often referred to as "the spirit of sport", it is the essence of Olympism; it is how we play true. The spirit of sport is the celebration of the human spirit, body and mind, and is characterized by the following values: Ethics, Fair play and honesty, Health, Excellence in performance, Character and education, Fun and joy, Teamwork, Dedication and commitment, Respect for rules and laws, Respect for self and other Participants, Courage, Community and solidarity. Doping is fundamentally contrary to the spirit of sport. (WADA 2009)

According to WADA the spirit of sport is thus the celebration of the human spirit, body and mind. The reasons advanced against doping are that it skews a level playing field, can threaten the health of the athlete, is against the spirit of sport, and incompatible with the concept of the athlete as a role model. All these arguments are problematic as the first author and several co-authors have explained before (Kayser et al. 2005, 2007; Kayser and Smith 2008). Elite sport is by definition a non-level playing field since it is about the celebration of differences. The protection of the health of the athlete argument is paternalistic and neglects the health hazards of sport itself while the distinction of avoidable and unavoidable risk is flawed. Moreover, the spirit of sport argument is fuzzy and fraught with problems, and the role-model argument is out of perspective as compared with any other role model in society. Anti-doping policies in sports have created an image of an idealized 'perfect' human. Obliging athletes to correspond to this ideal appears unfair compared to what is asked of other citizens.

15.10 What if Current Zero-Tolerance Anti-doping Policy Continues?

What can be foreseen over the next decades if the central tenet of the anti-doping movement – eradication of doping – continues to drive a global agenda of surveillance and suppression of doping and doping-like behaviour? We expect that, in the short term, continued pressure from the WADA and the IOC, backed by the international UNESCO convention, will oblige an increasing number of nations to adopt specific anti-doping legislation, especially those who aspire to organize Olympic Games. There is a international tendency to combat doping and related activities like trafficking through criminal law (e.g. in Italy, France and Slovenia, see McNamee and Tarasti 2010), quite akin to what happened in the field of psychotropic drugs, thus criminalizing the use, possession, traffic and commerce of doping substances. This development will be accompanied by increasingly repressive measures worldwide. Increasing numbers of citizens will have to comply with compulsory drug testing for an increasingly long list of substances. As mentioned before ordinary non-competitive fitness club members may have to comply with testing procedures for steroids and other illicit drugs, a procedure already introduced in Denmark and considered in several other countries (Christiansen 2011). This may be followed by testing of students for cognitive performance enhancing substances and other drugs and possibly others, like teachers, trainers, coaches, police personnel, etc. (Cakic 2009). Increased surveillance and testing will lead to increased numbers of convictions with an important burden imposed on the judicial system and the families of the convicted. For simple reasons of stochastic and procedural error frequency, a greater number of tests will lead to a greater number of false positives wrongly accusing innocent citizens. The prospect of such a development has worrying characteristics of a dystopia of Orwellian kind. The mentioned compulsory testing of non-competitive gym users in Denmark, applying the WADA code to non-athletes, clearly points in such a direction (Christiansen 2011). It appears paradoxical that gym users, generally conscious about their health and complying with general preventive principles like regular exercise and a healthy diet, making a balanced decision on steroid use to aid them in attaining their aspired body form, are punished for steroid use, while the general population can freely engage in dangerous behaviour combining bad nutrition, lack of exercise, tobacco and alcohol use without much of a constraint.

15.11 An Alternative Way to Deal with Doping

Full liberalization of substance use seems currently politically inconceivable while harsh repression has repeatedly been shown to induce more harm to society than it prevents (GCDP 2011). As a compromise between these extremes, regulation and harm minimisation is a pragmatic way to dynamically deal with such behaviour

(EMCDDA 2010; D'Angelo and Tamburrini 2010). It is dynamic because it has to be constantly adapted to changes in population behaviour. The difficulties of applying such a model to sports are of course huge, but more in line with anthropological generalizations of a socio-culturally moulded, but evolutionary defined 'human nature'. The choice between fighting doping by all means and regulation and harm minimisation is difficult, since neither will solve the problem; no ultimate solution exists. In our view, regulation and harm minimisation may come with less cost to society and the individual, as compared to a zero-tolerance approach, and therefore merits to be considered. We do not have a ready-made blueprint to offer; if an easy way existed it would already have been in place. As the Global Commission has proposed for the problem of psychotropic drugs, we should start experimenting with the regulation of doping instead of striving for its eradication.

We would like to make a series of suggestions for possible strategies. To begin with, the concept of performance enhancement by means of methods or substances, including pharmacology, should be seen as part of normal elite sports endeavour and not be negated by a utopic ideological 'spirit of sport' concept. Second, the health of elite athletes should still be protected, but taking into account the specificities of this risky profession (some sports come with a level of risk not acceptable in other professions). This can be done by continuing some form of testing, without going all the way as in today's testing. For example, a no-starting rule for a haematocrit above a given level, however the way it got to that level, is a pragmatic way to prevent excess use of red cell mass stimulation regimes that lead to a health hazard. Sure enough, athletes will find ways to cheat a bit around such strategies, but that would be part of the game while keeping the problem within acceptable boundaries. The argument that it would change sports into an arena akin to Formula 1 where the best engineering team wins is only partly correct. It will still take talent, a lot of hard work and some luck to become a champion. And then, it is likely that such a scenario is already in place anyway; today well-assisted athletes may engage in complex training regimes and strategic doping while remaining undetected. Third, the list of forbidden substances can be significantly shortened, leaving on the list only those substances with actually proven performance enhancing effects and major health hazards. For example, cannabis derivatives can be taken off the list, allowing athletes to be dealt with in the same way as the general population. The current arguments to keep cannabis on the list are flawed. There are no well-controlled trials that show any performance enhancing effect, while there is evidence for performance decreasing effects (Saugy et al. 2006; Campos et al. 2003).

With regard to the general population, instead of a crackdown on steroid users in gyms and fitness clubs with compulsory testing as in Denmark (Christiansen 2011), a harm reduction approach is very likely better (D'Angelo and Tamburrini 2010). This has already been shown in the UK where so-called steroid clinics, giving out clean syringes and thus lowering the threshold to medical care, have led to the number of syringes handed out now outnumbering those exchanged for injection for psychotropic drugs (McVeigh and Beynon 2003). These clinics, offering mostly free and anonymous services, make it possible to reach a previously hidden population. Potential advantages of providing harm minimisation measures, besides health benefits, include the personal and direct contact with a hidden population allowing them to be informed of the risks and dangers of doping substances, and helping them to take well informed decisions whether to continue use and if so in what way. These services show promise but need to be well evaluated (Jones et al. 2010).

In Switzerland the federal commission on drug-related affairs developed a conceptual model ('the cube') considering that for every substance, with different risk profiles, different levels of use exist (non-problematic, problematic, dependence), needing different levels of intervention at the level of prevention, treatment, harm minimisation and regulation. Such a model might help to conceptualise alternative policies on performance enhancing substances respecting public health and ethical principles (SEKD 2006).

15.12 What Problems Would Arise?

As Donohue et al. (2011) have noted with regard to the effect of alternative psychoactive drug policies, the two contentious questions here are also: "*By how much would the prevalence and intensity of doping rise under a different regime*?" and "*Would reduction in other costs outweigh the risks of increased doping*?". One would have to distinguish between elite athletes, amateur athletes, minors, gym users and the public in general, since it can be expected that the answers to these questions would vary between groups. It is impossible to predict what would happen in these different groups. In elite athletes one might expect limited harm since medical supervision and health oriented testing would constrain the possibilities. In amateur athletes the possibility of refraining from sourcing substances from the black market and having access to proper methods of use, might perhaps have a positive effect. It is of course difficult to propose how to regulate for minors, and developing children specifically. Since athletic careers often start very early, the protection of these young talents would be mandatory. Alternative policies should of course be continuously and extensively evaluated for desired outcomes and unintended negative consequences, carefully balancing the two.

15.13 What Are the Barriers to Change?

WADA has been successful in giving universal value to the code by having the UNESCO formulate an International Convention against Doping in Sport, which has now been signed by sufficient member states of the UN (165 as of January 2012) to have universal value. The Convention's intention is the elimination of doping in sports and it refers to WADA's code, thus leaving it up to WADA to define doping and anti-doping. As for illicit drugs, this makes policy changes moving away from the hard line of zero-tolerance politically difficult to accept, with the added difficulty that the International Olympic Committee uses the desire of participation in the Olympic Games as a lever to force nations to implement anti-doping legislation in accordance with the WADA Code. It seems that the ideal of sport as promoted by the International Olympic Committee is attaining universal value with a legal status that tends to supersede national law. We are worried that the inertia of this system is such that in the next 10–20 years little change can be expected. We hope that in the meantime the side effects of the 'war on doping' will not aggravate further, and that the tendency for a fusion of the 'war on drugs' with the 'war on doping', with excessive surveillance and harsh repression of a dystopian nature will be limited before it achieves truly Orwellian dimensions.

15.14 Conclusions

Many of us would probably welcome a world without wars, drugs or doping. Still, for many reasons, perhaps partly because of our perhaps 'innate imperfect human nature', the daily reality is quite different. The use of psychoactive substances is among the leading causes of preventable death across cultures and continents. Although prevalence of illicit substance use is lower than prevalence of licit substance use (e.g. alcohol), 50 years of 'war on drugs' have had little effect on this prevalence but have had many negative consequences. As Room and Reuter note:

> The system's emphasis on criminalisation of drug use has contributed to the spread of HIV, increased imprisonment for minor offences, encouraged nation states to adopt punitive policies (including executions, extra-judicial killings, imprisonment as a form of treatment, and widespread violations of UN-recognised human rights of drug users), and impaired the collection of data on the extent of use and harm of illicit drugs, all of which have caused harm to drug users and their families (Room and Reuter 2012, p. 84).

The Global Commission and many scientists, public health and law specialists have called for a transformation of the global drug prohibition regime, with experimentation with and evaluation of alternative regulation models, and access to evidence-based drug treatment and harm minimization services for those in need. The Commission's report asks us to "[*b*]*reak the taboo on debate and reform. The time for action is now*" (GCDP 2011, p. 3).

The current anti-doping policy, in place for just over 10 years, resembles the 'war on drugs' in several aspects, including a zero-tolerance approach, ideology encroaching on human rights and public health principles, high cost using public money for repression and control (and little for prevention or harm minimization), and attempts to shape internationally harmonized legal frameworks to attain its aim. The 'war on drugs' and the 'war on doping' are already tending to converge, as exemplified by the categorization of steroids as class III drugs in the USA, the compulsory drug testing for Danish gym users and the presence of cannabis on the list of forbidden doping substances.

Even if sufficiently complete and accurate data on the negative aspects of the anti-doping policy are still lacking, we suggest that the taboo on debate and reform also be broken in this field, now and not in 40 years. We cannot ignore the side effects of current anti-doping policy for society in general. We suggest experimentation with and evaluation of alternative ways of dealing with doping in elite (and amateur) sports, inspired by the experience gained with alternative drug policies, that are scientifically sound, respect human rights and public health, and treat athletes as humans and not as potential criminals. Regulation and harm minimization can be perceived as 'messy', but thanks to their pragmatic nature they might allow overall harm to society to be reduced below its current levels. The modern globalised sport entertainment industry with its almost unlimited financial means should not be allowed to hijack worldwide legal frameworks and orient society towards a zero-tolerance approach to both psychotropic and performance enhancing substances. Sports should remain what it is: play and entertainment, defined by a set of rules. Today there is perhaps only one valid reason against doping; it is against the rules and is therefore cheating. But in sports the rules can be changed and are, if there is a good reason to do so. We suggest that there are good reasons, and that it is time to change the anti-doping rule.

References

Aitken, C., C. Delalande, and K. Stanton. 2002. Pumping iron, risking infection? Exposure to hepatitis C, hepatitis B and HIV among anabolic-androgenic steroid injectors in Victoria, Australia. *Drug and Alcohol Dependence* 65(3): 303–308.

Amos, A. 2008. *Anti-doping policy: Rationale or rationalisation*. PhD thesis. University Library, Sydney.

Banfi, G. 2011. Limits and pitfalls of athlete's biological passport. *Clinical Chemistry and Laboratory Medicine* 49(9): 1417–1421.

Buss, D.M. 2001. Human nature and culture: an evolutionary psychological perspective. *Journal of Personality* 69(6): 955–978.

Cakic, V. 2009. Smart drugs for cognitive enhancement: ethical and pragmatic considerations in the era of cosmetic neurology." *Journal of Medical Ethics* 35(10): 611–615.

Campos, D.R., M. Yonamine, and R. de Moraes Moreau. 2003. Marijuana as doping in sports. *Sports Medicine (Auckland, N.Z.)* 33(6): 395–399.

Carlsohn, A., M. Cassel, K. Linné, and F. Mayer. 2011. How much is too much? A case report of nutritional supplement use of a high-performance athlete. *British Journal of Nutrition* 105(12): 1724–1728.

Christiansen, A.V. 2011. Bodily violations testing citizens training recreationally in gyms. In *Doping and anti-doping policy in sport*, eds. M. McNamee and V. Moller. Andover: Routledge.

Cole, C., L. Jones, J. McVeigh, A. Kicman, Q. Syed, and M. Bellis. 2010. Adulterants in illicit drugs: A review of empirical evidence. *Drug Testing and Analysis* 3(2): 89–96.

Currell, K., D.R. Moore, and P. Peeling. 2012. A–Z of nutritional supplements: Dietary supplements, sports nutrition foods and ergogenic aids for health and performance–Part 28. *British Journal of Sports Medicine* 46: 75–76.

D'Angelo, C., and C. Tamburrini. 2010. Addict to win? A different approach to doping. *Journal of Medical Ethics* 36(11): 700–707.

de Mondenard, J.-P. 2000. Historique et Évolution du Dopage. *Annales De Toxicologie Analytique* 12(1): 5–18.

de Mondenard, J.-P. 2011. *Tour De France, 33 Vainqueurs Face Au Dopage, Entre 1947 Et 2010 – Historique de L'évolution du Dopage dans le Cyclisme*. Paris: Hugo et Compagnie.

Degenhardt, L., and W. Hall. 2012. Extent of illicit drug use and dependence, and their contribution to the global burden of disease. *Lancet* 379(9810): 55–70.

Degenhardt, L., W.T. Chiu, N. Sampson, and R.C. Kessler. 2008. Toward a global view of alcohol, tobacco, cannabis, and cocaine use: Findings from the WHO world mental health surveys. *PLoS Medicine* 5(7): 1053–1067.

Donohue, J.J., B. Ewing, and D. Peloquin. 2011. *Rethinking America's illegal drug policy*. Cambridge, MA: National Bureau of Economic Research.

EMCDDA. 2010. *Harm reduction: Evidence, impacts and challenges*. Luxembourg: Publications Office of the European Union.

Fincoeur, B. 2009. Lutte Antidopage et Cyclisme à Deux Vitesses: Évolution du Rapport au Dopage chez les Cyclistes Belges depuis l'Affaire Festina. *Revue Internationale De Criminologie Et De Police Technique Et Scientifique* 3(9): 207–220.

Frederick, D.A., J. Lever, and L.A. Peplau. 2007. Interest in cosmetic surgery and image: Views of men and women the lifespan. *Plastic and Reconstructive Surgery* 120(5): 1407–1415.

Garland, D. 2008. On the concept of moral panic. *Crime, Media, Culture* 4(1): 9–30.

GCDP. 2011. *Report of the global commission on drug policy*. www.Globalcommissionondrugs.org.

Jelsma, M. 2003. Drugs in the UN system: The unwritten history of the 1998 United Nations General Assembly special session on drugs. *The International Journal on Drug Policy* 14(2): 181–195.

Jones, L., L. Pickering, H. Sumnall, J. McVeigh, and M.A. Bellis. 2010. Optimal provision of needle and syringe programmes for injecting drug users: A systematic review. *The International Journal on Drug Policy* 21(5): 335–342.

Kanayama, G., J.I. Hudson, and H.G. Pope. 2010. Illicit anabolic-androgenic steroid use. *Hormones and Behavior* 58(1): 111–121.

Kayser, B., and A. Smith. 2008. Globalisation of anti-doping: The reverse side of the medal. *British Medical Journal (Clinical Research Edition)* 337: a584.

Kayser, B., A. Mauron, and A. Miah. 2005. Viewpoint: Legalisation of performance-enhancing drugs. *Lancet* 366(Suppl 1): S21.

Kayser, B., A. Mauron, and A. Miah. 2007. Current anti-doping policy: A critical appraisal. *BMC Medical Ethics* 8: 2.

Kohler, M., A. Thomas, H. Geyer, M. Petrou, W. Schänzer, and M. Thevis. 2010. Confiscated black market products and nutritional supplements with non-approved ingredients analyzed in the Cologne doping control laboratory 2009. *Drug Testing and Analysis* 2(11–12): 533–537.

Kraska, P.B., C.R. Bussard, and J.J. Brent. 2010. Trafficking in bodily perfection: Examining the late-modern steroid marketplace and its criminalization. *Justice Quarterly* 27(2): 159–185.

Larance, B., L. Degenhardt, J. Copeland, and P. Dillon. 2008. Injecting risk behaviour and related harm among men who use performance- and image-enhancing drugs. *Drug and Alcohol Review* 27(6): 679–686.

Lentillon-Kaestner, V., and F. Ohl. 2011. Can we measure accurately the prevalence of doping? *Scandinavian Journal of Medicine & Science in Sports* 21(6): e132–e142.

Lentillon-Kaestner, V., M. S. Hagger, and S. Hardcastle. 2011. Health and doping in elite-level cycling. *Scandinavian Journal of Medicine & Science in Sports. Health and Doping in Elite-Level Cycling* (March 10): e-pub ahead of print.

Lippi, G., and M. Plebani. 2011. Athlete's biological passport: To test or not to test? *Clinical Chemistry and Laboratory Medicine: CCLM/FESCC* 49(9): 1393–1395.

Loland, S., and H. Hoppeler. 2011. Justifying anti-doping: The fair opportunity principle and the biology of performance enhancement. *European Journal of Sport Science* 12: 347–353.

McDowall, J.A. 2007. Supplement use by young athletes. *Journal of Sports Science and Medicine* 6(3): 337–342.

McNamee, M.J., and L. Tarasti. 2010. Juridical and ethical peculiarities in doping policy. *Journal of Medical Ethics* 36(3): 165–169.

McVeigh, J., and C. Beynon. 2003. New challenges for agency based syringe exchange schemes: Analysis of 11 years of data (1991–2001) in Merseyside and Cheshire, United Kingdom. *The International Journal on Drug Policy* 14(5–6): 399–405.

Melnik, B.C. 2009. Androgen abuse in the community. *Current Opinion in Endocrinology, Diabetes, and Obesity* 16(3): 218–223.

Mendoza, J. 2002. The war on drugs in sport: A perspective from the front-line. *Clinical Journal of Sport Medicine* 12(4): 254–258.

Nature. 2007. A sporting chance. 448(7153): 512.

Parkinson, A.B., and N.A. Evans. 2006. Anabolic androgenic steroids: A survey of 500 users. *Medicine and Science in Sports and Exercise* 38(4): 644–651.

Perneger, T. 2010. Speed trends of major cycling races: Does slower mean cleaner? *International Journal of Sports Medicine* 31(04): 261–264.

Petróczi, A. 2007. Attitudes and doping: A structural equation analysis of the relationship between Athletes' attitudes, sport orientation and doping behaviour. *Substance Abuse Treatment, Prevention, and Policy* 2: 34.

Petróczi, A., and T. Nepusz. 2011. Methodological considerations regarding response bias effect in substance use research: Is correlation between the measured variables sufficient? *Substance Abuse Treatment, Prevention, and Policy* 6(1): 1.

Petróczi, A., G. Taylor, and D.P. Naughton. 2011a. Mission impossible? Regulatory and enforcement issues to ensure safety of dietary supplements. *Food and Chemical Toxicology: An International Journal Published for the British Industrial Biological Research Association* 49(2): 393–402.

Petróczi, A., J. Mazanov, and D.P. Naughton. 2011b. Inside athletes' minds: Preliminary results from a pilot study on mental representation of doping and potential implications for anti-doping. *Substance Abuse Treatment, Prevention, and Policy* 6: 10.

Pitsch, W. 2009. 'The science of doping' revisited: Fallacies of the current anti-doping regime. *European Journal of Sport Science* 9(2): 87–95.

Pitsch, W., and E. Emrich. 2011. The frequency of doping in elite sport: Results of a replication study. *International Review for the Sociology of Sport* (August 23). E-pub before print.

Room, R., and P. Reuter. 2012. How well do international drug conventions protect public health? *Lancet* 379(9810): 84–91.

Sanchis-Gomar, F., V.E. Martinez-Bello, M.-C. Gomez-Cabrera, and J. Viña. 2011. Current limitations of the athlete's biological passport use in sports. *Clinical Chemistry and Laboratory Medicine* 49(9): 1413–1415.

Saugy, M., L. Avois, C. Saudan, N. Robinson, C. Giroud, P. Mangin, and J. Dvorak. 2006. Cannabis and sport. *British Journal of Sports Medicine* 40(Suppl 1): i13–i15.

Schweizerische Eidgenössische Kommission für Drogenfragen (SEKD). 2006. *Von Der Politik Der Illegalen Drogen Zur Politik Der Psychoaktiven Substanzen*. Bern: Huber.

Smith, M.E., and M.J. Farah. 2011. Are prescription stimulants 'smart pills'? the epidemiology and cognitive neuroscience of prescription stimulant use by normal healthy individuals. *Psychological Bulletin* 137(5): 717–741.

Sottas, P.E., N. Robinson, G. Fischetto, G. Dolle, J.M. Alonso, and M. Saugy. 2011. Prevalence of blood doping in samples collected from elite track and field athletes. *Clinical Chemistry* 57(5): 762–769.

Striegel, H., P. Simon, S. Frisch, K. Roecker, K. Dietz, H.-H. Dickhuth, and R. Ulrich. 2006. Anabolic ergogenic substance users in fitness-sports: A distinct group supported by the health care system. *Drug and Alcohol Dependence* 81(1): 11–19.

Striegel, H., R. Ulrich, and P. Simon. 2010. Randomized response estimates for doping and illicit drug use in elite athletes. *Drug and Alcohol Dependence* 106(2–3): 230–232.

Sullivan, R.J., and E.H. Hagen. 2002. Psychotropic substance-seeking: Evolutionary pathology or adaptation? *Addiction* 97(4): 389–400.

Teter, C.J., S.E. McCabe, K. LaGrange, J.A. Cranford, and C.J. Boyd. 2006. Illicit use of specific prescription stimulants among college students: Prevalence, motives, and routes of administration. *Pharmacotherapy* 26(10): 1501–1510.

WADA. 2009. *World anti-doping code*. www.wada-ama.org.

WADA. 2012. *International standard of testing*. www.wada-ama.org.

Wiesing, U. 2011. Should performance-enhancing drugs in sport be legalized under medical supervision? *Sports Medicine (Auckland, N.Z.)* 41(2): 167–176.

Wood, E., D. Werb, B. Marshall, J. Montaner, and T. Kerr. 2009. The war on drugs: A devastating public-policy disaster. *Lancet* 373(9668): 989–990.

Yesalis, C.E., and M.S. Bahrke. 2002. History of doping in sport. In *Performance-enhancing substances in sport and exercise*, ed. Charles E. Yesalis and Michael S. Bahrke. Champaign: Human Kinetics.

Zorzoli, M. 2011. The athlete biological passport from the perspective of an anti-doping organization. *Clinical Chemistry and Laboratory Medicine* 49(9): 1423–1425.

Chapter 16
A Simple Regulatory Principle for Performance-Enhancing Technologies: Too Good to Be True?

Roger Brownsword

16.1 Introduction

It is hard to draw meaningful comparisons between the 2012 Summer Olympics and the two previous London Olympics. In the 1908 Olympics, for instance, Tom Longboat, the Canadian favourite for the marathon event, collapsed after being given champagne to restore his flagging fortunes (evidently, although drugs were banned, the rules permitted various stimulants, including champagne); and the American tug-of-war team complained that the British team, who wore regulation police boots, had cheated by doing so (the rules prohibited only the wearing of footwear with projecting nails or tips) (Hampton 2011). By 1948, for the first time, the athletes had the benefit of starting blocks but, in the austere post-War years, their preparation was less than ideal. For example, the 'British athletes had to train on 2,600 cal/day, which included one slice of bacon, one ounce of cheese and one egg a week. This was increased to a docker's rations of 3,900 cal only when they were selected for the Olympic team. For some, this was only days before the games started.'(ibid: 27) Although the 'Austerity' Olympics had its stars, notably the legendary Fanny Blankers-Koen, we can reasonably expect that, in 2012, the performances, on both track and field, will be significantly better.

However, no one wants better performances to be achieved in breach of the rules; no one wants to see another Ben Johnson, stripped of his gold medal and sent home in disgrace after a positive drugs test. Even if it is not always entirely clear why the rules draw the line between what is prohibited and what is permitted in the particular places that they do—for example, why prosthetic limbs might be prohibited but not fibre-glass flexible poles for pole-vaulting, why anabolic steroids are prohibited but not altitude chambers, FastSkin swim-suits, or the Lotus bike

R. Brownsword (✉)
The Dickson Poon School of Law, King's College London, University of Sheffield, Room Number SW3.09, Strand, London, WC2R 2LS, UK
email: Roger.brownsword@kcl.ac.uk

J. Tolleneer et al. (eds.), *Athletic Enhancement, Human Nature and Ethics*, International Library of Ethics, Law, and the New Medicine 52, DOI 10.1007/978-94-007-5101-9_16,

used to such effect by Chris Boardman at the Barcelona Olympics—the spirit of the Olympics is at least one thread of continuity. Whether it is London in 1908, 1948, or 2012, competitors should observe the basic principles of fair play—and, when all is said and done, cheating is cheating.

Nevertheless, when there is the prospect of an ever-increasing array of performance-enhancing drugs together with technologies that are designed to extend the horizons of human capability, the background question of why some enhancers, but not others, are permitted assumes more prominence. In response to this question, in this paper, I will articulate an apparently simple, governing baseline principle, namely that *it is permissible for competent agents (such as Olympic competitors) to use enhancers unless either (i) use causes harm to others (who have not consented to this risk) or (ii) the user has freely agreed to act on a no-enhancement basis.* For agents who are not competent, the governing regulatory principle would be paternalistic.

On the face of it, the proposed principle has plenty going for it. First, it purports to be of general application, governing the use of enhancers not only in sports but also in medical or educational or employment settings; and, secondly, it means that we no longer need to agonise about whether an intervention is to be classified as 'therapeutic' or as an 'enhancement', as well as whether a situation is competitive or not.[1] However, like a drug-enhanced performance, is this principle too good to be true? Is it really as simple as it seems? And, even if it permits those who like their sport to be clean (the 'Corinthians' as I term them)[2] to compete on a no-enhancement basis, how are the concerns of those who would categorically prohibit enhancers to be addressed and accommodated?

In this paper, I will pursue these questions, starting with the apparent simplicity of the master principle (and its key elements of harm and consent) and then considering the case for categorical prohibition. Stated shortly, my conclusions are that, although there is a good deal of devil in the detail of the master principle, it is on the right track; that the master principle needs to be anchored to an ethic of rights; and that the only circumstances in which categorical prohibition might be appropriate is where the use of enhancers jeopardises the possibility of agents acting morally, in the sense of trying to do the right thing for the right reason.

16.2 First Thoughts and a Simple Principle

We might try to get an angle on the questions that we have posed in our introductory remarks by thinking about the background principles that we use to determine whether drugs or other interventions are acceptably used in medical contexts (compare Tännsjö

---

[1] This was a distinction that troubled me in Brownsword (2009). However, if we adopt the proposed master principle, the critical question is not whether the context is competitive or non-competitive but whether the agent has freely accepted the restriction on the use of enhancers.

[2] I use the term 'Corinthian' because (at any rate, to my mind) it evokes a spirit of honest participation, of idealism, of 'naturalism', and of 'amateurism' in the best sense. However, I am grateful to the editors who have pointed out to me that Oscar Pistorius, arguably the most 'unnatural' of elite athletes, has a biblical line from Corinthians tattooed across his back (the line says: 'I do not run like a man running aimlessly').

2009: 315). In clinical settings, it is axiomatic that doctors should not harm their patients. However, most interventions involve some degree of risk and a regulatory framework needs to be established. One approach is to anchor regulation to the best interests of patients, enjoining doctors to act paternalistically in whatever way that they in good faith judge to be in the best interests of their patients. However, modern medicine is orientated to a rather different regulatory approach, namely that the physician should advise the patient as to the clinical options (explaining the risks attached to each) and should proceed with a particular intervention only with the patient's free and informed consent.

If we were to apply the paternalistic principle to the regulation of sporting events, we would advise the relevant governing bodies that they should ban the use of performance-enhancing drugs or enhancing technologies where they are contrary to the best interests of participants. So, for example, governing bodies might ban the use of anabolic steroids because they judge that they are harmful to the health of athletes but permit new pole vault, swim-suit, or racing bike technologies because they do not significantly increase the risk of physical injury (or harm to health) for those who use them. Whilst this might seem a plausible basis for regulation, it is contrary to the ruling view in medicine. In clinical settings, the paternalistic principle continues to be applied in relation to children and those who are not competent to make their own treatment decisions; however, the general rule is that the relationship between doctors and patients should be governed by the principle of informed consent—or, at any rate, this is the governing principle in relation to the use of *therapeutic* products and procedures. This prompts the question: how would sports governance be affected if informed consent became the guiding regulatory principle?

Let us suppose that, instead of drawing on paternalistic principles, sports governance relied primarily on the principle of informed consent. This would mean that the governing bodies would have a responsibility to inform competitors about the risks associated with various enhancing technologies (including performance-enhancing drugs) but, young competitors apart, the rules would permit the use of any drugs or technologies. So, for example, while the rules might still prohibit the use of anabolic steroids by teenage swimmers, they would permit adult swimmers to use any drugs that they wished.[3] While this would revolutionise the rule-book for sports, it is a matter of debate how far it would actually change existing practice.[4]

[3] According to the Article 4.3 of the World Anti-Doping Agency (2003), a substance or method shall be considered for inclusion on the World Anti-Doping Agency's Prohibited List where it meets any two of the following three criteria: (i) it has the potential to enhance, or it enhances, sport performance; (ii) it represents an actual or potential health risk to the athlete; and (iii) its use violates the spirit of sport. At page 3 of the Introduction to the Code, the spirit of sport is said to be the 'celebration of the human spirit, body and mind' and it is characterised by the following values: ethics, fair play and honesty; health; excellence in performance; character and education; fun and joy; teamwork; dedication and commitment; respect for rules and laws; respect for self and other participants; courage; and community and solidarity. In short, doping is 'fundamentally contrary to the spirit of sport' (ibid).

[4] In general, the effectiveness of regulatory control depends on a combination of regulator-side factors (such as corruption, capture, and resources) and regulatee attitudes (such as whether regulatees are minded to 'game' the system or to invest in avoidance or evasion). For an interesting insight, see Møller (2010: 160).

That said, it would impact on both the media presentation of sport and the experience of many spectators. For the media, there would no longer be drug scandals to report; and, for spectators, there would no longer be the lingering thought that a stellar performance might have been drug-assisted—rather, the thought would be that the performance almost certainly would have been so assisted.

If, however, the regulation of *enhancing* technologies in sport were to be put on the same footing as the regulation of *therapeutic* interventions in medicine, this might elicit the objection that it ignores the watershed between therapy and enhancement that many (particularly in the context of medical practice) treat as critical. Not only is the distinction between therapy and enhancement heavily contested (Harris 2007; Sandel 2007; Savulescu and Bostrom 2009) there is a strong lobby arguing against the acceptability of enhancing technologies in any context. If the governing bodies in sport were to take their lead from this lobby, all forms of enhancement would be prohibited. Clearly, this is an objection that in due course we must reckon with. At this stage, however, let me persist with the thought that the regulation of enhancement in sport should be put on the same basis as the regulation of medical interventions that are intended to have a therapeutic effect.

If we continue with this train of thought, the regulatory principles for sport are not yet complete. So far, we have focused only on the possibility of competitors harming themselves. In contact sports, however, there is the possibility that competitors might harm one another, and that the use of enhancing technologies will increase the probability of harm or the seriousness of the harm caused or both. What should we make of this? In the regulatory mainstream, conduct that is directly and tangibly harmful to others is likely to be prohibited—or, at any rate, it is likely to be prohibited unless the person who is 'harmed' has consented to the conduct. On the basis of this highly significant rider, even though competitors in some sports (paradigmatically, boxing) act in ways that they know are likely to harm one another, we treat this as permissible (subject to some limiting rules) because the parties have consented to such risks of participation. Accordingly, following this line of reasoning, if there are enhancements that increase these risks, they should be permitted so long as the participants have given their free and informed consent.

On the basis of these first thoughts, we seem to have three simple regulatory principles for the London Olympics and, indeed, for sporting competitions more generally. These principles are:

- the use of enhancers (drugs or other technologies) that are harmful to parties other than the user (principally fellow competitors) should be prohibited unless the relevant parties have given their free and informed consent to the use of such enhancers; where such consent has been given, even enhancers that are harmful to others should be permitted
- where enhancers are harmful only to those who use them, then their prospective users should be made aware of the risks of use but, otherwise, they should be permitted to use such enhancers

- where participants are not yet of an age to be left to make their own prudential judgments as to the use of enhancers, the rules should be designed paternalistically to promote the best interests of such persons.

For out-and-out prohibitionists, who categorically oppose any use of enhancers, these regulatory principles, even if apparently simple, will seem to be wholly misguided. As we have said, prohibitionists of this stripe advocate a quite different, and also simple, rule: namely, that enhancers, irrespective of whether anyone (and, if so, who) might be harmed, and irrespective of who has consented, should not be permitted. For the second time in these remarks, let me pend this objection; but, rest assured, it is not being forgotten. Instead, let me acknowledge that there is likely to be a different breed of prohibitionists who might be unhappy about the proposed regulatory principles. For this group—let us call them the 'Corinthians'—the proposed principles threaten their enjoyment of sport as they know it. The Corinthians want the London Olympics to be open to all-comers but they like their sport clean and technology-free; hard but fair is their motto for sporting competition. Whether we are contemplating the use of EPO (remember Floyd Landis who was stripped of his yellow jersey after the 2006 Tour), modern tennis rackets with their accommodating sweet spots, or goal-line camera technology, Corinthian sports-lovers like things the way that they used to be. How might the regulatory principles be supplemented to assuage the preferences of this group?

Let us suppose that, with the proposed regulatory principles, sport becomes even more high-tech than it already is. Enhancers of various kinds are widely used by both participants and officials. This so frustrates the Corinthians that they resolve to set up their own enhancer-free sporting associations. Assuming that they are to be permitted to do this (and, it would be distinctly illiberal to argue otherwise), the rules of Corinthian-governed sports will provide that enhancers are prohibited. In the Corinthian Olympics, bamboo poles will be back in fashion, there will be no anabolic steroids, no starting blocks, and so on. However, if (in breach of the rules) competitors at the Corinthian games use performance-enhancing drugs that are harmful only to themselves, on what basis can we accuse them of doing wrong? For sure, such competitors have cheated (relative to the Corinthian rules) but their conduct is in line with the general regulatory principles that we have outlined. How is this tension to be reconciled?

If we treat the proposed regulatory principles as setting the defaults, then liberals will accept that there is the possibility of the defaults being legitimately set aside in favour of different principles. Other things being equal, the only condition is that the parties who switch the defaults do so on the basis of their free and informed consent. In other words, what binds the Corinthian competitors to the rules against the use of enhancers is that they have freely consented to participate on this basis. Hence, our supplementary (Corinthian) regulatory principle is:

- Where the rules of particular sporting competitions prohibit the use of enhancers, competitors are bound by these rules provided that they have freely agreed to participate on this basis.

When we put this principle together with the others, we see that consent can function in both a permissive and prohibitive direction. On the one hand, according to the default principles, a competitor can consent to the risk of harmful enhancers being used that would otherwise be prohibited; and, on the other hand, according to the Corinthian principle, a competitor can agree that enhancers, that otherwise would be permitted, should not be used.

Before we draw these tentative remarks to a close, the thought occurs that the regulatory principles that we have articulated might form a set that would have general application. In other words, no matter what the context, whether it be in the hospital, in the university, at the workplace, or in the Olympic stadium, no matter whether the context is competitive or non-competitive, the permissibility of using enhancers is governed by these principles—principles that orbit around the two ideas of harm and consent. For competent agents, the rule simply is that enhancers may be used unless either (i) use causes harm to others (who have not consented to this risk) or (ii) the user (like the Corinthians) has freely agreed to act on a no-enhancement basis.

The question now is whether the principle will stand up when we take a harder look at it. First, we can see how robust the ideas of harm and consent are; and then we can see how the principle fares when attacked by those who categorically condemn enhancement in any context, whether sporting, medical, employment, or academic.

16.3 Harm and Consent

The putative master principle hinges on whether A's enhancement harms B, coupled with riders where either B consents to A's use of enhancers or A agrees not to use enhancers. Breaking this down, we have the following four propositions. First, if A, by doing x (using enhancers), harms B, then it is prima facie wrong for A to do x. Secondly, however, if B consents to A doing x, even if x is 'harmful' to B, then A does no wrong to B by doing x. Thirdly, if A, by doing x, does not harm B, then it is prima facie permissible (not wrong) for A to do x. And, fourthly, if A has agreed with B that A will not do x, even if x is not 'harmful' to B, then A does wrong by doing x. If the central ideas of harm and consent are unproblematic in their application, we seem to have a simple and robust general regulatory principle. In this section, we can stress-test both the idea of harm and that of consent (and agreement).

16.3.1 Harm

The proposed principle follows the regulatory practice of liberal cultures by seeking out common ground in the harm principle–that is, in the principle that it is a necessary condition (although not also a sufficient condition) for the legitimate prohibition (and punishment) of conduct that it should be harmful to others. If conduct is not harmful in this way, then it should not be prohibited; unless regulators are exercising a paternalistic jurisdiction for the benefit of the incompetent or immature, it is no justification to argue that the prohibition protects the person from harming themselves. According to this version of liberal thinking, we can harm ourselves as much as we like; and, if we like to harm ourselves, then we should be permitted to do so. It follows that, if we choose to use performance-enhancing drugs–for example, if university students choose to use cognition-enhancing drugs–then regulators should permit this provided that it does no harm to others and notwithstanding any harm that we might inflict upon ourselves by such use.

Famously, this simple principle disintegrates under close analysis (Brownsword 2006: 527). The first problem is that it assumes that there is agreement on what precisely counts as a 'harm' to others. No doubt, there are many instances of an act being quite clearly harmful to another–for example, where the act occasions physical injury, or damage to property, or loss of reputation, and so on. However, once we move beyond such clear cases, we run into familiar difficulties. For example, what would we say about a display of poor sporting manners, which occasions some annoyance and mild distress, or about conduct that fails to live up to the expectations of a role-model and thus causes some disappointment? Are we dealing with relevant 'harm' in such cases? Similarly, when there is a huge 'let-down' following a positive drugs test (as in Ben Johnson's case), when celebration turns to contempt, is this a relevant 'harm'? If we replace a free-standing notion of harm with acts that have a negative impact on some significant or legitimate interest of the other, the spotlight turns to the criterion of significance or legitimacy. What is it that determines the significance or legitimacy of one's interests? Is it the preferences or the pains and pleasures of another? Is it their basic needs or their rights? Already, we can sense that a utilitarian might apply the harm principle in a way that is not quite identical to that adopted by a proponent of human rights, let alone by communitarians (Brownsword 2008).

Even if the idea of harm can be stabilised, there are further questions. In particular, how directly or foreseeably must the conduct cause the alleged harm? Suppose that there was some evidence to suggest that the use of drugs in sport encourages recreational drug use more generally. Would this be a sufficient causal link? Moreover, there is the perennial problem of determining who counts as a relevant 'other'. Do, say, horses and greyhounds count just like humans? To be sure, with the spirit of liberalism in the background, we can reply to the effect that there is more to harming others than simply acting against their preferences or causing mild offence, that the causal relationship between act and harm must be direct and obvious, and that relevant others are born humans. However, if we have to invoke the liberal spirit in order to craft these responses, the game is up; it is the liberal spirit, not the letter of the harm principle, that is doing the regulatory work.[5] How damaging are

[5] It also needs to be borne in mind that regulators can constrain conduct by designing products (for example, tennis or golf balls designed to control their velocity) and places in a certain way (for example, think about the way that different tennis court surfaces, including the particular kind or length of grass used at Wimbledon, affect the pace of the game). In other words, the ingredients in the regulatory environment go beyond the governing rules (Brownsword and Somsen 2009).

these stock reservations when we consider the master principle as a basis for regulating enhancing drugs or technologies?

We can focus on whether the use of an enhancer is harmful. As we have said, if the enhancer has a negative impact on the health of the user or others, or if there is damage to property, this will constitute harm on almost anyone's view. However, what if the complaint is less straightforward? A good example is the objection that enhancers give their users a positional advantage (which translates as a positional *disadvantage* for others). What should we make of this? Is it a relevant harm?

One of the leading advocates of enhancement, John Harris, defines enhancement as necessarily bringing about some improvement in human functioning. Thus:

> Enhancements of course are good if and only if those things we call enhancements do good, and make us better, not perhaps simply by curing or ameliorating our ills, but because they make us better people. Enhancements will be *enhancements properly so-called* if they make us better at doing some of the things we want to do, better at experiencing the world through all of the senses, better at assimilating and processing what we experience, better at remembering and understanding things, stronger, more competent, more of everything we want to be. (Harris 2007: 2)

On this view, post-enhancement (properly so-called), there is an improvement in A's capacity to perform some task (that is, relative to A's pre-enhancement capacity). For example, post-enhancement, A has significantly improved vision or hearing; or A is able to concentrate better or stay awake longer or train harder than was the case pre-enhancement. Given that A has these improved capacities, it follows not only that 'post-enhancement A' can outperform 'pre-enhancement A' but also that A can now perform better relative to B, C, and D. This is not to say that A will be able to outperform others (this depends on the capacities of B, C, and D, and whether or not they, too, are enhanced in the relevant respect) but it does mean that, post-enhancement, A gains some positional advantage. In the context assumed by Harris, it is not A's intention to gain such a positional advantage; in other words, if there is a harm (in the shape of a positional disadvantage) arising from A's new capacities, it is not one that A intends to inflict on others. Indeed, on Harris' account, A can say, in good faith, that he or she would be perfectly happy if B, C, and D, too, were able to enjoy the benefit of a particular enhancement (an enhancement that extends the span of life might be a good example). This allows Harris to argue that A can sign up for enhancement with a clear conscience and that regulators should not stand in A's way simply because A's act has some secondary effects (this, after all, is true of many of our acts) (Harris 2007: 28). However, even if we concede this much to Harris, once we switch to sporting or to educational contexts, the problem is that enhancement is intended precisely to gain positional advantage. Students, like sportsmen, who use enhancers or enhancing technologies do so precisely for their competitive (positional) advantage; they cannot say, other than quite disingenuously, that they are happy for others to have access to the same enhancers. How does this sit with the harm principle?

In preparation for the London Olympics, it seems that team GB is making innovative use of new technologies (Slot 2011: 24). The boxing coaches, for instance, are using a programme called iBoxer, which 'involves a network of cameras [so that] after a bout of sparring, a boxer can immediately look at his performance and his trainer can analyse various aspects of it down to the angle of his knee in throwing a punch.'(ibid.) There is no suggestion that the use of such enhancing technologies in any way infringes the rules; the only question is whether it should be permitted relative to the proposed regulatory principle. While team GB would have no grounds for opposing the use of similar technologies by other teams, they surely would not welcome the uptake of such technologies elsewhere. This would defeat the object–the object being not simply to improve the performance of team GB but to improve the medal tally (more golds, more silvers, more bronze medals) at the London Games. In short, the use of these technologies is for positional advantage, both individually and for the team as a whole. But, is it harmful to others in a way that would justify its prohibition?

Unless we already know the answer that we want to give, we can keep on repeating the question (is it harmful?) without making any progress. Moreover, if we try to find a way through the puzzle by turning to considerations of what is fair and whether or not there are level playing fields, and the like, we will probably make even less progress. We have already seen, for example, how difficult the experts have found it to assess whether Oscar Pistorius' carbon-fibre blades give him an unfair advantage over fellow (able-bodied) competitors and, with the multiplication of enhancements (including possibly genetic tweaks) the problem will be compounded (Lawson 2011; Stevens 2011). For its supporters, this is not yet a reason to abandon the harm principle. However, two points needs to be recognised. First, once we move beyond the clear cases of harm (primarily to person, property, and reputation), we are in difficulty; and, secondly, most examples of enhancement in sporting contexts are not harmful in a straightforward sense but raise the difficult question of whether acting for positional advantage is nevertheless causing harm to others.

16.3.2 Consent

Let us suppose that some enhancing technologies are harmful to others in a straightforward way (as might be the case, for example, in contact sports). According to the proposed regulatory principle, their use can be authorised by the consent of relevant others. Is this any less problematic than the harm element of the principle?

Once we start to unpack the idea of consent, there are many elements to analyse (Beyleveld and Brownsword 2007). At the core of the idea, however, is the requirement that the consent will only serve as a valid authorisation (for what would otherwise be a wrongful act) if it is given on a free and informed basis. Immediately, this raises the questions: what constitutes a free or unforced act; and what constitutes an informed act?

As is well-known, in medical contexts, the doctrine of informed consent has become the governing principle for the relationship between clinicians and their (competent adult) patients (Faden and Beauchamp 1986). Historically, the governing principle was that clinicians should do no harm (to their patients); but the modern priority is to act only with the informed consent of the patient. While this principle is robustly defended by advocates of patients' rights, it has been much criticised, not least because of the difficulties of determining how much information patients should have before the test of informed consent is satisfied (Manson and O'Neill 2007). Moreover, the longer that the information sheets become, in a desperate attempt to inform patients, the more confused patients are likely to be. It follows that if, in sporting contexts, harmful enhancements are to be permitted, those who are invited to consent to their use should be quite clear about the nature and extent of the increased risks they are now running. Just as in medical contexts, so too in sporting contexts (and particularly if there is some uncertainty about the precise effects of the enhancement) the requirement for an informed consent is likely to prove a minefield.

Even if informed consent were not problematic, the requirement that consent should be free and unforced is likely to prove highly contentious in its application. While there is some appeal in drawing a clean distinction between a 'threat' (which renders the act unfree) and a 'promise' or an 'offer' (which does not affect whether the act is free), this perhaps does not capture the concern about the way in which incentives can distort judgment. So, in more than one context, we find policy-makers grappling with the concern that any inducement should not be undue (see for instance Nuffield Council on Bioethics 2011). In sporting contexts, there might be some questions about just how freely B consented to enter the arena with an enhanced A, but the more obvious question perhaps is whether (enhanced) A freely agreed to the enhancement (which might be potentially harmful to both A and B).

To avoid any misunderstanding, we need to clarify an implicit element of the putative governing principle. According to the principle, it is permissible for A to use enhancers or enhancing technologies where the only person that might be harmed thereby is A. To be sure, we can take it that regulators have some responsibilities to inform persons about the risks that they might be running if they use enhancers; but, if A proceeds to self-administer some enhancers, the proposed principle has it that no wrong is done. Let us suppose, though, that A is under some pressure to undergo an enhancing procedure. For example, let us suppose that A is a high-jumper, whose performances put him on the fringe of the national team. C, A's coach, tells A that, unless he undergoes a procedure for leg extension, he will not be selcted for the national team. The procedure is not risk free and A might be harmed. If A is a youngster, the governing principle is paternalistic and, to some extent, A is protected—I say 'to some extent' because there might be more than one view about what is in A's best interest in such circumstances (compare Hagger and Hagger Johnson 2011). However, if A is a competent adult, would we judge that his decision to undergo the procedure is 'free' when he does so only because of his coach's advice and when he would not agree to the procedure if he could make the national team without it?

On the basis of these remarks, we can conclude that operationalizing a requirement of free and informed consent is far from straightforward. Again, this does not mean that consent should be abandoned as a regulative principle; for, in practice, we rely on consent all the time as a dynamic that enables us to shift our position. However, it is complex.

If we turn from consenting to the use of enhancers to consenting to their non-use (as is the case with the Corinthians), are the problems less acute? On the face of it, they are. No doubt, there might be instances in which parties complain that they were not fully informed or that they were pressurised into agreeing to the particular rules; intuitively, though, one would expect an agreement to forego the use of enhancers to be less problematic. Still, even in this seemingly less pressurised context, there might be difficult cases (as well as a standing temptation to defect from the agreement).

16.4 Categorical Prohibition

Taking a much less nuanced approach, some will argue that there simply should be no enhancement—irrespective of whether anyone is 'harmed', irrespective of whether there has been any covering consent, enhancement should not be permitted. This is the view of the prohibitionists.

Before we consider the credentials of prohibitionism, we need to distinguish it from the Corinthians. The Corinthians have a preference for sport played without enhancers or the use of enhancing technologies. However, they do not believe that there is anything fundamentally wrong or immoral about the use of enhancers. The prohibitionists, by contrast, hold precisely that it is wrong to use enhancers. For the prohibitionists, the restrictions on enhancers do not rest on local conventions, such as those imposed by the Corinthians; rather, they are applicable in all contexts (medical, sporting, educational, employment, and so on), irrespective of whether they actually are recognised or accepted. In simple terms, for the prohibitionists, any attempt to use enhancers or enhancing technologies to improve human performance is plain wrong.

Why do prohibitionists object (and, indeed, object so deeply)? Often, the prohibitionists present their objections as though they are self-evident. For example, they might say that enhancement is 'unnatural' or that it leads to performances that are not 'authentic', or that humans should not try to 'play God'. In each case, though, the Corinthians, the liberals, and others will want further and better particulars. If notions such as harm and consent invite further analysis, this is so *a fortiori* with appeals to the 'natural' or the 'authentic'.[6] However, in the most sophisticated recent objection that has been presented, Michael Sandel asserts that, while we all know that the argument against human enhancement rests on human dignity, the challenge is to articulate the sense in which enhancement compromises human dignity.[7]

---

[6] On the meaning of 'natural', see, e.g., Christian Lenk (Chap. 3) in this volume; for 'authentic', see, e.g., Parens (2009); and, in the context of nanotechnologies, for a helpful analysis of 'playing God', see Peters (2007) and Coady (2009).

[7] "It is commonly said that enhancement, cloning, and genetic engineering pose a threat to human dignity. This is true enough. But the challenge is to say *how* these practices diminish our humanity" (Sandel 2007: 24).

Of all concepts, that of human dignity is the most protean and problematic (Beyleveld and Brownsword 2001). We can attempt to identify some of the key axioms of dignitarian prohibitionism by focusing on the idea of commodification (insisting that no human should be treated as a mere 'thing'). So, for example, in an interesting discussion of nanomedicine, Bert Gordijn (2006) sketches a scenario in which nanosensors in our bodies monitor our health; and, where problems are detected, nanometric drug release systems are activated. Such early warning systems and nano-responses seem to be good for our health, but they might have a downside when viewed through a dignitarian lens. Thus:

> [S]uch developments will contribute to a more technologically inspired image of the body as something very similar to a machine. The body will increasingly be regarded as a whole, made up of many different components that might be fixed, enhanced or replaced if necessary. Development, functions, and appearance of the body will seem less and less fixed by nature and less frequently accepted without change, and more frequently controllable by technology. Instead of being in charge of our own health we might increasingly trust technology to take over this responsibility. In the process however, the body will be treated almost like the inanimate material of a machine. Hence, the body might become increasingly de-hallowed and de-mystified. (Gordijn 2006: 729)

To the extent that nanomedicine adopts or encourages the functional view that is already evident in human genetics and the new brain sciences, this will compound dignitarian concerns about commodification. The promise of in vivo nanosensors and drug release systems, like the promise of regenerative medicine, sounds fine until it is set alongside the disaggregation of humans into their component parts. Is there really no distinction between humans and, say, a motor car or a computer—just so many parts, so many functions, so many models?

What, then, should we make of the objections raised by the dignitarian prohibitionists? First, some might simply dismiss these objections as unpersuasive, as just so much conservative dogma (in the way that John Harris is dismissive of Sandel's plea against perfection). But, then, the prohibitionists might retort that those who advocate the harm principle or its cognates are no more than apologists for a liberal dogma. Secondly, however, we might be more assertive in support of a liberal rights view. After all, by and large, the public regulatory culture in Europe is not dignitarian; rather, the prevailing baseline values recognised by European political and legal institutions are ones of respect for human rights and fundamental freedoms.[8] It follows

---

[8] Granted, there are some significant pockets of dignitarianism in European law—notably in the moral exclusions against patentability in the EC Directive on the Legal Protection of Biotechnological Inventions (Directive 1998/44/EC). For the bite of the dignitarian reading of the exclusions, see Brüstle v Greenpeace eV (Case C34-10) (judgment of the ECJ handed down on 18 October, 2011); and, similarly, see the WARF case (Case G 0002/06, 25 November 2008) where the Enlarged Board of Appeal at the European Patent Office read the same exclusion in the same way. Coincidentally, at almost the same time that the ECJ decided the *Brüstle* case, the Grand Chamber of the European Court of Human Rights upheld dignitarian Austrian legislation concerning access to IVF (see application no. 57813/00, Judgment given on 3 November 2011).

that, unless the prohibitionists can come up with some compelling arguments against these values, they are simply tilting at windmills. Thirdly, we might argue, along Corinthian lines, that there is nothing wrong with a group or a community agreeing to eschew enhancers, or even wishing to define itself in terms of a rejection of what Sandel presents as hyperagency, or transhumanism, or whatever. That is fine; the problem is trying to impose such a communitarian view on those who have different preferences and predilections. In the case of sporting prowess, for example, some might prefer to celebrate natural talent, others sheer effort and endeavour, and yet others the technological support (as in FI motor racing). On this Corinthian approach, plurality is not a problem—the problem is an imperialist version of communitarianism.

That said, there is one application of enhancing technologies that does give cause for concern, no matter where one stands in debates about the ethics of enhancement. In an excellent discussion, the President's Council on Bioethics expresses the relevant concern with regard to the routine administration of methylphenidate (Ritalin) and amphetamine (Adderall) to children whose conduct is outside the range of acceptability:

> Behavior-modifying agents circumvent that process [i.e. the process of self-control and progressive moral education], and act directly on the brain to affect the child's behavior without the intervening learning process. If what matters is only the child's outward behavior, then this is simply a more effective and efficient means of achieving the desired result. But because moral education is typically more about the shaping of the agent's character than about the outward act, the process of learning to behave appropriately matters most of all. If the development of character depends on effort to choose and act appropriately, often in the face of resisting desires and impulses, then the more direct pharmacological approach bypasses a crucial element.... By treating the restlessness of youth as a medical, rather than a moral, challenge, those resorting to behavior-modifying drugs might not only deprive [the] child of an essential part of this education. They might also encourage him to change his self-understanding as governed largely by chemical impulses and not by moral decisions grounded in some sense of what is right and appropriate. (President's Council on Bioethics 2003: 105–106)

In other words, once we take an interventionist biotechnological or neurotechnological approach, to respond to (or manage) our social problems, there is a danger that, as the President's Council puts it, "we may weaken our sense of responsibility and agency." (ibid: 106) No doubt, we might try instead to rely on a technological intervention that, rather than correcting for the worst of our anti-social tendencies, is designed to enhance our moral disposition. However, this, too, is far from problematic (Daniels 2007: 40–41; Pacholczyk 2011). In particular, unless we assume that a ruling regulatory elite is morally omniscient, a moral community must aspire to do the right thing but, at the same time, it must accept that what the right thing is in any particular case is an open question (precisely in the way that protagonists disagree about the ethics of enhancement). In other words, first, there must be the opportunity for moral development and, then, there must be ample room for moral debate (Brownsword 2011).

16.5 A Rights Perspective

Elsewhere, I have argued that we should judge the permissibility of using enhancers and enhancing technologies relative to a set of rights that protect our basic needs as agents.[9] From that standpoint, any proposed enhancement—in any context, sporting or otherwise—would need to be compatible with (i) the array of rights recognised by the community and (ii) the community's aspiration to operate as a moral community.

If we redraft the proposed governing principle in a way that embeds this rights perspective, it becomes the following:

it is permissible for competent agents (such as Olympic competitors) to use enhancers unless either (i) use violates the rights of others (who have not consented to this risk) or (ii) the user has freely agreed to act on a no-enhancement basis. For agents who are not competent, the governing regulatory principle would be paternalistic.

It should not be thought that, by substituting the violation of 'rights' of others for 'harm' to others, we have retrieved a truly simple principle. There is still plenty of analysis, interpretation, and application to be carried out.

To understand why this is so, we need a sketch of some of the salient characteristics of the community of rights that I have in mind. First, each member of the community enjoys the protection of a set of generic rights, these being rights that focus on the basic needs of agency. The idea is that each agent will have his or her own distinctive preferences, plans, and projects; but, no matter what one's particular purposes, a set of (non-distinctive) generic conditions is presupposed. So, for example, while one agent might aspire to be an Olympic champion, another might prefer to be a concert pianist, and another a legal scholar—three agents, each with very different dispositions, but each with the same generic needs. It follows that the regulatory environment in such a community should start by assuring the conditions for viable agency (most basically for life itself) coupled with creating a context for flourishing community life. As individual members of the community, each agent will enjoy the guarantee of the generic rights but this will not guarantee that they have all that they require to achieve their particular purposes. Secondly, there is no assumption of moral omniscience in such a community and nor is it assumed that there is a mechanical specification of the particulars of the generic conditions. Accordingly, there might be disagreement as to whether a particular act does violate a generic right. Thirdly, members have the right to participate in the political and public life of their community. Political decisions cannot derogate from the generic rights but resources might be allocated in ways that favour particular purposes. If the community prefers to invest in sporting achievement rather than musical prowess, that is likely to be good for the aspirant Olympians but not so good for the would-be concert pianist.

---

[9] See Brownsword 2009. The seminal thinking that underlies this approach is found in Alan Gewirth (1978).

In such a community, what kind of public debate might there be with regard to the use of performance enhancers? Suppose it were proposed that there should be an across-the-board prohibition on the use of enhancers. Opponents might argue against such a prohibition in two ways. First, they might argue that members of the community have a *generic* right to use enhancers or enhancing technologies; and, secondly, they might simply express their preference for the availability of enhancers and cast their vote accordingly.

With regard to the first argument, if it were to be sustained, the proposal would fall (unless every member of the community agreed to waive standing on the right—which, ex hypothesi, is not the case). However, this begs the question of whether the argument would be sustained. For opponents to present the use of enhancers as covered by a generic right, they would need to show that such use is so much part of the generic conditions that it merits this special protection. While some enhancements (for example, enhanced eyesight) seem to be general purpose (and in this sense within the generic conditions), others look much more specific to particular purposes (for example, leg extensions for prospective high-jumpers) (compare Buchanan et al. 2000). No doubt, there would be much debate about whether a particular enhancement should be classified as general or special purpose; and, no doubt, there would be some anomalies (for example, Olympic archers might be able to plead a generic right to make use of their enhanced eyesight but not to use enhanced equipment). In a community of rights, the fact that such arguments are never quite settled is not a cause for concern; for, it is characteristic of such communities that they are constantly trying to give effect to their best understanding of their rights commitments and, to this extent, treating their public rules as provisional.[10]

The other argument that the opponents might present is simply that they prefer to have the option of using enhancers. The right that matters here is the right to participate in public debate and decision-making and to have equal consideration given to one's preferences. If the majority prefer a regulatory environment that does not categorically prohibit the use of enhancers, then the proposal will again be unsuccessful. However, so long as the proposal is not overridden by a generic right, but simply runs up against rival preferences there is always a chance that attitudes might change.

If we assume that the proposal for prohibition failed, the regulatory environment would permit the use of enhancers so long as they did not violate the generic rights of others (subject to their consent). There would still be many questions to be raised and answered depending upon whether the protected use of (general purpose) enhancers was seen as resting on a generic right or simply the balance of preferences. As for the Corinthians, they could continue to opt for prohibition; and, even if Corinthians, by eschewing the use of enhancers, thereby agreed to waive the benefit of a protected generic right, this would be within the power of members (unless, quite remarkably, the community viewed this particular generic right as indefeasible).

[10] For a rather different illustration of the same underlying idea, see Brownsword and Earnshaw (2010).

These remarks about the public decisions concerning enhancement that might be made in a community of rights help to dispel the notion that a community of rights has no real collective identity—that such communities, although firmly committed against violations of rights, have no sense of what they are for (of their mission, or some such objective). So, for example, it might be said that such communities declare that they are against the use of technologies that infringe on protected rights but, otherwise, they are neither for nor against modern technologies. Now, while it is correct to say that the first concern for such communities is that rights are respected, it is not correct to say that they have no sense of whether they are for or against technology. Through routine political decision-making, such communities will express their preferences; some will be technophiles, some technophobes, some neither technophile nor technophobe. The point is that there is no necessity that such communities all express the same preferences (and public character) or that their preferences remain constant.

In a particularly insightful report for the Rathenau Institute, the fact that liberal communities do not apparently commit to being either pro-enhancement or anti-enhancement prompts the thought that such communities are projects that are waiting to be completed. According to the report,

> [T]he moral discourse of the public domain is limited: it is about guaranteeing safety and freedom of the citizens, preventing risks, harm and violence, warranting justice and rights and asserting duties of individuals. It is not about the good life, about how to become a good person, about the meaning of love, loss and suffering, how to restrict insatiable wishes for joy, health, richness and honour, or how one should live and behave.... (Staman et al. 2008)

Of course, this picture is a bit of a caricature. Nevertheless, it serves to make our main point, namely that notwithstanding the great benefits of having a liberal society, in its basic structure it might have insufficient space for public discourse to address questions on collective issues that cannot be dealt with on a satisfactory level by either state regulation or individual choice (ibid: 152–153).

The authors conclude:

> It is doubtful whether the two main levels for public private decision-making, the state and the individual, offer us the right forum to give these developments [i.e., developments in human enhancement] collective treatment. Developments in several fields of human enhancement, however, pose questions that demand such a wider, collective, approach. They force us to reflect on who we are, how we think about new scientific, medical and cultural norms, identities and practices. Human enhancement requires a more moral debate than regulation and risk-control by the state can offer. And it demands more public consideration than is possibly available by reflection from individual consumers. (ibid.: 156)

While this diagnosis might fit some liberal communities, in an ideal-typical community of rights, there would be a strong culture of public participation in political debates so that collective decisions would reflect the tastes, preferences, and culture of the community. The upshot might be a community that seems quite 'communitarian' (for example, the community might decide that it understands sport, medicine, and education in a way that precludes the use of enhancers) or one that celebrates difference (as where Corinthian zones of privately ordered non-enhancement co-exist with a general permission on the use of enhancers). For categorical prohibitionists, such expressions of difference will seem pathological; but, where it is rights rather than duties that regulate, this is simply how the legitimate expression of individual preferences registers in the public and private ordering of the community.

16.6 Conclusion

In their contribution to this collection, Jan Tolleneer and Paul Schotsmans identify five reasons why one might argue against enhancement in sports. These are: (i) that the enhancer is harmful to the health of the user; (2) that it reduces the chances of other competitors; (3) that it undermines the spirit of sport; (4) that it creates negative role models; and (5) that it desecrates human nature. Although my paper has not been written with these particular reasons in mind, it is clear that they have figured in the analysis; and it might be helpful to conclude by seeing how these reasons are treated by the simple regulatory principle that has been the focus of the discussion and then, alongside this, how they might be viewed in a community of rights.

With regard to the first reason, the simple principle has a clause for the paternalistic protection of children and other vulnerable persons but, otherwise, it does not regard harm to the user as a good reason for prohibiting enhancers. To be sure, an individual might decide that the harm caused by enhancement is likely to outweigh any benefit and decide against use. That's fine. However, it is for individuals to decide for themselves, not for regulators to decide for them. In a community of rights, regulators will have a responsibility to promote decision-making that is both free and informed; but, in general, the philosophy and the approach would be much the same as that represented by the simple principle.

Secondly, the objection that enhancers reduce the chances of others is a difficult one for the simple principle. In sport, as in education, the use of enhancers is for positional advantage; and it is arguable that this is a relevant 'harm' to others. If so, unless those others consent to the use, it should not be permitted. In a community of rights, the strongest objection would be that using enhancers for positional advantage violates the generic rights of others. This, I suggest, looks less plausible than the counter-argument that it is actually the use of enhancers (or, at any rate, of general purpose enhancers) that is supported by generic rights. But, it might be that generic rights are not implicated and that the legitimacy of using enhancers for positional advantage turns on: (i) the expressed public preferences of the community and the rules that have resulted; and (ii) any applicable rules of Corinthian private ordering.

The third objection is that the use of enhancers undermines the spirit of sport. While the simple principle does not accommodate the categorical prohibitionists, it can license the Corinthians to compete on enhancer-free terms. Provided that agents freely sign up for the Corinthian rules, that is absolutely fine. And, much the same philosophy will inform the approach in communities of rights.

Fourthly, there is the objection that the use of enhancers by sports people creates negative role models. To some extent, this is uncontroversial: if using enhancers is cheating, then this is clearly a poor image to project (just like unsporting conduct, or any of its cognates). However, the real question is whether, without it being a case of cheating, the use of enhancers would set a poor example. For the simple principle, this presents another difficult case. Is this a case of relevant harm to others? The answer is not clear. In a community of rights, the question is whether others have a right that certain agents, such as sports people, should act as role models; and, then, whether using enhancers fails to meet whatever responsibility they have. If there is a generic right to use enhancers, the objection looks weak; and, if there is no generic right, but the balance of public preferences permits it, there is no reason to yield to the objection (unless one is a declared Corinthian, in which case one freely assumes special responsibilities).

Fifthly, the objection that human nature is desecrated is akin to the categorical prohibitionists' objection that human dignity is compromised. Supporters of both the simple principle and the rights approach unite in rejecting dignitarian conservatism.[11] If the dignitarians can show why their arguments should be privileged in ethical and regulatory debates, there will need to be a major rethink in those communities that start with liberal premises. However, until that time, the simple principle, for all its difficulties, points in the right direction; and an agency-based theory of rights holds the key to understanding how the use of enhancers should be regulated as a matter of both public and private (e.g., Corinthian) ordering.

Finally, what can we expect to be the legacy of the London Games? Will it be the memory of fine sporting achievements, or the regeneration of East London, or the renewal of debates about the acceptable limits to enhancing technologies? Quite likely, it will be all of this. However, by the time that the Games open in Rio in 2016, in most respects, the show will have moved on. As the torch is carried from one venue to another, the one element that will persist will be the continuing doubts that we have about the limits and legitimacy of human enhancement–for communities round the world it will take more than a few Olympics before we have clarity, let alone consensus, about these matters.

References

Application no. 57813/00 (Judgment given on 3 November 2011).

Beyleveld, D., and R. Brownsword. 2001. *Human dignity in bioethics and biolaw*. Oxford: Oxford University Press.

Beyleveld, D., and R. Brownsword. 2007. *Consent in the law*. Oxford: Hart.

Brownsword, R. 2003. Bioethics today, bioethics tomorrow: Stem cell research and the 'dignitarian alliance'. *University of Notre Dame Journal of Law, Ethics and Public Policy* 17(1): 15–51.

[11] Elsewhere, I have characterised this as an alliance of dignitarian views (Brownsword 2003).

Brownsword, R. 2006. Cloning, zoning, and the harm principle. In *First, do no harm (festschrift for Ken Mason)*, ed. Sheila McLean, 527–542. Aldershot: Ashgate.
Brownsword, R. 2008. *Rights, regulation and the technological revolution*. Oxford: Oxford University Press.
Brownsword, R. 2009. Regulating human enhancement: Things can only get better? *Law Innovation and Technology* 1(1): 125–152.
Brownsword, R. 2011. Lost in translation: Legality, regulatory margins, and technological management. *Berkeley Technology Law Journal* 26(3): 1321–1366.
Brownsword, R., and J. Earnshaw. 2010. Controversy: The ethics of screening for abdominal aortic aneurysm. *Journal of Medical Ethics* 6(12): 827–830.
Brownsword, R., and H. Law Somsen. 2009. Innovation and technology: Before we fast forward—a forum for debate. *Law Innovation and Technology* 1(1): 1–73.
Brüstle v Greenpeace eV (Case C34-10) (judgment of the ECJ handed down on 18 October, 2011).
Buchanan, A., D. Brock, N. Daniels, and D. Wikler. 2000. *From chance to choice: Genetics and justice*. Cambridge: Cambridge University Press.
Coady, C.A.J. 2009. Playing god. In *Human enhancement*, ed. J. Savulescu and N. Bostrom, 155–180. Oxford: Oxford University Press.
Daniels, N. 2007. Can anyone really be talking about ethically modifying human nature? In *Human enhancement*, ed. J. Savulescu and N. Bostrom, 25–42. Oxford: Oxford University Press.
EC Directive on the Legal Protection of Biotechnological Inventions (Directive 1998/44/EC).
Faden, R., and R. Beauchamp. 1986. *A history and theory of informed consent*. New York: Oxford University Press.
Gewirth, A. 1978. *Reason and morality*. Chicago: University of Chicago Press.
Gordijn, B. 2006. Converging NBIC technologies for improving human performance: A critical assessment of the novelty and prospects of the project. *The Journal of Law, Medicine & Ethics* 34(4): 726–732.
Hagger, L., and G. Hagger Johnson. 2011. 'Super kids' regulating the use of cognitive and psychological enhancement in children. *Law, Innovation and Technology* 3(1): 137–166.
Hampton, J. 2011. *London Olympics 1908 and 1948*. Oxford: Shire Publications.
Harris, J. 2007. *Enhancing evolution*. Princeton: Princeton University Press.
Lawson, D. 30 August 2011. Who gets to run in top-flight athletics? *The Independent*, 13.
Manson, N., and O. O'Neill. 2007. *Rethinking informed consent in bioethics*. Cambridge: Cambridge University Press.
Møller, V. 2010. The athletes' viewpoint. In *The ethics of sport. A reader*, ed. M. McNamee. London: Routledge.
Nuffield Council on Bioethics. 2011. *Human bodies: Donation for medicine and research*. London
Pacholczyk, A. 2011. Moral enhancement – What is it and do we want it? *Law, Innovation and Technology* 3(2): 251–277.
Parens, E. 2009. Towards a more fruitful debate about enhancement. In *Human enhancement*, ed. J. Savulescu and N. Bostrom, 71–90. Oxford: Oxford University Press.
Peters, T. 2007. Are we playing god with nanoenhancement? In *Nanoethics*, ed. F. Allhoff, P. Lin, J. Moor, and J. Weckert. Hoboken: Wiley. chapter 13.
President's Council on Bioethics. 2003. *Beyond therapy. Biotechnology and the pursuit of happiness*. Washington, DC: Dana Press.
Sandel, M. 2007. *The case against perfection*. Cambridge: Harvard University Press.
Savulescu, J., and N. Bostrom (eds.). 2009. *Human enhancement*. Oxford: Oxford University Press.
Slot, O. 13 August 2011. Generation Y's gold rush will be sparked by technology. *The Times*, 24.
Staman, J., H. Dijstelbloem, and M. Smits. 2008. Afterword. In *Reshaping the human condition: Exploring human enhancement*, ed. L. Zonneveld, H. Dijstelbloem, and D. Ringoir. The Hague: Rathenau Institute.

Stevens, D. 2011. Medalling in sport. *Newsletter of the ESRC Genomics Network*, 10.
Tännsjö, T. 2009. Medical enhancement and the ethos of elite sport. In *Human enhancement*, ed. J. Savulescu and N. Bostrom. Oxford: Oxford University Press.
WARF case (Case G 0002/06, 25 November 2008).
World Anti-Doping Agency. 2003. *World Anti-Doping Code*. http://www.wada-ama.org/rtecontent/document/code_v3.pdf. Accessed 30 Mar 2012.

Biographical Notes

Pieter Bonte studied philosophy at Ghent University and law at the Vrije Universiteit Brussel and works as a doctoral researcher at the Bioethics Institute Ghent, Department of Philosophy and Moral Sciences, Ghent University. His research project starts from a detailed study of intrinsic arguments for and against self-creation and self-enhancement via biotechnology and is aimed at the articulation of an enriched, rational and useable conception of 'human dignity'. His main interests lay in combining existentialist and evolutionary understandings of human nature and ensuing issues of personal identity, responsibility, human dignity and the art of living. He enjoys binding himself to the rules and rituals of basketball as well as cycling haphazardly through the country.

Barbara Broers, M.D., M.Sc., directs the Unit for Dependency in Primary Care at the Department of Community Health and Primary Care of the University Hospitals of Geneva. She is member of the Swiss Federal Commission for Drug Questions. Her research interests include public health aspects of psychotropic drug use, therapeutic compliance of drug users and harm reduction strategies.

Roger Brownsword is Professor of Law at King's College London, where he was the founding director of TELOS (a research centre that focuses on technology, ethics, law and society); he is an honorary Professor in Law at the University of Sheffield; and he is a visiting professor at Singapore Management University. His interests span legal theory, regulation and technology, and the common law. His more recent books include (with Deryck Beyleveld) *Human Dignity in Bioethics and Biolaw* (2001) and *Consent in the Law* (2007); *Contract Law: Themes for the Twenty-First Century* (2006); *Rights, Regulation and the Technological Revolution* (2008); and a co-edited collection on *The Foundations of European Private Law* (2011). His latest book (co-authored with Morag Goodwin) is *Law and the Technologies of the Twenty-First Century*. In total, he has published some 200 papers.

Andreas De Block is Assistant Professor of Philosophy at the KU Leuven. He studied philosophy and psychology in Leuven and Ghent, and worked as a postdoctoral researcher at Radboud University Nijmegen and the University of

J. Tolleneer et al. (eds.), *Athletic Enhancement, Human Nature and Ethics*, International Library of Ethics, Law, and the New Medicine 52, DOI 10.1007/978-94-007-5101-9,

Michigan. He is the author and editor of several books on psychoanalysis, philosophy of psychiatry, and philosophy of biology. His main research interests are cultural evolution, the philosophical assumptions of the nature-nurture debate, sexual orientation, and philosophy of psychiatry.

Trijsje Marie Franssen recently obtained a Ph.D. in Philosophy from the University of Exeter, Department of Sociology and Philosophy. Her research concentrates on the role of the ancient myth of Prometheus in today's debate on human enhancement and its relationship to normativity and arguments on human nature. She engages with the myth both to analyse the enhancement debate and its (implicit) moral and ontological claims and to reflect upon the particular characteristics that have given this myth its prominence in the history of philosophy in general.

Denis Hauw is Professor of Sport Psychology at the Institute of Sport Sciences of the University of Lausane. His research focuses on doping and substances uses in elite sports with a situated activity approach. He is the president of the French national call centre *Ecoute Dopage* and also is a sport psychologist consultant for diverse elite teams and athletes. From August 1, 2012 onwards, he will hold the position of Professor of Sport Psychology at the University of Lausanne.

John Hoberman is Professor of Germanic Studies, University of Texas at Austin. He has been active in sports studies and sports journalism over the past 35 years. He has a special interest in human enhancements in general and sports doping in particular. He is the author of *Mortal Engines* (1992) and *Testosterone Dreams* (2005). His current research concerns the use of anabolic steroids by police officers in the United States.

M. Andrew Holowchak teaches philosophy at Rider University in Lawrence, New Jersey. He has published over 60 peer-reviewed papers in areas such as ethics, psychoanalysis, ancient philosophy and science, philosophy of sport, and social and political philosophy and has authored 18 books including *Freud: From Individual to Group Psychology, Freud and Utopia, Happiness and Greek Ethics, Critical Reasoning & Philosophy, Ancient Science and Dreams, Philosophy of Sport, Aretism, The Stoics*, as well as two philosophical novels, *Life of a Jellyfish* and *Hotel Bob*. His current research is on philosophical elements in the writings of Thomas Jefferson. When not teaching or writing, he enjoys strength training (former super-heavyweight powerlifting champion), biking, gardening, travel, cooking, brewing beer, and polite conversation. He lives in Lindenwold, New Jersey, with his wife, Angela, and their several cats.

Marianne Raakilde Jespersen is post-doctoral researcher and lecturer at the Department of Sport Science, Public Health, Aarhus University, Denmark and the Department of Global Nutrition and Health at VIA University College, Aarhus. Her research focuses on the use of performance enhancing drugs in and out of sports, medicalization and body culture in general.

Eric T. Juengst directs the UNC Centre for Bioethics which serves as a hub for bioethics teaching, consultation, and research at the University of North Carolina, Chapel Hill. His own research and publications focus on ethical and conceptual issues in human genetics and genomic research, and his current project is a study of the ways that the goals of "personalized genomic medicine" are interpreted by its promoters and providers. He is also a co-founder of the "ELSI 2.0 Collaboratory," a new international network of scholars interested in the intersections of genomics and society.

Bengt Kayser, M.D., Ph.D., is director of the Institute of Movement Sciences and Sports Medicine at the University of Geneva. His research interests include the relationship between physical activity and health, the effect of hypoxia on physical performance and human enhancement.

Christian Lenk is a Privatdozent (senior lecturer) at the Institute for History, Philosophy & Ethics of Medicine of Ulm University. He studied philosophy, political sciences and social anthropology at the University of Hamburg. He concluded his Ph.D. thesis at the University of Münster in the field of medical ethics in 2002 with a work on *Therapy and Enhancement* and he has undertaken international research stays at the Universities of Swansea/UK (2008/09) and Zurich/CH (2011). Since 2008, he is Privatdozent at the University of Göttingen, and since 2011 acting director of Ulm University's research ethics committee.

Tara Magdalinski works at the UCD Centre for Sports Studies, School of Public Health, Physiotherapy and Population Science, University College Dublin. She has published widely in the area of sports studies focussing most recently on the cultural construction of performance enhancement. She co-edited (with Timothy Chandler), *With God on their Side: Sport in the Service of Religion* (Routledge, 2002), and published *Sport, Technology and the Body: The Nature of Performance* (Routledge, 2009). Her current projects include an examination of Olympic ephemera such as pins and mascots.

Michael J. McNamee is Professor of Applied Ethics in the College of Human and Health Sciences, Swansea University Wales. He has authored and edited many books including *Sports, Virtues and Vices* (2008), *Sports Ethics* (2010) and *Doping and Anti Doping Policy* (2011). He is Editor of the international journal *Sport, Ethics and Philosophy*, and is a former President of the International Association for the Philosophy of Sport. He still runs (slowly) and plays tennis (when not injured!).

Darian Meacham is Senior Lecturer in Philosophy in the Department of Arts at the University of the West of England, Bristol. An expert in Phenomenology and the Philosophy of the Body, he has particular interests in the phenomenology and politics of human enhancement. He is currently working on a project focusing on the fundamental philosophical issues surrounding cognitive enhancement.

Paul Schotsmans is Professor at the Centre for Biomedical Ethics and Law, Department of Public Health, Faculty of Medicine, KU Leuven University. He is the honorary president of the European Association of Centres of Medical Ethics (EACME) and former board member of the International Association of Bioethics. His research interest is mainly the application of 'personalism' as an ethical model for the ethical integration of reproductive technologies, pre-implantation and prenatal diagnosis, clinical genetics, stem cell research, organ transplantation and end of life decision making. He is the author and co-editor of several books and papers in leading journals in bioethics.

Sigrid Sterckx is Professor of Ethics at the Department of Philosophy & Moral Sciences of Ghent University and a part-time Professor of Ethics at the Department of Philosophy & Moral Sciences of the Vrije Universiteit Brussel. She lectures courses in theoretical ethics, methods in ethics and political philosophy. Her publications include three books, numerous book chapters and articles in legal and (bio) ethics journals on issues including end-of-life decisions and the patentability of human body material. Sigrid is a member of the Belgian Advisory Committee on Bioethics.

Jan Tolleneer is Professor at both the Department of Kinesiology, Faculty of Kinesiology and Rehabilitation Sciences, KU Leuven University, and the Department of Movement and Sport Sciences, Faculty of Medicine and Health Sciences, Ghent University. He also teaches at the National Training Academy of Flanders, Belgium. He is a former board member of the International Society for Comparative Physical Education and Sport (ISCPES) and of the International Society for the History of Physical Education and Sport (ISHPES). He teaches foundations of physical education, comparative studies, sports history and sports ethics. He coordinates the Interfaculty Research Group Sport and Ethics (KU Leuven). He published several books and articles in leading journals in sport studies.

Author Index

J. Tolleneer et al. (eds.), *Athletic Enhancement, Human Nature and Ethics*, International Library of Ethics, Law, and the New Medicine 52, DOI 10.1007/978-94-007-5101-9,